T0340993

Nonlinear Optics

Optical Sciences and Applications of Light

Series Editor
James C. Wyant
University of Arizona

Charged Particle Optics Theory: An Introduction, *Timothy R. Groves*

Photonics Modelling and Design, *Slawomir Sujecki*

Nonlinear Optics: Principles and Applications, *Karsten Rottwitt and Peter Tidemand-Lichtenberg*

Numerical Methods in Photonics, *Andrei V. Lavrinenko, Jesper Lægsgaard, Niels Gregersen, Frank Schmidt, and Thomas Søndergaard*

Please visit our website ***www.crcpress.com*** *for a full list of titles*

Nonlinear Optics

Principles and Applications

Karsten Rottwitt
Technical University of Denmark, DTU Fotonik

Peter Tidemand-Lichtenberg
Technical University of Denmark, DTU Fotonik

CRC Press is an imprint of the
Taylor & Francis Group, an **informa** business

CRC Press
Taylor & Francis Group
6000 Broken Sound Parkway NW, Suite 300
Boca Raton, FL 33487-2742

Printed on acid-free paper
Version Date: 20140922

International Standard Book Number-13: 978-1-4665-6582-1 (Hardback)

Visit the Taylor & Francis Web site at
http://www.taylorandfrancis.com

and the CRC Press Web site at
http://www.crcpress.com

Contents

Series Preface: Optical Sciences and Applications of Light

Optics and photonics are an enabling technology in many fields of science and engineering. The purpose of this book series is to present the state-of-the-art of the basic science of optics, applied optics, and optical engineering, and the applications of optics and photonics in a variety of fields including health care and life sciences, lighting, energy, manufacturing, information technology, telecommunications, sensors, metrology, defense, and education. This new and exciting material will be presented at a level that makes it useful to the practicing scientist and engineer working in a variety of fields

Volumes in this series cover topics that are a part of the rapid expansion of optics and photonics in a variety of fields, all over the world. The technologies discussed impact numerous real-world applications including new displays in smart phones, computers and television, new imaging systems in consumer cameras, biomedical imaging for disease diagnosis and treatment, and adaptive optics for space systems for defense and scientific exploration. Other applications include optical technology for providing clean and renewable energy, optical sensors for more accurate weather prediction, solutions for more cost-effective manufacturing and ultra-high capacity optical fiber communications technologies that will enable the future growth of the Internet. The complete list of areas optics and photonics are involved in is very long and always expanding.

Preface

Vision is one of the most important senses of humans. As a result of this, optics was developed as a physical science by the ancient Egyptians and Mesopotamians. Polished lenses made of quartz have been found dating back as early as 700 BC. The ancient Romans and Greeks used glass spheres filled with water to make lenses. The practical developments were followed by the development of theories of light and vision by ancient Greek and Indian philosophers, leading to the development of geometrical optics. The word optics comes from the ancient Greek meaning "appearance, look".

One of the first experimental demonstrations of a nonlinear optics phenomena may have been Raman scattering, more specifically in-elastic scattering of a photon against a molecule. When photons are scattered by a molecule, most photons are elastically scattered (Rayleigh scattering), that is the scattered photons have the same frequency (energy) as the incident photons. However, a small fraction of the scattered photons are in-elastically scattered, that is the scattered photons have a frequency different from, usually lower than, the incident photons. The effect was described experimentally and theoretically in the decade 1920 to 1930, and the phenomena was named after Sir C. V. Raman, who was also awarded the Nobel Prize in 1930 for his contribution to the work on understanding scattering of light.

During the last millennium more and more detailed and sophisticated light theories have been developed, allowing one to describe both the classical and quantum mechanical properties of light. As a result of the development of optical theories and the technological development of more and more sophisticated devices, optics today has a huge impact on our daily life. Particularly, the development of the laser back in the early 1960's which paved the way for many applications of light in society. The development of the laser also made it possible to do nonlinear optics, i.e. observe optical phenomena that depend on the electric field, be it the propagating field or an external field applied to a crystal through which light propagate, in a nonlinear manner. The first experimental demonstration of nonlinear optics using laser light followed very shortly after the first demonstration of the laser.

The development of nonlinear optics has led to the discovery of many new phenomena, some of which can be exploited for optical switching, frequency conversion, amplification of signals and spectroscopy. However, other nonlinear phenomena may limit the performance of optical devices, i.e. power transmission through fibers limited by nonlinear effects, as for example cross talk between communication channels in a transmission line.

Acknowledgements

As pointed out, this book started out as lecture notes. During the many years it has taken to develop the material, many colleagues and students have provided invaluable feedback and found many mistakes. Consequently, many colleagues, and students should be thanked for their inputs and for many discussions. To list every one is not an option even though it would be well deserved. However, we would like to mention especially Thorkild B. Hansen, Johan R. Ott, Martin E. V. Pedersen, Søren M.M. Friis, Lasse M. Andersen, Sidsel R. Petersen, Kristian R. Hansen, Lasse Høgstedt, Simon Lehnskov Lange, and Christian Agger.

Scope and structure

To write a textbook on nonlinear optics is a challenge, especially since there already exist many excellent textbooks in the topic. However, in the opinion of the authors of this textbook, the existing books fall into two categories, either very detailed leaning heavily toward basic physics as for example R. W. Boyd, *Nonlinear Optics*, Academic Press [33] and P.N. Butcher and D. Cotter, *The Elements of Nonlinear Optics*, Cambridge Studies in Modern Optics 9 [2] or very much directed toward applications as for example nonlinear optics used in optical fibers G. P. Agrawal *Nonlinear Fiber Optics*, Academic Press [3] or nonlinear optics in crystals by A. Yariv and P. Yeh, *Optical Waves in Crystals*, Wiley [4]. These books are all excellent books, and we strongly recommend readers of this book also read the other books. However, it is the ambition of this book to bridge the very thorough physics and mathematics in [33] and [2] with the more applied material in [3] and [4].

In addition, while teaching the subject of Nonlinear Optics, we found it difficult to find just one textbook covering the general theory as well as going into more detail on selected topics like frequency conversion and electro optical effect in crystals, Raman and Brillouin scattering in optical fibers and optical Kerr effect and four wave mixing in parametric devices. Consequently, this book started out as a lecture note, as a supplement to textbooks as for example Butcher and Cotter [2] which has a very detailed mathematical foundation, and R. Boyd which has a more physical approach to the field [1].

The book is structured so that the first five chapters of the book are dedicated to the description of the fundamental formalism of nonlinear optics, whereas the last five chapters of the book are devoted to a description of practical devices based on nonlinear phenomena. It has been the intention to discuss nonlinear optics using classical electromagnetism rather than quantum mechanics. Sections/subsections indicated with a star $*$ are intended to be background material and may contain quantum mechanics. These sections/subsections need not to be read as thoroughly as other sections/subsections to gain useful insight into nonlinear optics. Readers of the book, and students in particular, may enjoy further material including examples and problems, which is provided on the homepage www.fotonik.dtu.dk/english/Education/NLO.

In order to keep this book up to date, various data will be uploaded to this site, including:

- **Theoretical exercises**
- **Numerical exercises**
- **Lecture slides**
- **Important corrections**

Important constants

Table 1: Table of important constants

Constant	Symbol	Value	[Units]
Planck constant	h	$6.63 \cdot 10^{-34}$	[m/W]
Boltzmann constant	k_B	$1.381 \cdot 10^{-23}$	[J/K]
Vacuum permittivity	ε_0	$8.854{\cdot}10^{-12}$	[F/m]
Vacuum permeability	μ_0	$4\pi \cdot 10^{-7}$	[H/m]
Bohr radius	a_0	$0.529{\cdot}10^{-10}$	[m]
Vacuum speed of light	c	$2.99792458{\cdot}10^{8}$	[m/s]
Atomic mass unit	u	$1.660{\cdot}10^{-27}$	[kg]
Electron volt	e	$1.602{\cdot}10^{-19}$	[J]

Definition of coordinate systems

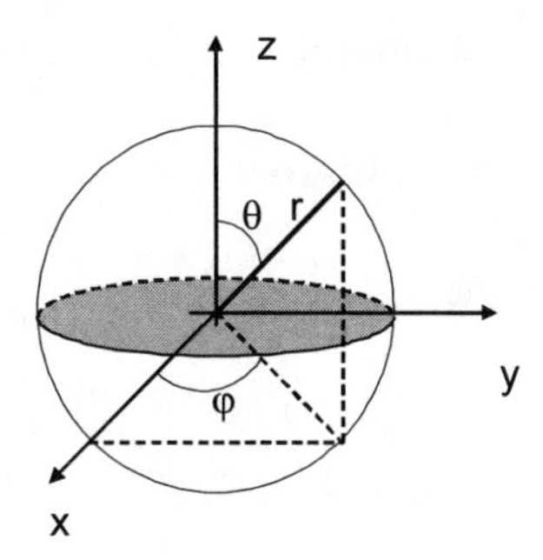

Figure 1: Coordinate systems

Cartesian and polar coordinates

$$x = r \sin\theta \cos\phi \tag{1a}$$
$$y = r \sin\theta \sin\phi \tag{1b}$$
$$z = r \cos\theta \tag{1c}$$

$$r = \sqrt{x^2 + y^2 + z^2} \tag{1d}$$
$$\phi = \arctan\left(y/x\right) \tag{1e}$$
$$\theta = \arccos\left(z/r\right) \tag{1f}$$

In standard spherical coordinates (r, θ, ϕ), the differential surface element perpendicular to a vector $\mathbf{r}$ is $da = r^2 \sin\theta d\theta d\phi$

$$\int\int \sin\theta d\theta d\phi. \tag{2}$$

V	Volume
dv	Volume element
A	Surface
da	Surface element

Definition of vectors and operators

$\mathbf{e}$	Unit vector (also $\mathbf{u}$ and $\mathbf{v}$ are used)
$\mathbf{n}$	Normal vector
$\mathbf{r}$	Vector from origin to observation point
ψ_p	The azimuthal angle on Poincaré sphere ($0 \leq 2\psi_p \leq 2\pi$)
θ_p	The polar angle on Poincaré sphere ($-\pi/2 \leq 2\psi_p \leq \pi/2$)
S_{ijk}	Symmetric tensor ($S_{ijk} = S_{ikj}$)
A_{ijk}	Antisymmetric tensor ($A_{ijk} = -A_{ikj}$)
$\hat{\mathcal{O}}$	Operator - $\hat{\mathcal{O}}^S$ (Schrödinger picture), $\hat{\mathcal{O}}^H$ (Heisenberg picture)
$\hat{U}(t)$	Unitary time evolution operator
$\hat{H}$	Hamiltonian operator - $\hat{H}_0$ (unperturbed), $\hat{H}_E$ (interaction)
$\hat{Q}$	Linear dipole moment operator
$\hat{\rho}$	Density matrix operator
$\hat{p}$	Dipole density matrix operator
$\langle\rangle$	Expectation value
$\|\psi_n\rangle$	Normalised state ($\langle\psi_n\|\psi_n\rangle = 1$)

$$\begin{pmatrix} x \\ y \end{pmatrix} = (x, y)^T \tag{3}$$

Dot product

$$\mathbf{A} \cdot \mathbf{B} = AB\cos(\theta) \tag{4}$$

Cross product

$$\mathbf{A} \times \mathbf{B} = (A_y B_z - A_z B_y)\,\mathbf{e}_x + -(A_z B_x - A_x B_z)\,\mathbf{e}_y + (A_x B_y - A_y B_x)\,\mathbf{e}_z \tag{5}$$

Gradient

$$\nabla \mathbf{A} = \frac{\partial A}{\partial x}\mathbf{e}_x + \frac{\partial A}{\partial y}\mathbf{e}_y + \frac{\partial A}{\partial z}\mathbf{e}_z \tag{6}$$

Divergence

$$\nabla \cdot \mathbf{A} = \frac{\partial A_x}{\partial x} + \frac{\partial A_y}{\partial y} + \frac{\partial A_z}{\partial z} \tag{7}$$

Curl

$$\nabla \times \mathbf{A} = \left(\frac{\partial A_z}{\partial y} - \frac{\partial A_y}{\partial z}\right)\mathbf{e}_x + \left(\frac{\partial A_x}{\partial z} - \frac{\partial A_z}{\partial x}\right)\mathbf{e}_y + \left(\frac{\partial A_y}{\partial x} - \frac{\partial A_x}{\partial y}\right)\mathbf{e}_z \tag{8}$$

Laplacian operator

$$\nabla^2 A = \nabla \cdot (\nabla A) = \frac{\partial^2 A}{\partial x^2} + \frac{\partial^2 A}{\partial y^2} + \frac{\partial^2 A}{\partial z^2} \tag{9}$$

$$\nabla_{\perp}^2 A = \frac{\partial^2 A_x}{\partial x^2} + \frac{\partial^2 A_y}{\partial y^2} \tag{10}$$

$$\nabla_z^2 A = \frac{\partial^2 A_z}{\partial z^2} \tag{11}$$

Important vector identity:

$$\nabla \times (\nabla \times \mathbf{A}) = \nabla (\nabla \cdot A) - \nabla^2 A \tag{12}$$

Transforms

Fourier transformation

$$E(\omega) = \frac{1}{2\pi} \int_{-\infty}^{\infty} E(t) \exp(i\omega t) dt. \tag{13}$$

Inverse Fourier-transform

$$E(t) = \int_{-\infty}^{\infty} E(\omega) \exp(-i\omega t) d\omega. \tag{14}$$

Miscellaneous notation

A	Variable or function
A^0	Amplitude
A_ω	Field oscillating at angular frequency ω
$(A_\omega)_x$	x-component of field oscillating at angular frequency ω
$\bar{A}$	Time average
A^*	Complex conjugate of variable or function
$(A^0)^*$	Complex conjugate of amplitude
$\mathbf{A}$	Vector
$\underline{\underline{A}}$	Matrix
$\det(\underline{\underline{A}})$	Determinant of matrix
$\mathrm{Tr}(\underline{\underline{A}})$	Trace of matrix
$\mathbf{A}_x$	x-coordinate of vector
i	Imaginary unit ($i = \sqrt{-1}$)
ζ	Complex number - $\zeta_i = \mathrm{Im}[\zeta]$ (imaginary part), $\zeta_r = \mathrm{Re}[\zeta]$ (real part)
$\wp$	Probability
$*$	Convolution
$\mathcal{F}$	Fourier transform
p.v.	Principal value

Subscripts

B	Brillouin
K	Kerr
R	Raman
coh	Coherence parameter
dfg	Difference-frequency generation parameter
eff	Effective value of parameter
sfg	Sum-frequency generation parameter
sol	Soliton parameter
e	*extraordinary*-component of parameter
i	*Idler*-component of parameter
o	*ordinary*-component of parameter
p	*Pump*-component of parameter
s	*Signal*-component of parameter
x	x-component of parameter
y	y-component of parameter
z	z-component of parameter
ω	Parameter oscillating at angular frequency ω
FOPA	Fiber optical parametric amplifier
FWHM	Full with half maximum
NCPH	Non-critical phase matching
PIA	Phase insensitive amplifier
PSA	Phase sensitive amplifier
QPM	Quasi-phase matching
RIN	Relative intensity noise
RMS	Root mean square
SK	Skewness
SNR	Signal to noise ratio

List of symbols

Symbol	Description	Unit	Page
$\mathbf{B}$	Magnetic induction vector	[Wb/m^2]	4
$\mathbf{D}$	Displacement field vector	[C/m^2]	4
$\mathbf{E}$	Electric field vector	[V/m]	4
$\mathbf{H}$	Magnetic field vector	[A/m]	4
$\mathbf{J}$	Current density vector - $\mathbf{J}_b$ (bound), $\mathbf{J}_f$ (free)	[A/m^2]	4
$\mathbf{M}$	Magnetization vector	[A/m]	5
$\mathbf{p}$	Momentum	[kg m/s]	190
$\mathbf{p}$	Microscopic induced electric polarization vector	[C/m^2]	23
$\mathbf{P}$	Induced electric polarization - $\mathbf{P}^{(\mathrm{NL})}$ (nonlinear)	[C/m^2]	5
$\mathbf{S}$	Pointings vector	[W/m^2]	12
A	Envelope distribution		120
D	Group velocity dispersion	[s/m^2]	219
E_L	The launched electric field	[v/m]	165
F	Force	[N]	24
H_R	Raman response function in the frequency domain	[m^2/w]	83
I	The moment of inertia	[kg m^2]	36
I	Intensity - I_0 (on-axis)	[W/m^2]	136
K	Surface current	[A/m]	7
K	Prefactor for nonlinear optical phenomena		72
L	Length - L_b (beat)	[m]	19
L	Angular momentum	[kg m^2/s]	36
N	Integer number		23
N_R	Raman contribution to intensity dep. refraction index	[m^{2}w]	83
P	Power - P_{wr} (re-radiated)	[W]	25
P_{dip}	The radiated power		165
R	Response tensor in the time domain		41
R	Phase front radius of curvature	[m]	154
R	Radial distribution		120
S_B	Recapture power fraction		166
T	Material response in the time domain	[s$^-$1]	41
V	Volume	[m^3]	46
a	Square root photon flux	[$m^{-1}s^{-1/2}$]	136
a_0	The Bohr radius	[m]	32
b	Confocal parameter	[m]	152
c	Speed of light in vacuum	[m/s]	8
d	Second-order nonlinear susceptibility	[m/V]	106
f_R	Raman fraction to electric Kerr response		60
g	Parametric gain parameter	[$\sqrt{s}$]	136

Symbol	Description	Unit	Page
g_B	Brillouin gain		204
g_R	Raman gain	[m/w]	82
g_{sfg}	Gain parameter	$[m^{-1}]$	147
g_{dfg}	Gain parameter	$[m^{-1}]$	150
h	Boyd-Kleinman factor		152
h_R	Raman response function in the time domain	$[\mathrm{s}^{-1}]$	60
k	Wavenumber	$[\mathrm{m}^{-1}]$	8
l	1 if generated frequency is $\omega_\sigma = 0$ else equal to zero		72
m	Mass - m_e (electron)	[kg]	26
m	Number of frequencies equal to zero		72
n	The refractive index of a material		7
n	Order of nonlinearity		72
p	Number of permutation of distinct frequencies		72
q	Point charge	[C]	22
r	Electro optic tensor		156
s	Quadratic electro optic tensor		156
t	Time	[s]	8
v	Velocity - v_p (phase), v_g (group)	[m/s]	35
w	Energy - w_e (electrical), w_m (magnetic)	[W]	12
w	Beam size - w_0 (waist)	[m]	152
$\lvert\wp\rvert^2$	The dipole strength		165
$\boldsymbol{\sigma}$	Circular polarized light - $\boldsymbol{\sigma}^+$ (left), $\boldsymbol{\sigma}^-$ (right)		21
Γ	Damping constant - Γ_e (electron)		26
Ω	Resonance frequency - Ω_e (electron)	$[\mathrm{s}^{-1}]$	26
α	Polarizability	$[\mathrm{cm}^2/\mathrm{v}]$	20
β	Propagation constant	$[\mathrm{m}^{-1}]$	10
β	Third-order polarizability	$[\mathrm{cm}^4\mathrm{v}^3]$	32
γ	Second-order polarizability	$[\mathrm{cm}^3\mathrm{v}^2]$	32
γ	Conversion efficiency		146
δ	Phase difference		15
ε	Permittivity - ε_0 (vacuum), ε_r (relative)	[F/m]	4
ζ	Gouy phase of Gaussian beam		152
η	Impedance of dielectric medium - η_0 (vacuum)	$[\Omega]$	136
λ	Wavelength	[m]	9
μ	Permeability - μ_0 (vacuum), μ_r (relative)	[H/m]	4
μ	Position of beam waist		152
ν	Optical frequency	$[\mathrm{s}^{-1}]$	149
ξ	Focusing parameter		152
ρ	Walk-off angle		152

Symbol	Description	Unit	Page
ρ	Charge density - ρ_b (bound) ρ_f (free)	[C/m^3]	4
σ	Charge density - σ_s (surface)	[C/m^2]	7
σ	Phase matching parameter	[m^{-1}]	152
τ	Time	[s]	35
ϕ	Photon flux	[$m^{-2}s^{-1}$]	136
χ	Susceptibility tensor - $\chi^{(n)}$ (nth-order)	[m/V]$^{(n-1)}$	8
$\chi^{(2)}_{\text{eff}}$	The effective second-order Susceptibility	[m/v]	136
ω	Angular frequency	[s^{-1}]	8

Authors

Karsten Rottwitt has been working with fiber optics, more specifically fiber optical amplifiers and application of nonlinear effects in optical fibers for more than 20 years. Karsten finished his PhD in 1993 from the Technical University of Denmark (DTU). His PhD studies focused on propagation of solitons through distributed erbium doped fiber amplifiers. From 1993 to 1995 Karsten continued as a Post doc at DTU. A total of half a year was spent at the femtosecond optics group at Imperial College, London. The Post doc was focused on stability of solitons while being amplified. From 1995 to 2000 Karsten continued his career within Bell Labs, New Jersey, USA, first within the submarine systems department of AT&T, and later within the fiber research department of Lucent Technologies. Throughout the five years, Karsten's research was directed toward Raman scattering in optical fibers. In 2000 Karsten moved back to Denmark in a research project at the university of Copenhagen related to near field optics. In 2002, Dr. Rottwitt moved to DTU Fotonik, where he is now a full professor within the field of nonlinear fiber optics. Since 2002 the research of Karsten Rottwitt has concentrated on Raman scattering, four-wave mixing and nonlinear interactions among higher order modes in optical fibers. The research has applications within bio-photonics, sensing and optical signal processing for example within optical communication.

Peter Tidemand-Lichtenberg has been working in the development of novel light sources and detection systems targeted for specific applications for 20 years. Peter completed his PhD in 1996 from the Technical University of Denmark. The PhD focused on compact coherent light sources based on solid-state laser and nonlinear frequency conversion technologies. After completion of the PhD, Peter spent 6 years in a small start-up company developing light source for various industrial applications. In 2002 Peter Tidemand-Lichtenberg returned to DTU to continue the development of coherent light sources in the UV and visible spectral region. The light sources were designed and tailored for specific applications, e.g. in terms of wavelength, tuneability, pulse duration, and pulse energy. Applications were mainly within bio-medicine; e.g. as excitation sources for fluorescence spectroscopy diagnostics and dermatology. The past 5 years Peter has mainly focused on extending the spectral coverage toward the mid-IR region, allowing for development of efficient light sources as well as low noise detection systems in the 2–12 um range based on frequency mixing.

Chapter 1

Introduction

Contents

A simple answer to the question: What is nonlinear optics ? is that nonlinear optics relates to propagation of an electromagnetic field, where the propagation depends on the electromagnetic field itself, for example the intensity of the field.

Examples include intensity dependent refractive index, intensity dependent absorption, mutual interaction among different frequencies in the propagating electromagnetic field, electro- and magneto-optical effects and scattering phenomena.

Some might argue that an early demonstration of nonlinear optics goes back to 1928 when several research groups reported scattering of light through mutual interaction of two frequencies with phonons, including Sir C.V. Raman and his coworkers who demonstrated the effect, and after whom the effect is named the Raman effect. Others may argue that some of the early research in nonlinear optics was done by P.A. Franken and coworkers who were the first to report observation of optical second harmonic generation [5]. However, it is without question that both results have contributed significantly, and it is very difficult, if possible at all, to define the most important result. In addition, there are many other results, not directly within nonlinear optics, that have made it possible to make significant advances within the field. These include the invention of the laser [6] and the development of low loss optical fibers in the late sixties. The latter led to the demonstration of a loss corresponding to 0.2 dB/km at 1.55 μm in 1979 [7]. This is perhaps one of the most important results, since it spurred an era within optical communication where the optical glass based fiber is the most important building block and nonlinear effects used for example to perform signal processing or optical amplification. [8].

The nonlinear interaction of light and matter leads to a broad range of exciting phenomena and has applications for example within optical communication, high power fiber lasers and optical sensors. These applications are all very important to society, and consequently, the field of nonlinear optics continues to expand as a field of research in electromagnetic wave propagation. A complete list of phenomena and applications would be enormous and serve no purpose. Instead, it is just noted here that in this text the main applications include

Optical signal processing: parametric amplification, modulators.

Transmission of optical signals: optical solitons, cross-phase modulation, four-wave mixing, phase conjugation, Raman scattering.

Sensing: temperature sensors, spectroscopy and imaging.

Lasers: pulse compression, and generation of super continuum.

In some of these applications it is necessary to understand nonlinear optics to avoid the nonlinear effects, whereas in other applications it is necessary to understand nonlinear optics to take maximum advantage of the nonlinear behavior.

This text is intended as an introduction to nonlinear optics, with focus on basic principles and applications. There are many excellent textbooks on nonlinear optics and fiber optics. Examples of the first include Butcher & Cotter [2] and R. W. Boyd [1], whereas the latter include Agrawal [3] and Snyder & Love [9]. It is the intention of this text to bridge the gap between these books. The description includes examples of propagation through crystals, optical waveguides and optical fibers, and the aim of the text is to bring the reader sufficient knowledge to understand and evaluate nonlinear optical phenomena. The text covers:

- a theoretical description of nonlinear interaction between light and matter.
- a description of optical material response functions in the time and frequency domain, the latter in the form of susceptibility tensors.
- a description of nonlinear wave propagation in bulk and waveguiding structures.
- specific examples of applied nonlinear wave propagation.

It is furthermore emphasized that the text focuses on effects related to non-resonant nonlinearities in non-absorbing media. Thus, we are able to restrict our analysis to classical electromagnetic descriptions. However, before we move on to a description of nonlinear optics in detail, we review basic linear optics in Section 1.2 and the classical harmonic oscillator model of the material polarization induced by an electric field in Sections 1.3 and 1.1.3.

1.1 Review of linear optics

Light is electromagnetic radiation, that is: a self-propagating wave in space with electric and magnetic field components. As opposed to sound waves, light can propagate in vacuum. Four basic properties of light include: amplitude (intensity), phase, frequency (wavelength), and state of polarization. Light exhibits properties of both waves and particles. The latter is photons, that carry energy and momentum, which may be imparted when they interact with matter. Typically, light is classified according to its frequency, for example in order of increasing frequency: Terahertz (THz), infrared radiation (IR), visible, and ultraviolet light (UV), see Figure 1.1.

As mentioned previously, nonlinear optics deals with phenomena where conventional linear optical properties such as the refractive index change as a function of the intensity of the light propagating through the media. Thus, nonlinear optics deals with propagation of electrical fields through different materials, i.e. interaction of light and matter. Consequently, we need to be able to describe how an electromagnetic field changes during propagation through matter. In this text we mainly do this by using classical electromagnetic theory, and propagation is described through Maxwell's equations.

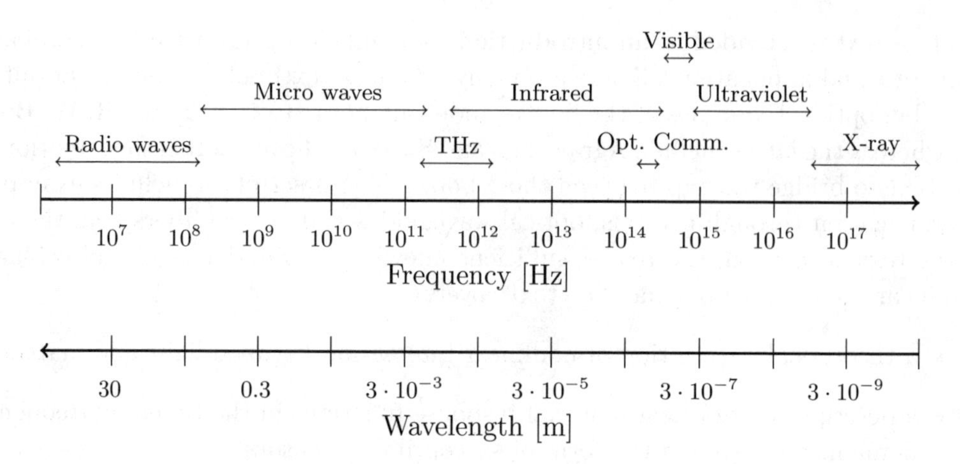

Figure 1.1: Frequencies and corresponding vacuum wavelengths of named radiation.

1.1.1 Maxwell's equations

The representation of the field of nonlinear optics given here is based on classical physics and electromagnetism. This means that we wish to describe propagation of the electric field vector $\mathbf{E}$, [V/m], and the magnetic induction vector $\mathbf{B}$, [Wb/m^2], or most often rather the (auxiliary) magnetic field vector $\mathbf{H}$, [A/m], as this propagates through a material with which it interacts in a nonlinear way. Consequently, we use Maxwell's equations to describe propagation of electromagnetic waves through a dielectric medium. We use Maxwell's equations in differential form, which in general exist in two formulations: a macroscopic and a microscopic form.

Microscopic formulation is in terms of total charge density ρ, [C/m^3], including free and bound charges, and total current density vector $\mathbf{J}$, [A/m^2], which similarly includes free and bound currents, including the charges and currents at the atomic level.

$$\nabla \cdot \mathbf{E} = \rho/\varepsilon_0, \tag{1.1a}$$

$$\nabla \times \mathbf{E} = -\frac{\partial \mathbf{B}}{\partial t}, \tag{1.1b}$$

$$\nabla \times \mathbf{B} = \mu_0 \varepsilon_0 \frac{\partial \mathbf{E}}{\partial t} + \mu_0 \mathbf{J}, \tag{1.1c}$$

$$\nabla \cdot \mathbf{B} = 0. \tag{1.1d}$$

Macroscopic formulation—also known as Maxwell's equations in matter—is in terms of only free charge density ρ_f, [C/m^3] and free current density $\mathbf{J}_f$, [A/m^2], i.e. bound charges and currents have been factored out. To do this, additional fields need to be defined: the displacement field vector $\mathbf{D}$, [C/m^2], and the magnetic field vector $\mathbf{H}$.

$$\nabla \cdot \mathbf{D} = \rho_f, \tag{1.2a}$$

$$\nabla \times \mathbf{E} = -\frac{\partial \mathbf{B}}{\partial t}, \tag{1.2b}$$

$$\nabla \times \mathbf{H} = \frac{\partial \mathbf{D}}{\partial t} + \mathbf{J}_f, \tag{1.2c}$$

$$\nabla \cdot \mathbf{B} = 0. \tag{1.2d}$$

The microscopic and the macroscopic formulations are equivalent using $\mathbf{J} = \mathbf{J}_b + \mathbf{J}_f$, where $\mathbf{J}_b$ is the bound current density and $\rho = \rho_b + \rho_f$, where ρ_b is the bound charge density, in addition to the relations

$$\mathbf{J}_b = \nabla \times \mathbf{M} + \frac{\partial \mathbf{P}}{\partial t}, \tag{1.3a}$$

$$\rho_b = -\nabla \cdot \mathbf{P}, \tag{1.3b}$$

$$\mathbf{D} = \varepsilon_0 \mathbf{E} + \mathbf{P}, \tag{1.3c}$$

$$\mathbf{B} = \mu_0(\mathbf{H} + \mathbf{M}), \tag{1.3d}$$

where $\mathbf{P}$, [C/m^2] is the induced electric polarization vector, and $\mathbf{M}$, [A/m] is the magnetization vector. $\varepsilon_0 = 8.854 \cdot 10^{-12}$ [F/m] is the vacuum permittivity. When describing electromagnetic fields in materials, the permittivity, ε, of the material, is often written as a product of the so-called relative permittivity ε_r and the vacuum permittivity, $\varepsilon = \varepsilon_r \varepsilon_0$. $\mu_0 = 4\pi \cdot 10^{-7}$ [H/m] is the vacuum permeability. Here we do not worry about magnetic materials and consequently do not discuss any materials having a relative permeability.

When an electric field is applied to a dielectric material its molecules respond by forming microscopic electric dipoles as their atomic nuclei move a tiny distance in the direction of the field, while their electrons move a tiny distance in the opposite direction. This produces a macroscopic bound charge density in the material even though all of the charges involved are bound to individual molecules and the total charge remain zero. For example, if every molecule responds the same, similar to that shown in Figure 1.2, these tiny movements of charge combine to produce a layer of positive bound charge on one side of the material and a layer of negative charge on the other side. The bound charge is most conveniently described in terms

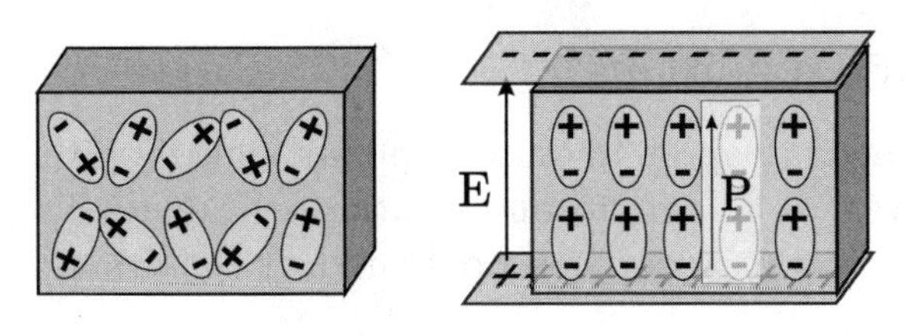

Figure 1.2: Induced polarization in material.

of a polarization, **P**, in the material. If **P** is uniform, a macroscopic separation of charge is produced only at the surfaces where **P** enters and leaves the material.

Somewhat similarly, in all materials the constituent atoms exhibit magnetic moments that are intrinsically linked to the angular momentum of the atoms' components, most notably their electrons. The connection to angular momentum suggests the picture of an assembly of microscopic current loops. Outside the material, an assembly of such microscopic current loops is no different from a macroscopic current circulating around the material's surface, despite the fact that no individual current flows a large distance. These bound currents may be described using the magnetization **M**.

The very complicated and granular bound charges and bound currents, may be represented on the macroscopic scale in terms of **P** and **M** which average these charges and currents on a sufficiently large scale so as not to see the granularity of individual atoms, but also sufficiently small that they vary with location in the material. As such, Maxwell's macroscopic equations ignore many details on a fine scale that are unimportant to understanding matters on a larger scale by calculating fields that are averaged over some suitably sized volume.

In this book we use the macroscopic formulation, that is Eqn. (1.2). In addition, we of will often be concerned with dielectrics, where there are no free charges and no free currents, hence $\mathbf{J}_f = 0$ and $\rho_f = 0$. In addition, most often the materials considered have no magnetization, i.e. $\mathbf{M} = 0$. Consequently, the set of Maxwell equations that is used in this book read:

$$\nabla \times \mathbf{E} = -\frac{\partial \mathbf{B}}{\partial t}, \tag{1.4a}$$

$$\nabla \times \mathbf{H} = \frac{\partial \mathbf{D}}{\partial t}, \tag{1.4b}$$

$$\nabla \cdot \mathbf{B} = 0, \tag{1.4c}$$

$$\nabla \cdot \mathbf{D} = 0. \tag{1.4d}$$

The description of nonlinear optics phenomena that follow in this text will to a large extent be based on the induced electric polarization **P**, which enters into Maxwells equations through the displacement vector field **D**, using Eqn. (1.3c).

1.1.2 Boundary conditions

Maxwell's equations only apply in discrete points where a given material does not have discontinuities. To be able to find solutions to Maxwell's equations across a junction of two different materials, a set of boundary conditions needs to be applied.

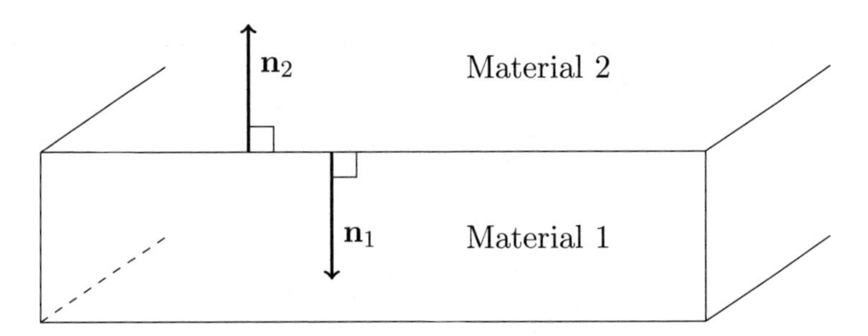

Figure 1.3: Boundary between two materials, material 1 with unit normal vector $\mathbf{n}_1$ and material 2 with unit normal vector $\mathbf{n}_2$.

Assuming two materials with unit normal vectors $\mathbf{n}_1$ and $\mathbf{n}_2$, as defined in Figure 1.3, the displacement field in the two materials are related through the surface-charge density σ_s as

$$\mathbf{n}_2 \cdot \mathbf{D}_2 + \mathbf{n}_1 \cdot \mathbf{D}_1 = \sigma_s. \tag{1.5}$$

That is, the difference between the normal components of the displacement current equals the surface-charge density. Note that since in some materials, $\mathbf{D} = n^2 \varepsilon_0 \mathbf{E}$, where n is the refractive index of the material, the boundary condition is often rewritten as a boundary condition for the electric field.

Sometimes only one normal vector is used, for example pointing from material 1 to 2, see [10]. In the notation used here this correspond to using only $\mathbf{n}_2$, i.e. replacing $\mathbf{n}_1$ by $-\mathbf{n}_2$

In analog, the normal components of the magnetic induction in the two materials go continuously across the boundary between the two materials

$$\mathbf{n}_2 \cdot \mathbf{B}_2 + \mathbf{n}_1 \cdot \mathbf{B}_1 = 0. \tag{1.6}$$

In addition to these two equations for the normal components, two other equations apply for the tangential components. The vectorial tangential components of the electric field are continuous across the junction of the two materials

$$\mathbf{n}_2 \times \mathbf{E}_2 + \mathbf{n}_1 \times \mathbf{E}_1 = 0, \tag{1.7}$$

and finally, the vectorial difference between the tangential components of the magnetic field equals the surface-current $\mathbf{K}$

$$\mathbf{n}_2 \times \mathbf{H}_2 + \mathbf{n}_1 \times \mathbf{H}_1 = \mathbf{K}. \tag{1.8}$$

1.1.3 Wave equation in lossless linear materials

Within nonlinear optics, propagation of light through a nonlinear material is an obvious core topic. Consequently, the wave equation is going to be an essential topic within the content of this text. In this section we review the linear wave equation. We start from Maxwell equations.

$$\nabla \times \mathbf{H} = \frac{\partial \mathbf{D}}{\partial t} \tag{1.9}$$

$$\nabla \times \mathbf{E} = -\mu_0 \frac{\partial \mathbf{H}}{\partial t}. \tag{1.10}$$

Taking the curl of Eqn. (1.10) and applying Eqn. (1.9) we arrive at

$$\nabla \times (\nabla \times \mathbf{E}) = -\mu_0 \frac{\partial^2 \mathbf{D}}{\partial t^2}. \tag{1.11}$$

Since $\mathbf{D} = \varepsilon_0 \mathbf{E} + \mathbf{P}$ where $\mathbf{P}$ is the induced polarization, which in linear optics is linearly proportional to the electric field with a proportionality factor given by the vacuum permittivity times a material parameter, the susceptibility, $\chi^{(1)}$, i.e. $\mathbf{P} = \varepsilon_0 \chi^{(1)} \mathbf{E}$ we get

$$\begin{aligned} -\nabla^2 \mathbf{E} &= -\mu_0 \varepsilon_0 \left(1 + \chi^{(1)}\right) \frac{\partial^2 \mathbf{E}}{\partial t^2} \\ &= -\frac{\left(1 + \chi^{(1)}\right)}{c^2} \frac{\partial^2 \mathbf{E}}{\partial t^2}, \end{aligned} \tag{1.12}$$

where it has been assumed that $\nabla \cdot \mathbf{E} = 0$. It is furthermore noted that by inserting the induced polarization into the expression for $\mathbf{D}$ it is common to define the relative permittivity ϵ_r as $\epsilon_r = \left(1 + \chi^{(1)}\right)$. Then Eqn. (1.12) has the solution

$$\mathbf{E} = \mathbf{E}^0 \exp\left[i(kz \pm \omega t)\right], \tag{1.13}$$

where $k^2 = \omega^2 \left(1 + \chi^{(1)}\right)/c^2$. This often referred to as the dispersion relation. Here and in the following, superscript index 0 is used to denote the amplitude. Finally, Eqn. (1.13) is the solution to the wave equation in complex notation. The physical field is the real part of Eqn. (1.13). In relation to nonlinear optics it is most convenient to use the complex notation since we will often have to find powers of the electrical field.

In **lossless** materials, $\chi^{(1)}$ is real and the wavenumber k is real and equals $k = \omega\sqrt{\varepsilon_r}/c$. In lossy materials or in leaky waveguides, the susceptibility, $\chi^{(1)}$, is complex and then the propagation constant k is complex [11].

Example 1.1. *Complex permittivity*
Typically a material is lossy and a complex susceptibility is required to describe propagation. If $\chi^{(1)}$ is complex then k is complex. Writing the real and imaginary parts of the complex susceptibility as $\chi_r^{(1)}$ and $\chi_i^{(1)}$, respectively, and the real and imaginary parts of the propagation constant as k_r and k_i, respectively, then

$$(k_r + ik_i)^2 = \omega^2 \left(1 + \chi_r^{(1)} + i\chi_i^{(1)}\right)/c^2, \tag{1.14}$$

which we may split into equations for the real and imaginary parts, i.e.

$$(k_r^2 - k_i^2) = \omega^2 \left(1 + \chi_r^{(1)}\right) / c^2 \tag{1.15}$$

and

$$2k_r k_i = \omega^2 \chi_i^{(1)} / c^2. \tag{1.16}$$

Solving this set of equations for the real and imaginary part of the propagation constant, we find the approximations

$$k_r \approx \frac{\omega}{c}\sqrt{1 + \chi_r^{(1)}} \tag{1.17}$$

and

$$k_i \approx \frac{1}{2}\frac{\omega}{c}\frac{\chi_i^{(1)}}{\sqrt{1 + \chi_r^{(1)}}}, \tag{1.18}$$

which are valid when $\left|\chi_i^{(1)}\right| \ll \left|1 + \chi_r^{(1)}\right|$.

—— ■ ——

The induced polarization of a material depends on the applied electric field. The physical quantity that relates induced polarization to the applied electric field is the permittivity ε. In a homogenous lossless linear material, i.e. a material where the permittivity is constant in space and time and independent of the electric field, the permittivity is $\varepsilon = \varepsilon_r \varepsilon_0$. If furthermore the material is non-magnetic, the permeability equals the vacuum permeability, i.e. $\mu = \mu_0$, and the squared refractive index equals the relative permittivity, $n^2 = \varepsilon_r$.

In general, the refractive index is a function that depends on various parameters, including the frequency of the applied electric field, often referred to as dispersion. If in addition there are no free charges and no free currents, then a harmonically time varying electric field $\mathbf{E}(\mathbf{r}, t) = 1/2\left(\mathbf{E}^0(\mathbf{r})e^{i(\mathbf{k}\cdot\mathbf{r}-\omega t)} + c.c.\right)$ satisfies the wave equation Eqn. (1.12) if the amplitude $\mathbf{E}^0(\mathbf{r})$ satisfies

$$\left(\nabla^2 + \frac{\omega^2}{c^2}\varepsilon_r\right)\mathbf{E}^0(\mathbf{r}) = 0, \tag{1.19}$$

where ω is the angular frequency of the electric field, $\mathbf{E}^0(\mathbf{r})$ is the amplitude, $\mathbf{r}$ the vector to the observation point, and $\mathbf{k}$ is the wavevector pointing in the direction of propagation. The magnitude of the wavevector equals $|\mathbf{k}| = 2\pi n/\lambda = \omega n/c$, where λ is the vacuum wavelength.

The amplitude vector $\mathbf{E}^0$ determines the state of polarization of the electrical field. In general the vector has three components, for example in cartesian coordinates an x-, a y- and a z-component, each of which are complex numbers. We return to a description of the state of polarization of the electric field in Section 1.1.6.

From Maxwell's equations $\mathbf{k}$ may be complex. In a lossless waveguide the propagation constant is real valued, whereas a leaky mode also has an imaginary contribution [11]. Furthermore, in a waveguide, it is not correct to express the size of the wavevector as a product of the vacuum wavenumber and the refractive index. Instead, a propagation constant β is introduced. Nevertheless, β is often decomposed into the vacuum wavenumber times an effective refractive index of the waveguide. We return to further details of this in Chapter 5.

The refractive index is often described as a real value. However, in a lossy material, the attenuation of the electric field is described through an imaginary part of the refractive index [11]. Furthermore, the refractive index is not simply a scalar but a tensor, that is, in linear optics, the refractive index is described by a matrix. This matrix (second rank tensor) describes that the electric field component along one axis may be affected by the electric field component along another axis. We return to applications of tensors in Chapter 2. The consequence is birefringence and is discussed in many textbooks, see e.g. [12].

Assuming that the electromagnetic field does not depend on x, y but only varies with z, a solution of particular interest is the superposition of two exponential functions, and may be written as

$$\mathbf{E} = \mathbf{E}^{+} e^{i\beta z - i\omega t} + \mathbf{E}^{-} e^{-i\beta z - i\omega t}, \tag{1.20}$$

where $\beta = k_0 n$, and where $\mathbf{E}^{+}$ and $\mathbf{E}^{-}$ are complex valued constant vectors that describe the amplitude of the right and left propagating electric fields, k_0 is the vacuum wavenumber $k_0 = \omega/c = 2\pi/\lambda$ where λ is the vacuum wavelength. This solution is a representation of two plane waves, in complex notation, that travel in the z-direction. The first term represents a wave traveling in the positive z-direction whereas the second term represents a wave traveling in the negative z-direction.

While Eqn. (1.20) is a solution to the wave equation that represents two plane waves traveling in opposite directions in complex notation, it is noted that the physical field, that is, the electric field that can be measured, is expressed for example as the real part of Eqn. (1.20) or as

$$\mathbf{E} = \frac{1}{2}\left(\mathbf{E}^{+} e^{i\beta z - i\omega t} + \mathbf{E}^{-} e^{-i\beta z - i\omega t} + c.c.\right), \tag{1.21}$$

where $c.c.$ denotes complex conjugate. This is important to remember when describing nonlinear optics since we will be using powers of the electrical field. In some cases a wave is propagating in one direction only, as for example a short pulse propagating in an optical fiber, or the wave in one direction is much stronger than the wave propagating in the opposite direction, as for example an optical fiber amplifier, whereas in other cases, as for example a laser, the electrical fields propagate in both directions.

Before we move on to a more detailed description of nonlinear optics, a few more fundamental aspects from linear optics should be reminded. Let us consider a wave that only propagates in the forward direction, i.e. $\mathbf{E}^- = 0$ in Eqn. (1.21).

One of the parameters that characterizes such a wave is the **phase velocity**, v_p, which is the rate at which the phase of the wave evolves. In the plane wave, for example from Eqn. (1.20) the phase of the forward propagating wave $\phi(z,t) = (\beta z + \omega t)$, is a simple function of time and position. The position at which the phase equals zero is then: $z_{\phi=0} = \omega t/\beta$. From this, the phase velocity equals

$$v_p = \frac{\partial}{\partial t} z_{\phi=0} = \frac{\omega}{\beta} = c/n. \tag{1.22}$$

Of equal importance is the **group velocity**, v_g, which is the velocity at which the variations in the shape of the wave's amplitude (known as the modulation or envelope of the wave) propagate through space

$$v_g = \frac{\partial \omega}{\partial \beta}, \tag{1.23}$$

where $\beta = \omega n/c$. From this the group velocity may also be expressed through the refractive index as $v_g = c/(\partial n/\partial \omega) \approx c/(n_0 + \omega(\partial n/\partial \omega))$. In general, if one pictures a wavepacket consisting of a carrier wave and an envelope, the phase velocity is the speed of the carrier wave whereas the group velocity is the speed of the envelope.

In addition to a propagating wave, a **standing wave** may also exist. This is achieved if the two amplitudes $\mathbf{E}^+$ and $\mathbf{E}^-$ are identical. Assuming $\mathbf{E}^+ = \mathbf{E}^- = \mathbf{A}^0$, the electric field equals

$$\mathbf{E} = 2\mathbf{A}^0 \cos(\beta z) \frac{1}{2} \left\{ e^{-i\omega t} + c.c \right\}. \tag{1.24}$$

1.1.4 Power

In nonlinear optics, we wish to describe changes in optical properties of various materials as a function of the intensity of the optical field that propagates through the material. The intensity is the power per unit area carried by the propagating electromagnetic wave.

The flow of intensity of the electromagnetic wave is found using Poyntings theorem. In general this states that the rate at which work is done on all charges in a volume V, i.e $\int_V \mathbf{E} \cdot \mathbf{J} dv$, equals the decrease in stored electromagnetic energy by the electric and magnetic field and the loss of energy that has propagated out through the surface A bounding the volume V. From Jackson [10], pg. 259,

$$\int_V (\mathbf{E} \cdot \mathbf{J}) dv = -\int_V \left(\mathbf{E} \cdot \frac{d\mathbf{D}}{dt} + \mathbf{H} \cdot \frac{d\mathbf{B}}{dt} \right) dv - \int_V \nabla \cdot (\mathbf{E} \times \mathbf{H}) dv. \tag{1.25}$$

Stored energy of an electric field may be pictured as the energy required to place a single charge in the electric field from infinety in analogy to the energy required to charge a capacitor by moving electrons from one plate to another. Rather than single charges one may put the electric field and the energy stored in a volume V of the electric field is

$$w_e = \int_V \frac{1}{2}\mathbf{E}\cdot\mathbf{D}dv.$$

In analogy, the stored magnetic energy may be pictured as the energy required to establish a current

$$w_m = \int_V \frac{1}{2}\mathbf{H}\cdot\mathbf{B}dv.$$

Using Gauss' divergence theorem, the last integral in Eqn. (1.25) of the divergence of the vector field inside the volume V may be replaced by an integral of the outward flux of the vector field $\mathbf{E}\times\mathbf{H}$ through the closed surface A. If, in addition, a situation is considered where there are no currents i.e. $\mathbf{J}_f = 0$, as for example in dielectrics such as glass, we then get

$$\begin{aligned}\oint_A (\mathbf{E}\times\mathbf{H})\cdot d\mathbf{a} &= -\int_V \left(\mathbf{E}\cdot\frac{d\mathbf{D}}{dt} + \mathbf{H}\cdot\frac{d\mathbf{B}}{dt}\right)dv \\ &= -\frac{d}{dt}\int_V \left(\frac{1}{2}\varepsilon(\mathbf{E}\cdot\mathbf{E}) + \frac{1}{2}\mu(\mathbf{H}\cdot\mathbf{H})\right)dv,\end{aligned} \tag{1.26}$$

by using the constitutive relations. In Eqn. (1.26), the final integral is recognized as the stored electromagnetic energy. That is, the flow of energy out through a surface equals the rate of change of stored energy in the electromagnetic field within that volume.

The vector field in the integral on the left-hand side represents the energy per unit time transported by the electromagnetic field and is called the Poynting vector

$$\mathbf{S} = \mathbf{E}\times\mathbf{H}. \tag{1.27}$$

1.1.5 Harmonically time varying fields

Harmonically time varying fields are essential for the treatment of linear as well as nonlinear optics, and for most practical cases only the time averaged power flow is of relevance. Thus, we consider a harmonically time varying field which in complex notation is written as: $\mathbf{E}(t) = \frac{1}{2}(\mathbf{E}^0 e^{-i\omega t} + c.c)$. Using this, Eqn. (1.26) may be rewritten as

$$\oint_A \left(\mathbf{E}^0\times(\mathbf{H}^0)^*\right)\cdot d\mathbf{a} = i\omega\int_V \left(\varepsilon\mathbf{E}^0\cdot(\mathbf{E}^0)^* - \mu\mathbf{H}^0\cdot(\mathbf{H}^0)^*\right)dv. \tag{1.28}$$

For harmonically time varying fields, the time average, in the following indicated by a bar, of the complex Poynting vector at frequency ω is expressed through

$$\bar{\mathbf{S}} = \frac{1}{2}\mathrm{Re}\left[\mathbf{E}^0\times(\mathbf{H}^0)^*\right]. \tag{1.29}$$

By noting that the average stored electric and magnetic energy density equals $\bar{w}_e = \frac{1}{4}\varepsilon\mathbf{E}^0 \cdot (\mathbf{E}^0)^*$ and $\bar{w}_m = \frac{1}{4}\mu\mathbf{H}^0 \cdot (\mathbf{H}^0)^*$, respectively, then Eqn. (1.28) may be rewritten as

$$\frac{1}{2}\oint_A \left(\mathbf{E}^0 \times (\mathbf{H}^0)^*\right) \cdot d\mathbf{a} = i2\omega \int_V (\bar{w}_e - \bar{w}_m) dv. \tag{1.30}$$

This is a complex equation. Separating this into its real and imaginary parts, the real part of Eqn. (1.30) gives

$$\oint_A \bar{\mathbf{S}} \cdot d\mathbf{a} = 0. \tag{1.31}$$

The volume integral on the right-hand side of Eqn. (1.30) is proportional to the difference between the average stored electric energy in the volume and the average stored magnetic energy. Taking into account a factor of 1/2 in the energy expressions and another 1/2 for the averaging of squares of the sinusoids we obtain from the imaginary part of Eqn. (1.30)

$$\mathrm{Im}\left[\oint_A \left(\mathbf{E}^0 \times (\mathbf{H}^0)^*\right) \cdot d\mathbf{a}\right] = 4\omega \int_V (\bar{w}_e - \bar{w}_m) dv. \tag{1.32}$$

Thus, the imaginary part of Poynting flow through the surface, Eqn. (1.32) may be thought of as reactive power flowing back and forth to supply the instantaneous changes in the net stored energy in the volume. This is the so-called reactive power, which is power that is not propagating.

Example 1.2. *Traveling plane wave*
Consider a plane wave which is linearly polarized along the x-axis, $\mathbf{E}^0 = E_x^0\mathbf{e}_x$. This is defined through

$$E_x(z,t) = \frac{1}{2}\left(E_x^0 e^{i\beta z} e^{-i\omega t} + c.c.\right). \tag{1.33}$$

The corresponding auxiliary magnetic field vector then consists only of a y-component and is given by

$$H_y(z,t) = \frac{\beta}{\omega\mu}\frac{1}{2}\left(E_x^0 e^{i\beta z} e^{-i\omega t} + c.c.\right). \tag{1.34}$$

From this, the time average Poynting vector only has a z-component, which equals

$$\bar{S}_z = \frac{1}{2}\frac{\beta}{\omega\mu}|E_x^0|^2, \tag{1.35}$$

which is purely real. From Eqn. (1.31)

$$\oint_A \frac{1}{2}\frac{\beta}{\omega\mu}|E_x^0|^2 \mathbf{e}_z \cdot d\mathbf{a} = 0 \tag{1.36}$$

stating that the average energy entering the volume bounded by A also leaves the volume.

To evaluate the imaginary part, i.e. Eqn. (1.32) we need to evaluate the average stored electric and magnetic energy. From the text below Eqn. (1.29), these equal

$$\bar{w}_e = \frac{1}{4}\varepsilon|E_x^0|^2, \quad \bar{w}_m = \frac{1}{4}\mu(\frac{\beta}{\omega\mu})^2|E_x^0|^2. \tag{1.37}$$

Since $\beta^2 = \omega^2\mu\varepsilon$, the average stored electric energy equals the average stored magnetic energy, i.e

$$(\bar{w}_e - \bar{w}_m) = 0 \tag{1.38}$$

as expected.

— ■ —

Example 1.3. *Standing wave*
A standing plane wave is defined as

$$E_x(z,t) = E^0\frac{1}{2}\left(e^{i\beta z} + e^{-i\beta z}\right)e^{-i\omega t} + c.c., \tag{1.39}$$

where the first term $\propto e^{i(\beta z - \omega t)}$ represents a forward propagating wave whereas the second term, $\propto e^{-i(\beta z + \omega t)}$ represents a backward propagating wave. Correspondingly, the auxiliary magnetic field equals

$$H_y(z,t) = H_y^0\left(e^{i\beta z} - e^{-i\beta z}\right)e^{-i\omega t} + c.c. = E_x^0\frac{\beta}{\omega\mu}\frac{1}{2}\left(e^{i\beta z} - e^{-i\beta z}\right)e^{-i\omega t} + c.c. \tag{1.40}$$

The time averaged Poynting vector, equals

$$\bar{\mathbf{S}}_z = \frac{1}{2}\mathbf{E}^0 \times (\mathbf{H}^0)^* = -i\sin(2\beta z)|E^0|^2\frac{\beta}{\omega\mu}\mathbf{e}_z. \tag{1.41}$$

Since the real part of Poynting's vector equals zero, no power is flowing in and out of a given volume encapsulating the standing wave. However, since the imaginary part of Poynting's vector differs from zero, electromagnetic energy exist in the volume. To find the energy stored in the electromagnetic field using $\bar{w}_e = \frac{1}{4}\varepsilon\mathbf{E}^0 \cdot (\mathbf{E}^0)^*$ and $\bar{w}_m = \frac{1}{4}\mu\mathbf{H}^0 \cdot (\mathbf{H}^0)^*$ it is necessary to rewrite Eqn. (1.39) and Eqn. (1.40). Following this, we find

$$\bar{w}_e = \frac{1}{2}\epsilon|E_x^0|^2(1 + \cos(2\beta z)) \tag{1.42}$$

and

$$\bar{w}_m = \frac{1}{2}\mu|H_y^0|^2(1 - \cos(2\beta z)) = \frac{1}{2}\mu\left(\frac{\beta}{\omega\mu}\right)^2|E_x^0|^2(1 - \cos(2\beta z)) \tag{1.43}$$

leading to

$$\begin{aligned}(\bar{w}_e - \bar{w}_m) &= \frac{1}{2}\epsilon|E_x^0|^2(1 + \cos(2\beta z)) - \frac{1}{2}\mu\left(\frac{\beta}{\omega\mu}\right)^2|E_x^0|^2(1 - \cos(2\beta z)) \\ &= \epsilon|E_x^0|^2\cos(2\beta z).\end{aligned} \tag{1.44}$$

The right hand side of Eqn. (1.32) equals

$$4\omega \int_V (\bar{w}_e - \bar{w}_m)dv = \int_V 4\omega\epsilon |E_x^0|^2 \cos(2\beta z)dv \tag{1.45}$$

and the left hand side of Eqn. (1.32) equals

$$\mathbf{E}^0 \times (\mathbf{H}^0)^* = -2\sin(2\beta z)|E_x^0|^2 \frac{\beta}{\omega\mu}, \tag{1.46}$$

implying that the energy is shifting between the electric and magnetic fields within the volume but is not propagating through the surface enclosing the volume.

— ■ —

1.1.6 Polarization state of the electric field

An electric field is characterized by its amplitude which is a complex vector, i.e. it has both a length, defining the intensity, and a direction, defining the state of polarization. Throughout most of this book we consider an electric wave that propagates in the z-direction, and the state of polarization is then characterized by the x- and y-component of the electrical field, more specifically the length of the two components and their relative phase. If the components are in phase then the wave is linearly polarized whereas the wave is elliptically polarized if the components are out of phase. A special case is when the components are of equal magnitude but $\pi/2$ out of phase, in which case the wave is circularly polarized. Light may be polarized, partially polarized or even non-polarized. Beside spontaneous emission which is unpolarized, we will be concerned only with electric fields with a well-defined state of polarization.

In linear optics the state of polarization is an important characteristic property, which has a huge impact on many applications; examples follow later in this section. In nonlinear optics the state of polarization is even more important since different components may contribute to the induced polarization and hence couple to each other. This coupling is described through the susceptibility tensors, that is one of the main topics of this book. In the following we provide a brief review of how the state of polarization is described, mapped and characterized.

The electric field vector of any arbitrary state of polarization may be written as

$$\mathbf{E}(z,t) = \frac{1}{2}\left(E_x^0 e^{i(\beta z - \omega t)}\mathbf{e}_x + E_y^0 e^{i(\beta z - \omega t)} e^{i\delta}\mathbf{e}_y + c.c.\right), \tag{1.47}$$

where $\mathbf{e}_x$ and $\mathbf{e}_y$ are unit vectors defining the x- and y-coordinates. When the two components E_x^0 and E_y^0 are chosen to be real, δ describes the phase difference between the x- and y-components of the electric field. Consequently, E_x^0, E_x^0 and δ define the state of polarization.

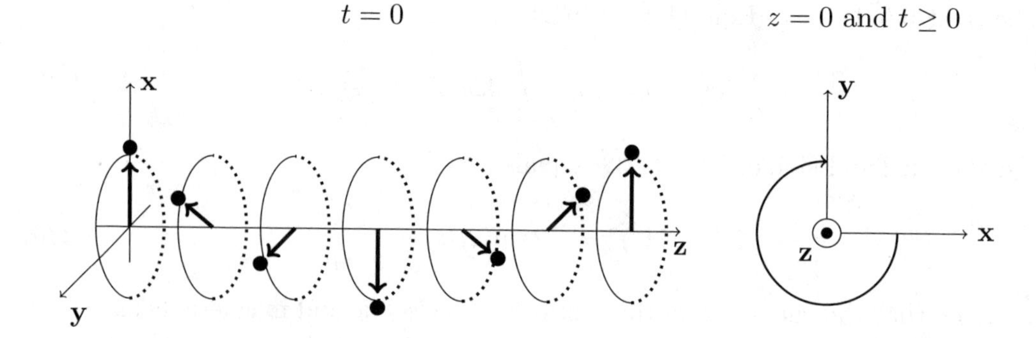

Figure 1.4: The electrical field is propagating along the positive z-axis. The state of polarization is defined by observing the endpoint of the electrical field vector, when looking towards the negative z-direction. Both the **left** and **right** figure illustrate the case when the y component is delayed by ($\delta = -\pi/2$) relative to the x component of the electrical field. In the **left** figure the time is fixed ($t = 0$) whereas in the **right** figure, the position is fixed ($z = 0$). From **left** figure: The handedness of the polarization is determined by pointing one's thumb in the direction that the wave is propagating, and matching the curling of one's fingers to the direction of the temporal rotation of the field at a given point in space - i.e. here right handedness polarization. From **right** figure: The field is considered clockwise (counter clockwise) circularly polarized when, from the point of view of the observer, looking toward the wave propagation, the field rotates in the clockwise (counter-clockwise) direction.

The three cases, linear, circular and elliptical, states of polarization may be characterized through

- $\delta = 0$ when the light is linearly polarized, note this also includes $E_x^0 = 0$ or $E_y^0 = 0$.
- $\delta = \pm\frac{\pi}{2}$ and $E_x^0 = E_y^0$ when the light is circularly polarized.
- Otherwise the light is elliptically polarized.

In a linearly polarized electric field, the amplitude vector is pointing in a fixed direction in the xy-plane during propagation, whereas the endpoint of the electric field vector traces out a circle or an ellipsoid when the electric field is circularly or elliptically polarized, respectively. Figure 1.4 illustrates a right-handed circularly polarized electric field propagating in the positive z-direction. For further details see [10].

In addition to Eqn. (1.47), there are several other methods used to map and characterize the state of polarization of an electric field. A comprehensive method is by using the so-called Stokes parameters which include the intensity of the wave, the degree of polarization, i.e. to which degree the state of polarization is well defined (does the electric field have a state of polarization or not or only partially).

In addition, the Stokes parameters also include information about what state of polarization the electric field is in.

Assuming that the light is fully polarized one may use a vector, i.e. two parameters rather than four Stokes parameters. The vector is referred to as the **Jones vector**, and it is simply: $(E_x^0, E_y^0)^T$, where it is noted that both components are complex numbers. To visualize the state of polarization the vector is mapped on a sphere, the so-called Poincaré sphere.

In the following we mainly deal with fully polarized light. Consequently, the state of polarization is visualized on the Poincaré sphere.

Consider an electric field

$$\mathbf{E}(\mathbf{r}, t) = \frac{1}{2}\left(\mathbf{E}^0(\mathbf{r}) \exp\left(-i\omega t\right) + c.c\right), \tag{1.48}$$

where the amplitude vector $\mathbf{E}^0(\mathbf{r})$ include the propagation constant β_x and β_y through an exponential term, that is

$$\mathbf{E}^0(\mathbf{r}) = \begin{pmatrix} E_x^0(x,y)e^{i\beta_x z} \\ E_y^0(x,y)e^{i\beta_y z} \end{pmatrix}. \tag{1.49}$$

In a birefringent material the propagation constant is different along the two axes of the material. The amplitude vector may be rewritten as

$$\mathbf{E}^0 = e^{i\beta_x z}\begin{pmatrix} E_x^0 \\ E_y^0 e^{i\Delta\beta z} \end{pmatrix}, \tag{1.50}$$

where $\Delta\beta = \beta_y - \beta_x$. To map $\mathbf{E}^0$ on the Poincaré sphere, we introduce the complex number ζ [12] [13] as

$$\zeta = (E_y^0/E_x^0)e^{i(\beta_y - \beta_x)z} \tag{1.51}$$

and from this we find the relation for the azimuthal angle $2\psi_p$ along the equatorial plane

$$\tan(2\psi_p) = \frac{2\mathrm{Re}[\zeta]}{1 - |\zeta|^2} \tag{1.52}$$

and the altitude

$$\sin(2\theta_p) = \frac{2\mathrm{Im}[\zeta]}{1 + |\zeta|^2}, \tag{1.53}$$

where $0 \leq 2\psi_p < 2\pi$ and $-\pi/2 \leq 2\theta_p \leq \pi/2$.

Note the the angles are defined different from standard spherical coordinates, especially the angle $2\theta_p$ which in spherical coordinates is defined from the z-axis.

Mapping the state of polarization on the Pioncaré sphere also enables us to trace the evolution of the state of polarization during propagating of the electric field.

This is for example relevant when monitoring a change in the state of polarization as light propagates through various components or as light propagates through a nonlinear material. Finally, by calculating the induced polarization corresponding to all states of polarization of the applied electric field, we may find the average induced polarization by averaging over all points on the Poincaré sphere.

Often it is useful to separate the amplitude vector, $\mathbf{E}^0$, into a product of its modulus value E^0 and a unit vector $\mathbf{e}$. Then the polarization of the electric field is described by

$$\mathbf{E}^0 = E^0\mathbf{e} = \begin{pmatrix} E_x^0 \\ E_y^0 \end{pmatrix} = E^0 \left(\cos(\theta_p) \begin{pmatrix} \cos(\psi_p) \\ \sin(\psi_p) \end{pmatrix} + i\sin(\theta_p) \begin{pmatrix} -\sin(\psi_p) \\ \cos(\psi_p) \end{pmatrix} \right). \tag{1.54}$$

Note that the two vectors $(\cos(\psi_p), \sin(\psi_p))^T$ and $(-\sin(\psi_p), \cos(\psi_p))^T$ form the basis of the equatorial plane in Poincaré sphere, see Figure 1.5.

In this notation, linear polarization is obtained for $\theta_p = 0$ and circular polarization for $\theta_p = \pm\pi/4$. This may be plotted on the Poincaré sphere, as shown in Figure 1.6, and using the Jones vector the electric field corresponding to the two states of polarization are written as

Circularly polarized light $\theta_p = \pm\pi/4$

$$\begin{pmatrix} E_x^0 \\ E_y^0 \end{pmatrix} = E^0\sqrt{2}/2 \begin{pmatrix} \cos(\psi_p) - i\sin(\psi_p) \\ \sin(\psi_p) + i\cos(\psi_p) \end{pmatrix} = E^0\sqrt{2}/2e^{-i\psi_p} \begin{pmatrix} 1 \\ \pm i \end{pmatrix}. \tag{1.55}$$

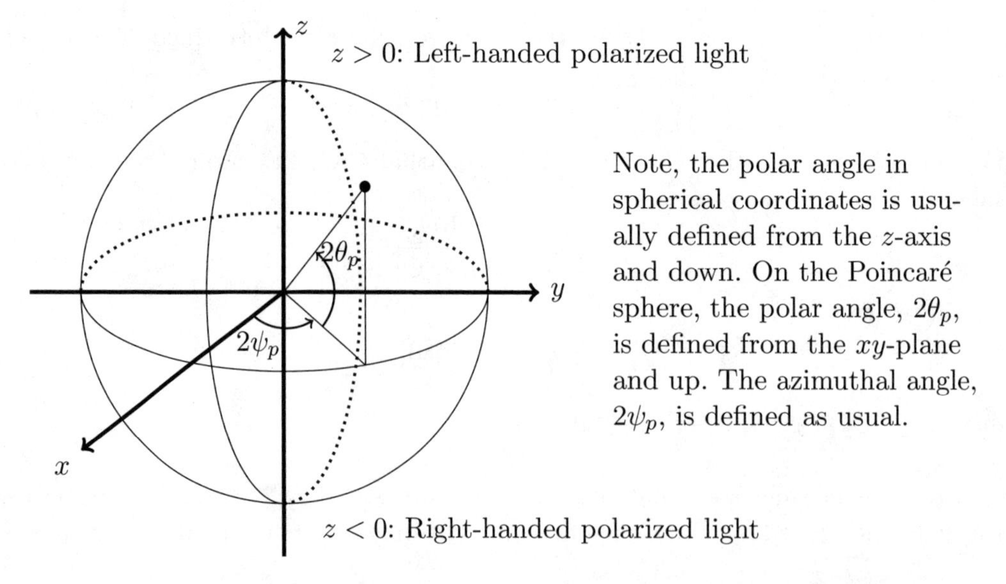

Figure 1.5: The polarization state of light is visualized on the Poincaré sphere.

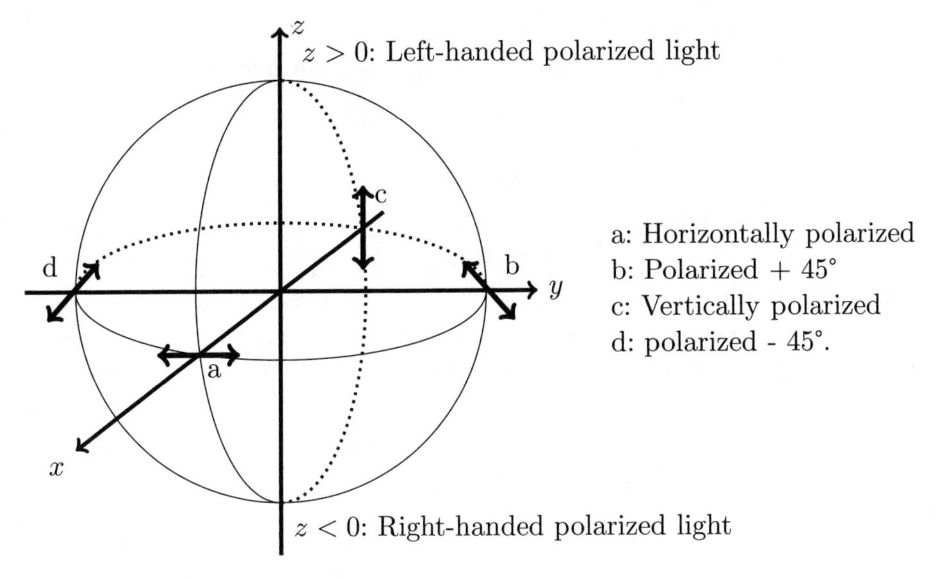

Figure 1.6: Linear states of polarization are found on the equatorial plane while circular polarizations are found on the poles of the Poincaré sphere.

Linearly polarized light $\theta_p = 0$

$$\begin{pmatrix} E_x^0 \\ E_y^0 \end{pmatrix} = E^0 \begin{pmatrix} \cos(\psi_p) \\ \sin(\psi_p) \end{pmatrix}. \tag{1.56}$$

Example 1.4. *Birefringence in fibers*
Typically, an optical fiber is considered as an example of a circular cylindrical geometry. However, this is only true in an ideal world, and for many reasons, including the fabrication process of fibers, any fiber has some degree of birefringence. Consequently, the propagation constant along the x- and y-axis are slightly different.

In some cases, for example within nonlinear optics, it is desirable to propagate a beam with a well defined state of polarization throughout propagation. For such cases optical fibers are made which have a large difference between β_x and β_y. These fibers are referred to as polarization maintaining fibers.

Assume a difference $\Delta\beta$ between β_x and β_y, then after a distance called the beat length, $L_b = 2\pi/\Delta\beta$, the beam has traveleved one great circle, Figure 1.6. In typical fibers this distance is in the order of few meters.

—— ■ ——

Example 1.5. *Averaged induced polarization*
The induced polarization in a material providing loss/gain may be written as

$$P_x = \varepsilon_0 \alpha_x E_x \tag{1.57a}$$
$$P_y = \varepsilon_0 \alpha_y E_y, \tag{1.57b}$$

where α_x and α_y are referred to as the polarizability, we return to a discussion of polarizability in Section 1.2 page 185. When $\alpha_x \neq \alpha_y$, we refer to this case as polarization dependent loss/gain. In the above it is assumed that there is no coupling between the y-component of the electric field and the x-component of the induced polarization, and likewise that there is no coupling between the x-component of the electric field and the y-component of the induced polarization. Under these assumptions the propagation equation for the two components may be written as

$$\frac{dE_x}{dz} = i\beta_x E_x - \frac{\alpha_x}{2} E_x \tag{1.58a}$$
$$\frac{dE_y}{dz} = i\beta_y E_y - \frac{\alpha_y}{2} E_y. \tag{1.58b}$$

As the components are amplified/attenuated differently, the ratio E_x/E_y is modified and the phase difference is modified by the difference $\beta_x - \beta_y$ which also varies as the beam propagates. However, we may evaluate the effective induced polarization as $P_{\text{eff}} = \mathbf{e}^* \cdot \mathbf{P}$, and we may then find the mean of the effective induced polarization by averaging P_{eff} over the Pioncaré sphere.

By using

$$\begin{pmatrix} E_x \\ E_y \end{pmatrix} = E^0 \left(\cos(\theta_p/2) \begin{pmatrix} \cos\psi_p \\ \sin\psi_p \end{pmatrix} + i\sin(\theta_p/2) \begin{pmatrix} -\sin\psi_p \\ \cos\psi_p \end{pmatrix} \right) \tag{1.59}$$

we find

$$P_{\text{eff}} = \varepsilon_0 E^0 \alpha_x (\cos^2\frac{\theta_p}{2}\cos^2\psi_p + \sin^2\frac{\theta_p}{2}\sin^2\psi_p) + \alpha_y(\cos^2\frac{\theta_p}{2}\sin^2\psi_p + \sin^2\frac{\theta_p}{2}\cos^2\psi_p). \tag{1.60}$$

The mean over all possible polarizations is then

$$\begin{aligned} \langle P_{\text{eff}} \rangle &= \frac{\varepsilon_0 E^0}{4\pi} \int_0^{2\pi} \int_{-\pi/2}^{\pi/2} P_{\text{eff}} \cos\theta_p d\theta_p d\psi_p \\ &= \varepsilon_0 E^0 \frac{\alpha_x + \alpha_y}{2}. \end{aligned} \tag{1.61}$$

In summary, the mean effective induced polarization is a simple average of the polarizability times the amplitude of the electric field times the vacuum permittivity.

— ■ —

1.1.7 Jones vector

As pointed out earlier, one of the advantages of using a Jones vector to describe the state of polarization is that it is much simpler to propagate the electric field through discrete elements, since these may be described using simple matrices. If the electric field before an optical element is $\mathbf{E}^0$, then after the element with a propagation matrix $\underline{\underline{A}}$ the electric field is simply $\underline{\underline{A}}\mathbf{E}^0$.

Below a few examples of the propagation matrix corresponding to typical components are highlighted.

A polarizer: As a simple example we choose an x-polarizer, which only allows x-polarized light to pass. This element is characterized by the matrix
$\begin{pmatrix} 1 & 0 \\ 0 & 0 \end{pmatrix}$.

Half wave plate: $\begin{pmatrix} -i & 0 \\ 0 & i \end{pmatrix}$

A half wave plate may be used to change the handedness of the propagating electric field. Consider for example a circularly polarized electric field given by

$$\mathbf{E}^0 = \frac{E^0}{\sqrt{2}} \begin{pmatrix} 1 \\ e^{i\pi/2} \end{pmatrix}. \tag{1.62}$$

By applying a half wave plate this field becomes

$$\mathbf{E}^0 = \frac{E^0}{\sqrt{2}} \begin{pmatrix} -i & 0 \\ 0 & i \end{pmatrix} \begin{pmatrix} 1 \\ e^{i\pi/2} \end{pmatrix} = -i\frac{E^0}{\sqrt{2}} \begin{pmatrix} 1 \\ e^{-i\pi/2} \end{pmatrix}. \tag{1.63}$$

Quarter wave plate: $\begin{pmatrix} 1 & 0 \\ 0 & i \end{pmatrix}$

Angled linear polarizer: $R(-\theta)T_x R(\theta)$

$$\begin{pmatrix} \cos\theta & -\sin\theta \\ \sin\theta & \cos\theta \end{pmatrix} \begin{pmatrix} 1 & 0 \\ 0 & 0 \end{pmatrix} \begin{pmatrix} \cos\theta & \sin\theta \\ -\sin\theta & \cos\theta \end{pmatrix} = \begin{pmatrix} \cos^2\theta & \cos\theta\sin\theta \\ \sin\theta\cos\theta & \sin^2\theta \end{pmatrix}$$

1.1.8 Superposition of circularly polarized light

Alternatively, the electric field in Eqn. (1.48) may also be written as a sum of a left- and right-hand circularly polarized light, i.e.

$$\mathbf{E}^0 = E^+\boldsymbol{\sigma}^+ + E^-\boldsymbol{\sigma}^-, \tag{1.64}$$

where $\boldsymbol{\sigma}^\pm$ are unit-vectors for right ($\boldsymbol{\sigma}^- = \frac{1}{\sqrt{2}}(\mathbf{e}_x - i\mathbf{e}_y)$) and left hand ($\boldsymbol{\sigma}^+ = \frac{1}{\sqrt{2}}(\mathbf{e}_x + i\mathbf{e}_y)$) circularly polarized light. These unit vectors are obtained from Eqn. (1.54) by letting $\delta = \pm\pi/2$. It is noted that

- $(\boldsymbol{\sigma}^{\pm})^* = \boldsymbol{\sigma}^{\mp}$,
- $\boldsymbol{\sigma}^+ \cdot \boldsymbol{\sigma}^- = \frac{1}{2}(\mathbf{e}_x + i\mathbf{e}_y) \cdot (\mathbf{e}_x - i\mathbf{e}_y) = 1\boldsymbol{\sigma}^- \cdot \boldsymbol{\sigma}^+$,
- $\boldsymbol{\sigma}^+ \cdot \boldsymbol{\sigma}^+ = \frac{1}{2}(\mathbf{e}_x + i\mathbf{e}_y) \cdot (\mathbf{e}_x + i\mathbf{e}_y) = 0 = \boldsymbol{\sigma}^- \cdot \boldsymbol{\sigma}^-$,
- $\boldsymbol{\sigma}^+ \cdot (\boldsymbol{\sigma}^-)^* = \frac{1}{2}(\mathbf{e}_x + i\mathbf{e}_y) \cdot (\mathbf{e}_x + i\mathbf{e}_y) = 0$.

Applying these dot products we find that $\mathbf{E}^0 \cdot \mathbf{E}^0 = 2E^+E^-$ and $(\mathbf{E}^0)^* \cdot \mathbf{E}^0 = |E^+|^2 + |E^-|^2$.

1.2 Induced polarization

In Section 1.1 we have reviewed basic properties of light and linear optics and we are now ready to introduce nonlinear optics. To do this we need to be able to describe the interaction between light and the material that the light propagates through. This interaction is described by the induced electric polarization.

The induced electric polarization, *not* to be confused with the polarization state of the electromagnetic field, is a vector that describes the extent to which a dielectric becomes polarized by an applied electric field.

In dielectric materials the electron cloud of individual atoms may be displaced from the nucleus if an electric field is applied to the material. For simplicity we start the following discussion by considering the simplest case i.e. a hydrogen atom. If an electric field is applied to a hydrogen atom the electron cloud of charge $-e$ is displaced from its equilibrium position around the nuclei of charge $+e$, i.e. the hydrogen atom has become an electric dipole, more specifically a system consisting of two equal but oppositely charged point charges separated by a distance. More generally, the two charges could be a charge $q_1 = q$ and a charge $q_2 = -q$. If q_1 is located at $\mathbf{r}_1$ and q_2 is located at $\mathbf{r}_2$, then the dipole is characterized by a dipole moment vector $\mathbf{p} = q_1\mathbf{r}_1 + q_2\mathbf{r}_2 = q\mathbf{l}$ where $\mathbf{l} = \mathbf{r}_1 - \mathbf{r}_2$ is the vector from q_2 to q_1. If the distance between the two charges approaches zero, for all practical purposes, i.e. when the length of the dipole is much smaller than the distance to an observation point, then the dipole is referred to as a *point dipole*. The dipole moment is in units Coulomb times meter [Cm]. The dipole moment per unit volume is called the induced polarization and is expressed in units of [C/m^2].

In a neutral material which may be considered to consist of many charges, the above definition of the dipole moment vector is extended to the sum

$$\mathbf{p} = \mathbf{r}_1 q_1 + \mathbf{r}_2 q_2 + \cdots + \mathbf{r}_n q_n, \tag{1.65}$$

where $\mathbf{r}_i$ is the vector from the origin to charge number i and q_i is the charge of charge number i. Consequently the dipole moment vector of the ensemble of charges is the sum of all dipole moment vectors.

When describing a material an often used description is to imagine the material as if it were made up of N dipoles per unit volume. Assuming that all dipoles are pointing in the same direction, the induced macroscopic electric polarization vector $\mathbf{P}$ is then

$$\mathbf{P} = N\mathbf{p}. \tag{1.66}$$

The dipole moment vector $\mathbf{p}$ depends on the applied electric field, i.e. $\mathbf{p}(\mathbf{E})$ and consequently so does the macroscopic polarization. If the applied external electric field is rather weak, the microscopic dipole moment vector may be described as a linear relation i.e. $\mathbf{p} = \alpha\mathbf{E}$, where α is denoted as the polarizability, and the macroscopic induced polarization is $\mathbf{P} = \varepsilon_0\chi^{(1)}\mathbf{E}$, where $\chi^{(1)}$ is the linear susceptibility. If the applied electric field is rather strong, the dipole moment and hence the macroscopic induced polarization becomes nonlinear in the electric field. The expansion of the induced polarization to include nonlinearities is the core content of this book.

In the following we are mainly concerned with the macroscopic induced electric polarization. To prevent confusion with the state of polarization of the electric field while at the same time keeping a simple notation, we make an effort to denote the macroscopic induced electric polarization as the induced polarization. The induced polarization may be expanded in a power series in the applied electric field where the term that depends linearly on the electric field is referred to as the linear contribution and all terms that depend on the electrical field to powers higher than one represents the nonlinear contributions. Though the expansion of the induced polarization into a power series in the electric field is merely a mathematical description, all nonlinear terms have physical origin.

There are many causes for a nonlinear response in the induced polarization. These include distortion of the electronic cloud around atoms in the material, distortions in the molecular network of the material, vibrational as well as rotational, reorientation of molecules, acoustic waves, etc. All of these may be described using simple theory known from the forced harmonic and an-harmonic oscillator. In the following we review the oscillator and later return to the magnitude and time response of the various contributions to the optical nonlinear response.

1.2.1 Dipole in external field

A dipole orients itself to align with an applied electric field. Consequently a material consisting of a large number of small dipoles, which all align with the applied electric field, becomes polarized. To describe this we consider a dipole placed in an external electric field $\mathbf{E}_e$, and we calculate the energy of the dipole in the electric field.

For simplicity we consider a stationary field pointing in the x-axis, $\mathbf{E}_e = E^0\mathbf{e}_x$. In addition a dipole is placed in the xy- plane with its center at origo and making an angle θ with the x-axis.

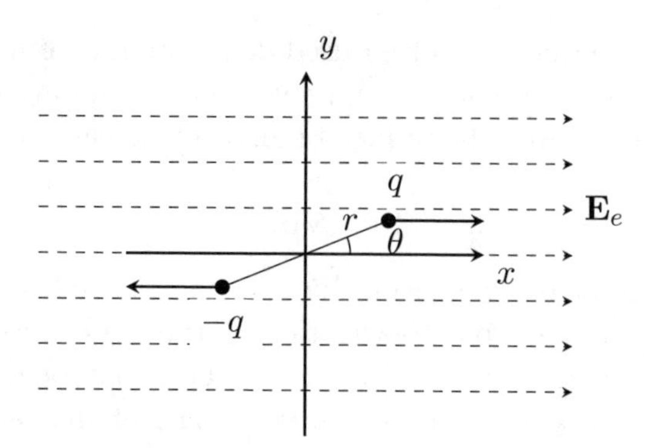

Figure 1.7: A dipole in an applied electric field. The dipole is defined by a charge $+q$ at position $(r\cos(\theta), r\sin(\theta))$, an a charge $-q$ at position $(-r\cos(\theta), -r\sin(\theta))$.

Since the potential that corresponds to the external field is calculated through $-\nabla V = \mathbf{E}_e$, we get in the case we consider: $-dV/dx = E^0$, hence $V(x) = -E^0(x - x_0)$, where $x - x_0$ is an increment along the x-axis defined by an arbitrary starting point with x-coordinate x_0 and endpoint x, and where we note that the potential only has an x-coordinate.

From this, and by using that the work done to move a charge q from position $(r, 0)$ to position $(r\cos(\theta), r\sin(\theta))$ is $W_{+q} = (+q)(-E^0)r\cos\theta$ and like-wise W_{-q} the work done to move a charge $-q$ from position $(-r, o)$ to $(-r\cos(\theta)_1 - r\sin(\theta))$ $W_{-q} = (-q)(-E^0 x_{-q}) = -qE^0 r\cos(\theta)$ adding W_{+q} and W_{-q} we get $W = -2qE^0 r\cos(\theta)$ since $2r = d$, i.e. the distance between the two charges, we may write the potential energy as $W = -\mathbf{p}\cdot\mathbf{E}_e$, where we have used $\mathbf{p} = q\mathbf{d}$. From this we conclude that the potential energy E_{pot} of a dipole in an external electric field is $E_{\text{pot}} = -\mathbf{p}\cdot\mathbf{E}_e$. It is clear that if the dipole moment is in the same direction as the applied electric field, then the dipole has no potential energy.

1.2.2 Torque on a dipole in external field

In general, the torque $\mathbf{M}$ or moment of force is $\mathbf{M} = \mathbf{r}\times\mathbf{F}$, where the displacement vector $\mathbf{r}$, is a vector from the the point where the torque is measured to the point where the force $\mathbf{F}$ is applied. Here the torque is $\mathbf{M} = \mathbf{p}\times\mathbf{E}$, which is a vector pointing in the direction orthogonal to the dipole moment vector and the applied electric field. One may imagine a nail punched through the center of the dipole. Remembering that the coulomb force of each charge due to the applied electric field is equal in magnitude $(= |q|E)$ but pointing in opposite directions, i.e. there is no net force to move the dipole in a certain direction, then the resulting effect of the external electric field is to rotate the dipole. This is of course under the assumption that the dipole is not elongated or in other ways impacted by the applied electric field. The work done by rotating the dipole equals the difference in potential energy between the initial and final position.

1.2.3 The electric field from a 'point' dipole

A dipole in itself will also set up an electric field. The electric field at a position $\mathbf{r}$ from a point dipole located at Origo is

$$\mathbf{E}(\mathbf{r}) = \frac{1}{4\pi\varepsilon_0}\left[\frac{(3\mathbf{r}\cdot\mathbf{p})\mathbf{r}}{r^5} - \frac{\mathbf{p}}{r^3}\right], \tag{1.67}$$

where $\mathbf{p}$ is the dipole moment vector and r is the distance to the observation point.

As opposed to a static dipole, an oscillating dipole, as for example a dipole created by an oscillating applied electric field, radiates. The electric field from such a dipole is given for example by P. Milonni and J. Eberly [14]. If the time dependent dipole moment vector is given as

$$\mathbf{p} = \mathbf{e}_p p(t), \tag{1.68}$$

and the position $\mathbf{r} = r\mathbf{e}_r$ is measured from the dipole, then the electric field is

$$\begin{aligned}\mathbf{E}(\mathbf{r},t) =& \frac{1}{4\pi\varepsilon_0}\left[(3\mathbf{e}_p\cdot\mathbf{e}_r)\mathbf{e}_r - \mathbf{e}_p\right]\left[\frac{1}{r^3}p(t-r/c) + \frac{1}{cr^2}\frac{d}{dt}p(t-r/c)\right] + \\ & \frac{1}{4\pi\varepsilon_0}\left[(\mathbf{e}_p\cdot\mathbf{e}_r)\mathbf{e}_r - \mathbf{e}_p\right]\left[\frac{1}{c^2r}\frac{d^2}{dt^2}p(t-r/c)\right].\end{aligned} \tag{1.69}$$

If the length of the dipole is very short, the dipole is referred to as a Hertzian dipole.

Example 1.6. *Rayleigh scattering*
Several linear as well as nonlinear optical phenomena may be described using radiation from a dipole. In the following we provide a short review of how radiation from a dipole may be used to describe Rayleigh scattering, i.e. elastic scattering, where the frequency of the launched and scattered light are equal. It is noted that Rayleigh scattering is caused by molecules, particles, etc., which are much smaller than the wavelength of light. In an optical fiber most of the loss at short wavelengths, 400 nm to 1600 nm, is caused by Rayleigh scattering.

As already described, an electric field does set up a dipole. The power re-radiated from the dipole, p_{wr}, is [14]

$$P_{wr} = \frac{2}{12\pi\varepsilon_0 c^3}\left[\frac{d^2}{dt^2}p(t-R/c)\right]^2, \tag{1.70}$$

where R is the radius at which the scattered light is measured. If the microscopic polarization $\mathbf{p}$ is $\mathbf{p} = \mathbf{e}_p\alpha E_0\cos(\omega(t-z/c))$, where α is the polarizability, then the radiated power is

$$P_{wr} = \frac{\omega^4}{6\pi\varepsilon_0 c^3}\alpha^2(E^0)^2\cos^2(\omega(t-z/c)). \tag{1.71}$$

It is noted that the scattered power depends on the frequency to the power of four. Thus it is much stronger at short wavelengths compared to long wavelengths.

— ■ —

1.3 Harmonic oscillator model

For dielectrics the calculations of the induced electric polarization may be performed using a spring model of the bound charges, as for example electrons bound to the nucleus of an atom. The force on charge q moving with velocity $\mathbf{v}$ in an electromagnetic field is known as the Lorentz force, $\mathbf{F} = q(\mathbf{E} + (\mathbf{v} \times \mathbf{B}))$, where $\mathbf{E}$ is the electric field, and $\mathbf{B}$ is the magnetic field. To evaluate the induced polarization, the contribution from the magnetic field is typically neglected. This is validated by comparing the contribution from the magnetic field relative to the contribution from the electric field. Inserting a plane wave, the contribution from the magnetic field is $(|\mathbf{v}|/c)E^0$, where E^0 is the amplitude of the electric field. Since the speed of the electron is much smaller than the speed of light, this approximation is justified. In a simple one-dimensional picture, the motion of an electron with mass m_e at displacement x from its equilibrium position, due to an applied electric field $E(t)$ is described as

$$m_e \left(\frac{d^2x}{dt^2} + 2\Gamma \frac{dx}{dt} + \Omega^2 x + ax^2 + bx^3 + \ldots \right) = -eE_x(t), \qquad (1.72)$$

where Γ is the damping constant, Ω is the natural frequency of the electron, and a, b the nonlinear material parameters. In Eqn. (1.72), the origin of the motion is chosen to coincide with the center of the nucleus. Below we apply this classical model to the derivation of the induced nonlinear polarization of a dielectric.

In the following we first evaluate the polarizability using the harmonic oscillator. Then we evaluate the nonlinearities to order of magnitude due to interactions between the electric field and electrons in Section 1.5.

In the one-electron oscillator model, we assume that the wavelength of the electromagnetic field is sufficiently large to neglect any spatial variations of the electric field over the spatial extent of the oscillator. In addition, we assume that the proton is fixed in space, with the electron free to oscillate around the nucleus. These are crude approximations, which may be relaxed by assuming instead

1. that the bound proton-electron pair is considered as constituting a two-body central force problem of classical mechanics in which one assumes a fixed center of mass of the system, around which the proton as well as the electron are free to oscillate.

2. that the center of mass is allowed to oscillate as well. In this case an equation of motion for the center of mass is added to the evolution of the dipole moment.

1.3.1 Simple classical harmonic oscillator

Consider an electron of mass m_e and charge $-e$ that is attached to an ion by a spring. When the displacement of the electron from its equilibrium position is denoted x, and the acceleration of the electron d^2x/dt^2, then Newton's second law provides $m_e d^2x/dt^2 = -k_s x$, where the spring force is $-k_s x$. If the system is subject to a damping force $F_d = -c \cdot dx/dt$, the motion of the electron is governed by

$$m_e \frac{d^2x}{dt^2} + 2\Gamma m_e \frac{dx}{dt} + k_s x = 0, \tag{1.73}$$

which is often written in the form [2]

$$m_e \left[\frac{d^2x}{dt^2} + 2\Gamma \frac{dx}{dt} + \Omega^2 x \right] = 0, \tag{1.74}$$

where Γ is a damping constant and $\Omega = \sqrt{k_s/m_e}$ is the resonance frequency.

In the case where the system is undamped, i.e. $\Gamma = 0$ the solution to the displacement is a simple harmonic oscillation with frequency Ω. When the system is damped ($\Gamma \neq 0$) the solution is divided into three cases

1. $\Gamma^2 > \Omega^2$ then the system is over-damped
   The solution is
$$x = e^{-\Gamma t} \left[c_1 e^{(\sqrt{\Gamma^2 - \Omega^2})t} + c_2 e^{(-\sqrt{\Gamma^2 - \Omega^2})t} \right], \quad (c_1, c_2) \in \Re \tag{1.75a}$$
2. $\Gamma^2 < \Omega^2$ then the system is under-damped
   The solution is
$$x = e^{-\Gamma t} \left[c_1 e^{i(\sqrt{\Omega^2 - \Gamma^2})t} + c_2 e^{-i(\sqrt{\Omega^2 - \Gamma^2})t} \right], \quad (c_1, c_2) \in \Re \tag{1.75b}$$
3. $\Gamma = \Omega$ then the system is critically-damped.
   The solution is
$$x = e^{-\Gamma t} \left[c_1 t + c_2 \right], \quad (c_1, c_2) \in \Re \tag{1.75c}$$

1.3.2 Forced classical harmonic oscillator

When now applying an electric field to the matter, one needs to include a driving term that drives the oscillator. This leads to the forced harmonic oscillator, which may have a so-called linear response

$$m_e \left[\frac{d^2x}{dt^2} + 2\Gamma \frac{dx}{dt} + \Omega^2 x \right] = -eE(t). \tag{1.76}$$

When the applied electric field is a simple harmonic wave i.e.

$$E(t) = E^0 \cos(\omega t) = \frac{1}{2} E^0 \left[e^{-i\omega t} + c.c. \right] \tag{1.77}$$

the displacement, due to the electric field, from equilibrium equals

$$x(t) = \frac{-eE^0}{2m_e} \left[\frac{e^{-i\omega t}}{\Omega^2 - \omega^2 - 2i\Gamma\omega} + c.c. \right]. \tag{1.78}$$

The complete solution to Eqn. (1.76) is then the sum of the solution to the homogeneous un-forced oscillator i.e. Eqn. (1.75a)-Eqn. (1.75c) and a single solution to the inhomogeneous differential equation, i.e. Eqn. (1.78). The induced polarization originating from the applied electric field is then

$$P = -eNx(t) = \frac{Ne^2E^0}{2m_e} \left[\frac{e^{-i\omega t}}{\Omega^2 - \omega^2 - 2i\Gamma\omega} + c.c. \right], \tag{1.79}$$

where N is the number of dipoles per unit volume. When the damping constant equals zero the induced polarization is linearly proportional to the electric field with a real constant of proportionality. However, when the damping force is non-zero, then the induced polarization is delayed relative to the applied electric field, and from Section 1.1.3 propagation is lossy. The real and imaginary parts of the susceptibility are found from Eqn. (1.78), and from this the real and imaginary parts of the propagation constant are determined.

The induced polarization may be rewritten as

$$\begin{aligned} P(t) &= N\frac{e^2E^0}{m_e} \frac{1}{(\Omega^2 - \omega^2)^2 + 4\Gamma^2\omega^2} \left[(\Omega^2 - \omega^2)\cos(\omega t) + (2\Gamma\omega)\sin(\omega t) \right] \\ &= N\frac{e^2E^0}{m_e} \frac{1}{((\Omega^2 - \omega^2)^2 + 4\Gamma^2\omega^2)^{1/2}} \cos(\omega t - \delta), \end{aligned} \tag{1.80}$$

where $\tan(\delta) = 2\Gamma\omega/(\Omega^2 - \omega^2)$. It is noted that the applied electric field $E(t)$ is real, hence the material response is also real. Figure 1.8 illustrates an example of the linear induced polarization.

1.3.3 Forced an-harmonic oscillator

If the amplitude of the applied electric field increases, the oscillator may have a nonlinear response as illustrated in Figure 1.9. To describe this, it is necessary to include so-called an-harmonic terms in the oscillator equation, compared to Eqn. (1.76),

$$\ddot{x} + 2\Gamma\dot{x} + \Omega^2 x + ax^2 + bx^3 = -eE(t)/m_e. \tag{1.81}$$

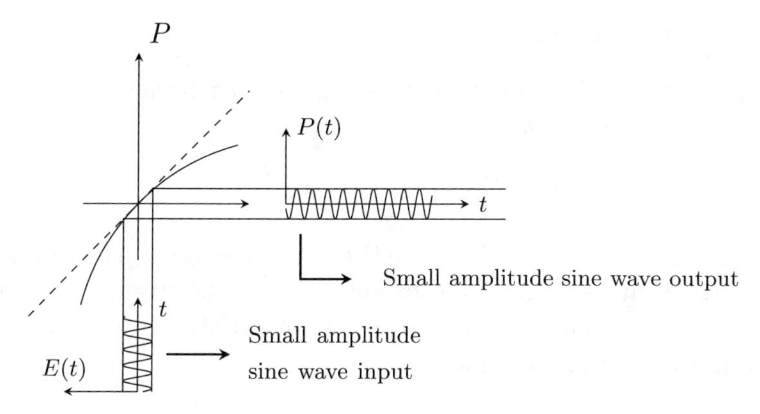

Figure 1.8: The induced polarization for small input fields, [2].

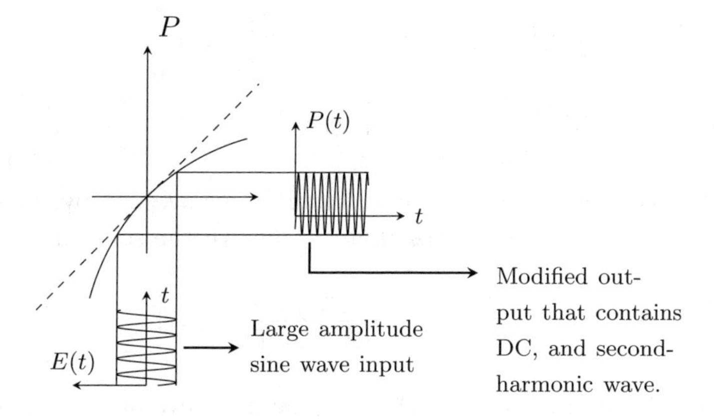

Figure 1.9: The induced polarization for large input fields compared to Figure 1.8 [2].

This leads to two cases. The one where $a \neq 0$ is the so-called non-centro-symmetric materials, that is materials where the potential energy [1] of the electron is non-symmetric around the equilibrium displacement. An example is $LiNbO_3$ used in for example high speed optical modulators. The other where $a = 0$ and $b \neq 0$ is so-called centro-symmetric materials. An example is fused silica as used in optical fibers. In this case the potential energy is a symmetric function around the equilibrium position of the electron. In both cases Eqn. (1.81) needs to be solved for example by means of a perturbation expansion of the displacement. This may be done under the assumption that the nonlinear terms are much smaller than the linear terms for any displacement.

[1] The potential energy stored in a spring is calculated by finding the work necessary to stretch the spring a distance x from its equilibrium or un-stretched length. From Hooke's Law, the spring force equals $F = -k_s x$, where k_s is the spring constant. Consequently, the work done (and therefore the stored potential energy) is

$$U_e = -\int F dx = \int k_s x dx = \frac{1}{2} k_s x^2.$$

Non-centro-symmetric case

The perturbation method begins by expressing the displacement as a power series in the perturbation strength λ, i.e.

$$x = \lambda x^{(1)} + \lambda^2 x^{(2)} + \lambda^3 x^{(3)} + \cdots \tag{1.82}$$

and by replacing the driving field with $\lambda E(t)$. In order for x as expressed in Eqn. (1.82) to be a solution to Eqn. (1.81) for any value of the strength λ, it is required that the terms proportional to λ,λ^2, λ^3 etc. each satisfy the equation separately. From collecting these terms, we find

$$\frac{d^2x^{(1)}}{dt^2} + 2\Gamma\frac{dx^{(1)}}{dt} + \Omega^2 x^{(1)} = -eE(t)/m_e \tag{1.83a}$$

$$\frac{d^2x^{(2)}}{dt^2} + 2\Gamma\frac{dx^{(2)}}{dt} + \Omega^2 x^{(2)} + a\left\{x^{(1)}\right\}^2 = 0 \tag{1.83b}$$

$$\frac{d^2x^{(3)}}{dt^2} + 2\Gamma\frac{dx^{(3)}}{dt} + \Omega^2 x^{(3)} + 2ax^{(1)}x^{(2)} = 0. \tag{1.83c}$$

From this, the lowest order in the perturbation, $x^{(1)}$, is governed by the same equation as the classical forced harmonic oscillator, i.e. with the solution

$$x^{(1)} = \frac{-eE^0}{2m_e}\left[\frac{e^{-i\omega t}}{\Omega^2 - \omega^2 - 2i\Gamma\omega} + c.c.\right]. \tag{1.84}$$

Inserting this into Eqn. (1.83a-b) for $x^{(2)}$, we may now evaluate $x^{(2)}$, however, we leave the detailed calculations to the reader. We note that the term $\left\{x^{(1)}\right\}^2$ equals

$$\left\{x^{(1)}\right\}^2 = \frac{e^2(E^0)^2}{4m_e^2}\left[\frac{e^{-i2\omega t}}{(\Omega^2 - \omega^2 - 2i\Gamma\omega)^2} + \frac{e^{i2\omega t}}{(\Omega^2 - \omega^2 + 2i\Gamma\omega)^2} + 2\frac{1}{(\Omega^2 - \omega^2)^2 + 4\left(\Gamma\omega\right)^2}\right]. \tag{1.85}$$

From this we see that the induced polarization, beside the contribution from $x^{(1)}$ at the same frequency as the electrical field, has two new contributions, one at twice the frequency of the electrical field and one at the DC component. This analysis can easily be continued to determine higher-order nonlinearities $x^{(3)}$, $x^{(4)}$, etc.

Centro-symmetric case

For the case of a centro-symmetric medium a similar analysis may be carried out. Such analysis shows that the lowest order nonlinearity originates from $x^{(3)}$, and only odd nonlinear terms are non-zero. Silica glass is an example of such a nonlinearity, and we return to this in much more detail later.

1.4 Local field corrections

In the previous description of light matter interaction, two effects were described separately i) an applied electric field would create a dipole ii) a dipole would create an electric field. However, the dipoles created are also impacted by the electric field created by surrounding dipoles.

In dense media neighboring molecules give rise to an internal field E_i in addition to the macroscopic field E_m. Thus the local field at a given molecule is $E_l = E_m + E_i$. The internal field may be approximated by the field inside a sphere with a uniform charge distribution [10], [15]. If P is the uniform macroscopic induced polarization defined by the macroscopic electric field E_m, i.e. ($P = \varepsilon_0(\epsilon_r - 1)E_m$), then Jackson [10] shows that the internal electric field is

$$E_i = P/3\varepsilon_0. \tag{1.86}$$

The average molecular dipole moment due to the local electric field is then

$$< p_{\text{mol}} >= \varepsilon_0 \alpha (E_m + E_i), \tag{1.87}$$

where α is the molecular realizability. From this, the local field corrected induced polarization is $P_c = N < p_{\text{mol}} >= N\varepsilon_0\alpha(E + P/3\varepsilon_0)$, where N is the average number of molecules per unit volume. Inserting the macroscopic induced polarization P, we find $P_c = N\varepsilon_0\alpha E_m(1 + (\epsilon_r - 1)/3)$. As per definition, and in terms of measurable quantities, we express this as: $\varepsilon_0(\epsilon_r - 1)E_m$ i.e.

$$\begin{aligned} N\varepsilon_0\alpha E_m(1 + (\epsilon_r - 1)/3) \equiv \varepsilon_0(\epsilon_r - 1)E_m \quad & \Longleftrightarrow \\ \frac{N\alpha}{3} = \frac{\epsilon_r - 1}{2 + \epsilon_r} \quad & \end{aligned} \tag{1.88}$$

This should be compared against $\epsilon_r = N\alpha + 1$, which is obtained if local field corrections are ignored.

1.5 Estimated nonlinear response

In previous sections we have modeled the electronic nonlinearity using the harmonic oscillator. In the following we evaluate the order of magnitude and response of the electronic nonlinearity based on [16] and [17].

1.5.1 Electronic response

Non-resonant electronic nonlinearities occur as the result of the nonlinear response of bound electrons to an applied optical field. These exist in all materials and are in some cases the dominant contribution to the nonlinear optical responses for example in silica glass.

If an electric field is applied to an atom, the electron cloud is displaced a distance relative to the nucleus which consists of positive charges. On the other hand, as the electron cloud and the nucleus are displaced, the electron cloud experiences a Coulomb force pulling it back toward the nucleus.

In a simple scalar model, the microscopic induced polarization can be expanded in a power series in the applied electric field E as

$$p(t) = \alpha E(t) + \beta E^2(t) + \gamma E^3(t), \tag{1.89}$$

where α is the linear polarizability and β and γ the nonlinear polarizabilities.

Since the linear atomic polarizability is a property of matter in equilibrium, the displacement adjusts itself so that the force on the electron cloud due to the internal field matches the force due to the external field.

To evaluate the nonlinear polarizabilities we follow the strategy outlined by R. W. Boyd [16] based on the assumption that the nonlinear terms $\beta E^2(t)$, $\gamma E^3(t)$ etc. all become comparable to the linear term if the applied field strength approaches or exceeds the Coulomb field responsible for binding the valence electrons to the target material, in the following denoted $E_{\mathrm{at}} = e/(4\pi\varepsilon_0 a_0^2)$, where the Bohr radius a_0 is ($a_0 = 4\pi\varepsilon_0\hbar^2/(m_e e^2) = 0.53\mathring{A}$) and m_e is the electron mass at rest. From this we find

$$\beta = \alpha/E_{\mathrm{at}} \quad \text{respectively} \quad \gamma = \alpha/E_{\mathrm{at}}^2. \tag{1.90}$$

To estimate the nonlinear polarizability we thus simply need to estimate the linear polarizability. In the following we model the linear polarizability, first using a static model and then a harmonic oscillator model

A) Static charge

If an electric field E is applied to an atom, the electron cloud, of charge $-e$ and extending a sphere with a radius which equals the Bohr radius a_0, is displaced a distance d relative to the nucleus of positive charge, of similar absolute value as the electron cloud. The absolute value of the force that is pushing the electron cloud from the nucleus is $F_E = |eE|$ and the direction is such that the electron cloud is pushed from the nucleus. This leads to a dipole moment of size $p = |e|d$. On the other hand, as the electron cloud and the nucleus are displaced, the electron cloud experiences a Coulomb force, F_C, pulling the nucleus and the electron cloud toward each other again. The latter Coulomb force and its corresponding electric field is often referred to as the internal field.

To evaluate the Coulomb force, we use Gauss' law to find the internal electric field at any position r from the center of the spherical electron cloud, i.e. we forget the nucleus for a moment. Choosing a Gaussian surface as a sphere with radius d, then the charge included by the Gaussian surface is $q = -e(4\pi d^3/3)/(4\pi a_0^3/3) =$

$-ed^3/a_0^3$. From this the electric field at any point along the Gaussian sphere has the numerical value $|E_{\text{at}}| = |q|/(\varepsilon_0 4\pi d^2)$. The Coulomb force F_C on the nucleus is then $F_C = |E_{\text{at}}e| = |e|d/(a_0^3\varepsilon_0 4\pi)$, and the direction is such that the nucleus and the electron cloud are pulled toward each other.

Since atomic polarizability is a property of matter in equilibrium, the force due to the external field matches that of the internal field, i.e: $|eE| = |e|^2 d/4\pi\varepsilon_0 a_0^3$. Thus the displacement $d = |E4\pi\varepsilon_0 a_0^3/e|$ is proportional to the applied electric field as expected. Hence, the electric dipole moment is now expressed as $p = 4\pi\varepsilon_0 a_0^3 E$, which yields the linear atomic polarizability

$$\alpha = 4\pi\varepsilon_0 a_0^3. \tag{1.91}$$

From Eqn. (1.90) we find, by using α from above and by evaluating the internal field as if the nucleus were separated from the core by a distance equal to the Bohr radius, that the second-order nonlinear polarizability is

$$\beta = 4\pi\varepsilon_0 a_0^3/E_{\text{at}} = \frac{(4\pi\varepsilon_0)^2 a_0^5}{e}. \tag{1.92}$$

B) The harmonic oscillator
Using the harmonic oscillator the atomic polarizability is from Eqn. (1.84) using $\Gamma = 0$

$$\alpha(\omega) = \sum_{i=1}^{Z} \frac{e^2/m_e}{\Omega_i^2 - \omega^2}, \tag{1.93}$$

where Ω_i is the natural frequency of the ith electron and the sum is over the Z electrons. m_e is the electron mass. This is a complicated expression since the natural frequency corresponding to each electron is undefined. Thus, for simplicity we model the system as a collection of hydrogen atoms and Ω_i reduces to the natural frequency, Ω_0, for the electron orbiting the nucleus. In this case the natural frequency is related to the Bohr radius a_0 through

$$\Omega_0 = \frac{E_f}{\hbar} = \frac{m_e e^4}{(4\pi\varepsilon_0)^2\hbar^3}, \tag{1.94}$$

where $E_f = m_e e^4/((4\pi\varepsilon_0)^2\hbar^2)$ is the fundamental energy level for the hydrogen atom. From this Ω_0 approximates $2{\cdot}10^{16}$ s^{-1}.

For low frequencies, i.e. for frequencies below 10^{16} s^{-1}, the polarizability approximates

$$\alpha \approx e^2/(m_e\Omega_0^2). \tag{1.95}$$

Inserting this together with the Coulomb field β reduces to

$$\beta = a_0^5\frac{(4\pi\varepsilon_0)^2}{e}, \tag{1.96}$$

which matches the value estimated using the static model.

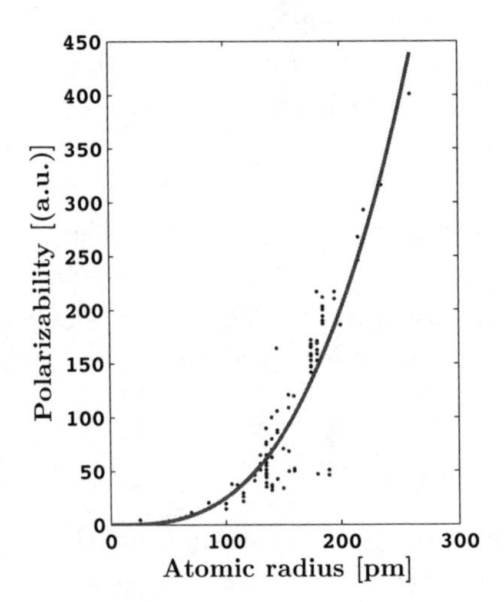

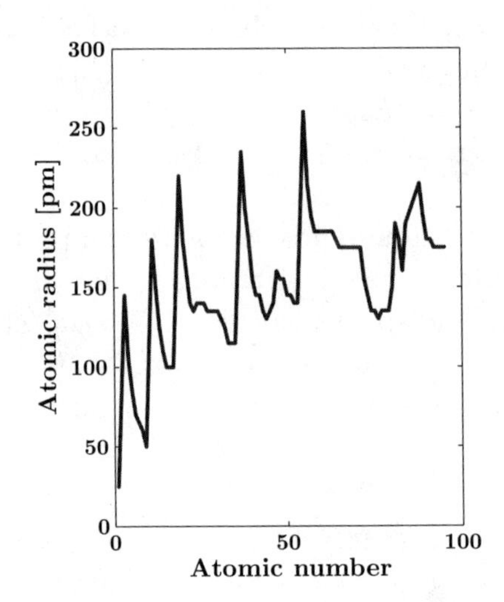

Figure 1.10: Atomic radius versus atomic number. **Left** atomic polarizability in atomic units versus atomic number [18] ($(a.u.) = 0.1648776^{-40}$ [Cm^2/V]). Solid curve is theoretical fit to a cubed dependence in atomic radius. **Right** atomic radius versus atomic number.

The two models agree, and demonstrate that the linear as well as the nonlinear polarization depends on the size of the atoms. Figure 1.10 displays the atomic radius for most elements in the periodic table.

1.5.2 Miller's rule

From the above, the first-order susceptibility is proportional to the atomic Bohr radius whereas the the third-order susceptibility is proportional to the Bohr radius to the power of five. Consequently, it is reasonably to expect that the larger the first-order susceptibility the larger the third-order susceptibility. This is confirmed by the so-called Miller's rule [19], according to which

$$\chi^{(3)} \propto (\chi^{(1)})^3, \tag{1.97}$$

that is, the more polarizable a material is, the more nonlinear the material is.

Since the refractive index is related to the first-order susceptibility and as we will show in Chapter 9 the nonlinear refractive index related to the third-order susceptibility one may also assume that the higher a refractive index a material has, the stronger its refractive index depend on the intensity of the electric field propagating through the material, i.e. the higher its nonlinear refractive index is. Monro and Ebendorff-Heidepriem have suggested the relation [20]

$$n_2 = \left(\frac{n^2 - 1}{4\pi}\right)^4 \frac{160\pi^2}{cn^2} 10^{-10}, \tag{1.98}$$

where n is the refractive index. Figure 1.11 shows predicted values from Eqn. (1.98), compared against experimental data.

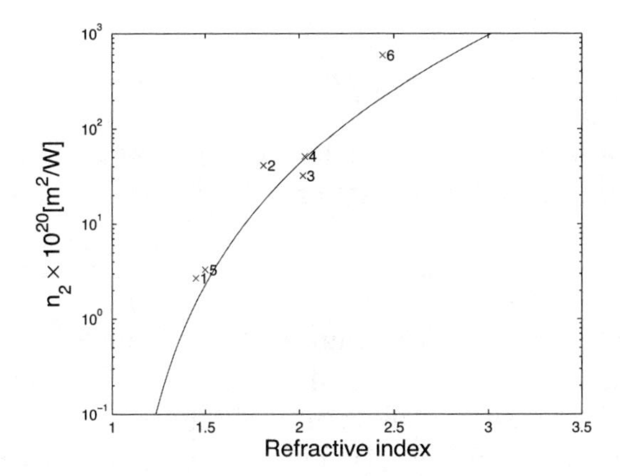

Figure 1.11: Solid curve from Eqn. (1.98), data from [2] and refs. therein. 1: silica, 2: lead silicate, 3: bismuthoxide, 4: tellurite, 5. flouride, 6: chalgonide.

Response time

The electronic response is extremely fast since it equals the time required for the electron cloud to become distorted in response to an applied optical field. This response time approximates the orbital period of the electron in its motion about the nucleus [1].

Consider an electron which is orbiting a nucleus. The centripetal force $m_e v^2/a_0$ to this motion is supplied by the Coulomb force $e^2/(4\pi\varepsilon_0 a_0^2)$ where a_0 is the radius of the orbit, i.e in the simplest case the Bohr radius. From this the electron speed equals

$$v = \frac{e^2}{2h\varepsilon_0} = 2.2 \cdot 10^6 \text{m/s}, \tag{1.99}$$

and consequently the response time equals

$$\tau = 2\pi a_0/v = 1.5 \cdot 10^{-16} \text{s}. \tag{1.100}$$

1.5.3 Intra-molecular response

In addition to the distortion of the electronic cloud, the polarization may also be distorted due to intra-molecular motion, meaning molecular vibrations and/or rotations, as for example in Raman scattering, as will be discussed in detail in Chapter 7.

Molecular vibrations

For order of magnitude we consider only a diatomic molecule. In this simple model the two nuclei may be pictured as being connected by a spring. The energy of the spring is

$$E = \left(n + \frac{1}{2}\right) h\nu \quad , \quad n \in \aleph_0. \tag{1.101}$$

Typical values of the energy range from 10 meV to 0.5 eV. This corresponds to a duration of the vibrational period in the range from around 10 fs to 400 fs.

Molecular rotations

For order of magnitude we consider again a diatomic molecule which can rotate about an axis through its center of mass. The kinetic energy of a rotating body may be expressed as $1/I\omega^2$ where I is the moment of inertia $I = mr^2$. The energy may also be expressed in terms of the angular momentum $L = I\omega$ as $E = L^2/(2I)$.

From a quantum mechanical description the energy eigenvalues are expressed as

$$E_l = \frac{l(l+1)\hbar^2}{2I}. \tag{1.102}$$

Typical values are $1.2 \cdot 10^{-5}$ eV to $3.7 \cdot 10^{-3}$ eV. From this the duration of the rotational period is thus on the order of 1 ps to 300 ps.

1.5.4 Other contributions

For some materials there may be many other contributions than the above mentioned. They may even be more significant, depending on the considered medium and the applied electric field. A complete list is out of the scope of this book however, a few contributions should be mentioned

Molecular reorientation

Examples include reorientation of molecules in a liquid, for example seen in liquid crystals. The response time depends on the viscosity of the molecules in the liquid but is typically slower than the intra-molecular contributions. In addition, the nonlinear strength also varies significantly from material to material but may be very strong.

Acoustic motion

Light may also cause acoustic motion or so-called electrostriction, as for example is known from Brillouin scattering. The magnitude of this nonlinearity may be very strong, however, the response time is typically very slow in the order of 10^{-9} s to 10^{-10} s.

Overview of nonlinear effects

A complete list of nonlinear optical effects and examples of applications of nonlinear optical effects would never be complete for a very long time, and in addition, such a list would exceed the ambitions of this text and not serve the purpose of learning nonlinear optics. Consequently, we have decided to focus on a few nonlinear effects and their applications.

To make it easier to read the text we provide a short description of the examples we use throughout the text, with more details in the following chapters. In general, the list may be divided into second-order nonlinear effects and third-order nonlinear effects, in accordance with

Second-order nonlinear effects The induced polarization depends quadraticly on the electric field

OPO and OPA Examples of second-order nonlinear processes includes optical parametric oscillators (OPOs) and optical parametric amplifiers (OPAs).

SHG Another example of a second-order nonlinear process is second-harmonic generation (SHG), for example used to change the frequency of a laser.

Third-order nonlinear effects The induced polarization depends on the electric field cubed.

Intensity dependent refractive index - Kerr effect Light at a single frequency may modify the refractive index that the light experiences. This may for example be due to a coupling between different vectorial components of the electric field or it may be due to a strong peak power in a narrow pulse as for example when used to form solitons. In glass, the intensity dependent refractive index is nearly frequency independent over a large wavelength range around 1550 nm. Consequently it is often used as the basis for parametric amplifiers, and wavelength converters in optical fibers as well as in super continuum generation.

Scattering An electric field consisting of two or more frequencies may interact with the material that the light is propagating through. This happens in a scattering process such as Brillouin or Raman scattering, where energy is transferred from the short wavelength component to the long wavelength component. Such inelastic scattering processes may be used to make amplifiers, to shift the center wavelength of a pulse, or to make an optical sensor.

1.6 Summary

- We use the macroscopic version of Maxwell's equations in Eqn.(1.4)

$$\nabla \times \mathbf{E} = -\frac{\partial \mathbf{B}}{\partial t}, \tag{1.4a}$$
$$\nabla \times \mathbf{H} = \frac{\partial \mathbf{D}}{\partial t} + \mathbf{J}_f, \tag{1.4b}$$
$$\nabla \cdot \mathbf{B} = 0, \tag{1.4c}$$
$$\nabla \cdot \mathbf{D} = \rho_f, \tag{1.4d}$$

and Eqn.(1.3)

$$\mathbf{D} = \varepsilon_0 \mathbf{E} + \mathbf{P} \tag{1.3c}$$
$$\mathbf{B} = \mu_0(\mathbf{H} + \mathbf{M}), \tag{1.3d}$$

where $\mathbf{H}$ and $\mathbf{D}$ are auxiliary fields. Typically, in this book, there are no free charges, $\rho_f = 0$, nor free currents, i.e. $\mathbf{J}_f = 0$.

- $\mathbf{P}$ is the induced polarization, which is introduced using a microscopic description and the dipole momentum $\mathbf{p} = \varepsilon_0 \alpha \mathbf{E}$.

- Linear induced polarization density $\mathbf{P} = \varepsilon_0 \chi^{(1)} \mathbf{E}$, where $\chi^{(1)}$ is the susceptibility. $\mathbf{D} = \varepsilon_0 \mathbf{E} + \mathbf{P} = \varepsilon_0(1 + \chi^{(1)})\mathbf{E} = \varepsilon_0 \varepsilon_r \mathbf{E}$.

- In lossless materials the refractive index n is defined through $\chi^{(1)}$ or equally well through the relative permittivity ε_r.

- Materials may be described by dipoles that interact mutually, this needs to be considered if a macroscopic quantity as the refractive index is predicted from microscopic quantities like the microscopic polarizability. This is done using local field correction

$$\frac{N\alpha}{3} = \frac{\epsilon_r - 1}{2 + \epsilon_r}. \tag{1.88}$$

- The power in the electromagnetic field may be evaluated through the Poynting vector

$$\mathbf{S} = \mathbf{E} \times \mathbf{H}. \tag{1.27}$$

For a harmonically time varying field Poynting vector may be evaluated using complex notation $\bar{\mathbf{S}} = \frac{1}{2}\mathrm{Re}\left[\mathbf{E}^0 \times (\mathbf{H}^0)^*\right]$.

- The state of polarization of a propagating electric field is essential. An electric field having a specific state of polarization may be written as

$$\begin{pmatrix} E_x^0 \\ E_y^0 \end{pmatrix} = E^0 \left(\cos(\theta_p) \begin{pmatrix} \cos(\psi_p) \\ \sin(\psi_p) \end{pmatrix} + i \sin(\theta_p) \begin{pmatrix} -\sin(\psi_p) \\ \cos(\psi_p) \end{pmatrix} \right), \tag{1.54}$$

where the angles ψ_p and θ_p define the state of polarization.

Chapter 2

Time-domain material response

Contents

A nonlinear material response may occur if for example either there is a resonance between the light and some natural oscillation mode of the medium, or if the intensity of the light is sufficiently strong to merit a nonlinear material response. Direct resonance can occur in isolated intervals of the electromagnetic spectrum at

- Ultraviolet and visible frequencies (10^{15} s^{-1}) where the oscillator corresponds to an electronic transition of the medium
- Infrared (10^{13} s^{-1}), where the medium has vibrational modes
- Far infrared-microwave range (10^{11} s^{-1}), where there are rotational modes.

These interactions are also called **one-photon processes**. An advantage of a resonant interaction between the light and matter is that the nonlinearity is significantly enhanced. However, a general drawback with the resonance-enhancement is that the response time of the induced polarization density of the material is slowed down, affecting applications such as optical switching or modulation where speed is of great importance. A proper description of resonant nonlinearities often involves quantum mechanical descriptions, and is not given here. We restrict ourselves to non-resonance effects.

Low frequency modes (phonons) of the material can be excited at optical frequencies (10^{15} s^{-1}) through indirect resonant processes in which the difference in frequency between the initial and final state of light and the wave vectors of the two light beams match the frequency and wave vector of the excited lower frequency mode of the material. In the case where the 'lower frequency mode' is a rotational or a vibrational mode, the process is called Raman scattering. If the lower frequency mode instead is an acoustic mode, the process is called Brillouin scattering. We return to these processes in later chapters of this text.

2.1 The polarization time-response function

Our starting point is a series expansion of the induced electric polarization

$$\mathbf{P}(\mathbf{r},t) = \mathbf{P}^{(0)}(\mathbf{r},t) + \mathbf{P}^{(1)}(\mathbf{r},t) + \mathbf{P}^{(2)}(\mathbf{r},t) + \ldots + \mathbf{P}^{(n)}(\mathbf{r},t) + \cdots, \tag{2.1}$$

where $\mathbf{P}^{(0)}$ is the static electric polarization, i.e. independent of the electric field, as for example found in some crystals. Consequently, the static polarization is normally independent of time. $\mathbf{P}^{(1)}$ is the linear induced polarization, i.e. linear in the electric field, $\mathbf{P}^{(2)}$ is the quadratic induced polarization, i.e. quadratic in the electric field, etc.

Throughout this text we consider only the local response, where it is assumed that the induced polarization is determined solely by the electric field at the actual observation point in space. This implies that we can avoid the complication of an extra integral over all points in space when evaluating the induced polarization. It is noted that nonlocal optics is an important field for example in nano-photonics [21][22][23].

In addition, we assume that the material parameters are invariant in time, such that $\mathbf{P}(\mathbf{r},t)$ is identical to the polarization obtained if it were induced by a time-displaced electric field, only with the output, i.e. the resultant induced polarization, shifted in time in accordance with the applied electric field. This is referred to as time invariance.

2.1.1 The linear response function

The linear induced polarization $\mathbf{P}^{(1)}(\mathbf{r},t)$ is linear in the electric field, and the μth component of the induced polarization is related to the αth component of the electric field through

$$P_{\mu}^{(1)}(\mathbf{r},t) = \varepsilon_0 \int_{-\infty}^{\infty} T_{\mu\alpha}^{(1)}(t;\tau) E_{\alpha}(\mathbf{r},\tau) d\tau, \tag{2.2}$$

where $T^{(1)}_{\mu\alpha}(t;\tau)$ is a rank-two tensor, i.e. a 3×3 matrix in units of [s^{-1}] that describes the linear material response at observation time t due to an applied electric field at time τ, and (μ, α) are typically a subset of the Cartesian coordinates x, y or z. Note that a summation over index α is implicitly implied; for completeness, this notation is referred to as the Einstein notation. More details on tensors are given in Appendix A. By performing the integration, all contributions in time from the electric field of the light are weighted to take into account that the material may have a delayed response. It is required that $T^{(1)}_{\mu\alpha}(t;\tau)$ is causal, i.e. that no optically induced polarization occurs before the field is applied, that is $T^{(1)}_{\mu\alpha}(t;\tau) = 0$ for $t \leq \tau$. In addition to being causal, it is also noted that since both the induced polarization as well as the applied electric field are measurable, i.e. real quantities, the response function must be real.

By applying time invariance and causality to Eqn. (2.2), it can be shown that the linear induced polarization only depends on the difference between the observation time and the time of the action of the applied electric field [2]. Starting from Eqn. (2.2) and applying time invariance

$$T^{(1)}(t;\tau) \equiv R^{(1)}(t-\tau), \tag{2.3}$$

where $R^{(1)}(t-\tau)$ is a second-rank tensor depending only on the time difference $(t-\tau)$. Through variable change $(\tau \rightarrow t-\tau')$ the polarization response tensor $R^{(1)}(\tau)$ is introduced. In analog to the constraints discussed for $T^{(1)}_{\mu\alpha}(t;\tau)$, $R^{(1)}(t-\tau)$ is subject to the two following restrictions

- Causality: The response must equal zero for $\tau < 0$, $R^{(1)}(\tau) \equiv 0$ for $\tau < 0$.
- Reality: The response function must be real.

From Eqn. (2.2) and Eqn. (2.3), the induced polarization is evaluated through a convolution of the material response and the applied electric field. That is

$$\begin{aligned} P^{(1)}_{\mu}(\mathbf{r},t) &= \varepsilon_0 \int_{-\infty}^{\infty} R^{(1)}_{\mu\alpha}(t-\tau) E_{\alpha}(\mathbf{r},\tau) d\tau \\ &= \varepsilon_0 \int_{-\infty}^{\infty} R^{(1)}_{\mu\alpha}(\tau) E_{\alpha}(\mathbf{r},t-\tau) d\tau. \end{aligned} \tag{2.4}$$

It is noted that some materials, or some phenomena, have a very fast response time and some have a very slow response time. This is of course due to the physical mechanisms responsible for the response; some processes are fast, some are slow. In relation to this, one has to recall that in optics there are often two timescales, the timescale of the carrier wave, 10^{-14} s, and the timescale of an optical envelope pulse 10^{-12} s to 10^{-13} s. This is illustrated in Fig. 2.1, which shows the electrical field of a 100 fs Gaussian pulse modulated onto a carrier wave at 1555 nm. From the figure, the fast harmonic oscillation of the carrier wave is visible, and, in addition, there is also a much slower envelope, which in this case is a Gaussian envelope, superimposed on the carrier wave.

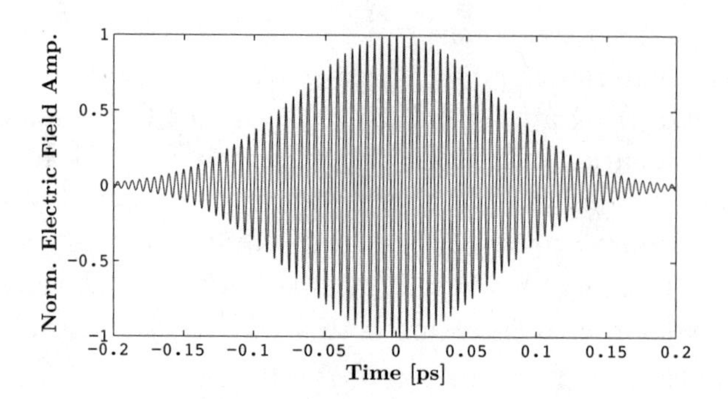

Figure 2.1: A Gaussian envelope on a 1555 nm carrier wave. The envelope is $f(t) = e^{-(t/T_0)^2}$, $T_0 = 100$ fs. The number of cycles in a Gaussian envelope counted to the e^{-1} width equals $2T_0\omega/(2\pi)$ where ω is the radial frequency of the light. Here, this corresponds to 40 cycles. In the figure, the pulse is assumed to propagate in vacuum.

2.1.2 The quadratic response function

Similar to the linear induced polarization, the quadratic response is a sum of all infinitesimal contributions in time. In the quadratic induced polarization though, not all contributions originate from the same time scale, and hence the proper formulation of the second-order induced polarization density is a double integral,

$$P_\mu^{(2)}(\mathbf{r}, t) = \varepsilon_0 \int_{-\infty}^{\infty} \int_{-\infty}^{\infty} T_{\mu\alpha\beta}^{(2)}(t; \tau_1, \tau_2) E_\alpha(\mathbf{r}, \tau_1) E_\beta(\mathbf{r}, \tau_2) d\tau_1 d\tau_2, \tag{2.5}$$

where again a summation over α and β is implied. The tensor $T_{\mu\alpha\beta}^{(2)}(t; \tau_1, \tau_2)$ uniquely determines the quadratic, second-order induced polarization of the medium, in units of [m/(Vs^2)]. However, the tensor $T_{\mu\alpha\beta}^{(2)}(t; \tau_1, \tau_2)$ is not unique in itself, implying that contributions to the right-hand side from $E_\alpha(\mathbf{r}, \tau_1)E_\beta(\mathbf{r}, \tau_2)$ and $E_\beta(\mathbf{r}, \tau_2)E_\alpha(\mathbf{r}, \tau_1)$ that correspond to two different tensor elements are not the same. However, we may choose $T_{\mu\alpha\beta}^{(2)}(t; \tau_1, \tau_2)$ to be unique and symmetric. To see this, the tensor is expressed as a sum of a symmetric and antisymmetric part under interchange of the index and time variable pairs (α, τ_1) and (β, τ_2)

$$T_{\mu\alpha\beta}^{(2)}(t; \tau_1, \tau_2) = S_{\mu\alpha\beta}^{(2)}(t; \tau_1, \tau_2) + A_{\mu\alpha\beta}^{(2)}(t; \tau_1, \tau_2), \tag{2.6}$$

where

$$S_{\mu\alpha\beta}^{(2)}(t; \tau_1, \tau_2) = \frac{1}{2}\left[T_{\mu\alpha\beta}^{(2)}(t; \tau_1, \tau_2) + T_{\mu\beta\alpha}^{(2)}(t; \tau_2, \tau_1)\right] \tag{2.7}$$

and

$$A_{\mu\alpha\beta}^{(2)}(t; \tau_1, \tau_2) = \frac{1}{2}\left[T_{\mu\alpha\beta}^{(2)}(t; \tau_1, \tau_2) - T_{\mu\beta\alpha}^{(2)}(t; \tau_2, \tau_1)\right]. \tag{2.8}$$

By definition the antisymmetric tensor satisfies $A_{\mu\alpha\beta}^{(2)}(t; \tau_1, \tau_2) = -A_{\mu\beta\alpha}^{(2)}(t; \tau_2, \tau_1)$, and the symmetric tensor satisfies $S_{\mu\alpha\beta}^{(2)}(t; \tau_1, \tau_2) = S_{\mu\beta\alpha}^{(2)}(t; \tau_2, \tau_1)$. Hence, the second-order polarization is

$$P_{\mu}^{(2)}(\mathbf{r},t) = \varepsilon_0 \int_{-\infty}^{\infty}\int_{-\infty}^{\infty} T_{\mu\alpha\beta}^{(2)}(t;\tau_1,\tau_2)E_{\alpha}(\mathbf{r},\tau_1)E_{\beta}(\mathbf{r},\tau_2)d\tau_1 d\tau_2 \tag{2.9}$$

$$= \varepsilon_0 \int_{-\infty}^{\infty}\int_{-\infty}^{\infty} \left[S_{\mu\alpha\beta}^{(2)}(t;\tau_1,\tau_2) + A_{\mu\alpha\beta}^{(2)}(t;\tau_1,\tau_2)\right] E_{\alpha}(\mathbf{r},\tau_1)E_{\beta}(\mathbf{r},\tau_2)d\tau_1 d\tau_2.$$

When evaluating the sum over $\alpha\beta$ the contributions from $A_{\mu\alpha\beta}^{(2)}(t;\tau_1,\tau_2)E_{\alpha}(\mathbf{r},\tau_1)$ $E_{\beta}(\mathbf{r},\tau_2)$ and $A_{\mu\beta}\alpha^{(2)}(t;\tau_1,\tau_2)E_{\beta}(\mathbf{r},\tau_1)E_{\alpha}(\mathbf{r},\tau_2)$, for all $\alpha\beta$ counterbalance each other because of the tensor A being antisymmetric. Thus, the response tensor may now be chosen uniquely as symmetric, i.e. $T_{\mu\alpha\beta}^{(2)}(t;\tau_1,\tau_2) = T_{\mu\beta\alpha}^{(2)}(t;\tau_2,\tau_1)$. This property and its generalization to higher rank tensors is a very important result, known as **intrinsic permutation symmetry**. The intrinsic permutation symmetry holds whether the nonlinear interaction under analysis is highly resonant or far from resonance. We return to this symmetry and its implications in the following sections.

By again applying the arguments of time invariance, as previously for the linear response function, we find for the quadratic response

$$T_{\mu\alpha\beta}^{(2)}(t+t_0;\tau_1,\tau_2) = T_{\mu\alpha\beta}^{(2)}(t;\tau_1 - t_0,\tau_2 - t_0) \tag{2.10}$$

for all t, t_0 τ_1 and τ_2. By setting $t = 0$, and replacing t_0 with t, it is shown that the tensor depends only on the differences $t - \tau_1$ and $t - \tau_2$. This allows the introduction of the quadratic polarization response function $R_{\mu\alpha\beta}^{(2)}(\tau_1,\tau_2)$, which has the following properties

- causality requirement: $R_{\mu\alpha\beta}^{(2)}(\tau_1,\tau_2) = 0$ if either $\tau_1 < 0$ or $\tau_2 < 0$.
- reality condition: $R_{\mu\alpha\beta}^{(2)}(\tau_1,\tau_2)$ is real.
- intrinsic permutation symmetry: $R_{\mu\alpha\beta}^{(2)}(\tau_1,\tau_2)$ is invariant under the change of pairs (α,τ_1) and (β,τ_2).

The second-order induced polarization is then expressed as

$$P_{\mu}^{(2)}(\mathbf{r},t) = \varepsilon_0 \int_{-\infty}^{\infty}\int_{-\infty}^{\infty} R_{\mu\alpha\beta}^{(2)}(t-\tau_1,t-\tau_2)E_{\alpha}(\mathbf{r},\tau_1)E_{\beta}(\mathbf{r},\tau_2)d\tau_1 d\tau_2 \tag{2.11}$$

$$= \varepsilon_0 \int_{-\infty}^{\infty}\int_{-\infty}^{\infty} R_{\mu\alpha\beta}^{(2)}(\tau_1',\tau_2')E_{\alpha}(\mathbf{r},t-\tau_1')E_{\beta}(\mathbf{r},t-\tau_2')d\tau_1' d\tau_2'.$$

Example 2.1. *Second-order response functions*
A second-order material response function is given by

$$T_{\mu\alpha\beta}^{(2)}(t;\tau_1,\tau_2) = T_{iii}^{(2)}(t;\tau_1,\tau_2) = A\exp\left[-(t-\tau_1)^2/T_1^2\right]\exp\left[-(t-\tau_2)^2/T_2^2\right], \tag{2.12}$$

for $t > \tau_1$, $t > \tau_2$, $(i = x, y, z)$ and otherwise $T^{(2)}_{\mu\alpha\beta}(t; \tau_1, \tau_2) = 0$, where A is an amplitude, T_1 and T_2 two time constants and τ_1 and τ_2 two different times in the applied electric field, and t the time of the observation. Typically A is chosen such that the response is normalized i.e. by setting the observation time t equal zero

$$\int_{-\infty}^{\infty}\int_{-\infty}^{\infty} A \exp\left[-(\tau_1)^2/T_1^2\right] \exp\left[-(\tau_2)^2/T_2^2\right] d\tau_1 d\tau_2 \equiv 1. \tag{2.13}$$

In the considered case the double integral decouples and the integration is simple. By performing the integration we get $A = (\pi T_1 T_2)^{-1}$.

It is noted that when T_1 and T_2 approach zero, then the response becomes

$$T^{(2)}_{iii}(t; \tau_1, \tau_2) = A\delta(t - \tau_1)\delta(t - \tau_2). \tag{2.14}$$

The induced polarization from this response equals

$$P_i^{(2)}(\mathbf{r}, t) = \varepsilon_0 \int_{-\infty}^{\infty}\int_{-\infty}^{\infty} A\delta(t - \tau_1)\delta(t - \tau_2) E_i(\mathbf{r}, \tau_1) E_i(\mathbf{r}, \tau_2) d\tau_1 d\tau_2. \tag{2.15}$$

By integrating over τ_1 we get

$$P_i^{(2)}(\mathbf{r}, t) = \varepsilon_0 E_i(t) \int_{-\infty}^{\infty} A\delta(t - \tau_2) E_i(\mathbf{r}, \tau_2) d\tau_2, \tag{2.16}$$

and after the final integration over τ_2 we arrive at the final result

$$P_i^{(2)}(\mathbf{r}, t) = \varepsilon_0 A E_i^2(\mathbf{r}, t). \tag{2.17}$$

—— ■ ——

2.1.3 Higher-order response function

The nth-order induced polarization is in general given by

$$P_\mu^{(n)}(\mathbf{r}, t) = \varepsilon_0 \int_{-\infty}^{\infty} \cdots \int_{-\infty}^{\infty} T^{(n)}_{\mu\alpha_1\cdots\alpha_n}(t; \tau_1, \ldots, \tau_n) E_{\alpha_1}(\mathbf{r}, \tau_1) \cdots E_{\alpha_n}(\mathbf{r}, \tau_n) d\tau_1 \cdots d\tau_n, \tag{2.18}$$

where $T^{(n)}_{\mu\alpha_1\cdots\alpha_n}(t; \tau_1, \ldots, \tau_n)$ is a tensor of rank $n + 1$ in units $[\mathrm{s}^{-1}(\mathrm{m}/(\mathrm{Vs}))^{(n-1)}]$. Similar to the quadratic response the nth-order response may be expressed as a symmetric response and an antisymmetric response. The symmetric response is obtained by specifying a symmetric tensor

$$S^{(n)}_{\mu\alpha_1\cdots\alpha_n}(t; \tau_1, \ldots, \tau_n) = \frac{1}{n!} \sum_{\text{permutations}} T^{(n)}_{\mu\alpha_1\cdots\alpha_n}(t; \tau_1, \ldots, \tau_n), \tag{2.19}$$

where the sum is over $n!$ permutations of the n pairs (α_1, τ_1), $(\alpha_2, \tau_2) \cdots (\alpha_n, \tau_n)$. In analogy to the quadratic polarization, the antisymmetric contribution may be set to zero leaving only the symmetric nth-order polarization. Applying the principles of time-invariance and causality, an nth-order polarization response tensor $R^{(n)}$ of rank $n+1$ is derived. The response equals zero if any τ_i is negative, and is invariant under any of the $n!$ permutations of the n pairs (α_1, τ_1), (α_2, τ_2) and (α_n, τ_n).

The nth-order induced polarization is then expressed as

$$\begin{aligned} P_\mu^{(n)}(\mathbf{r},t) &= \varepsilon_0 \int_{-\infty}^{\infty} \cdots \int_{-\infty}^{\infty} R^{(n)}_{\mu\alpha_1\cdots\alpha_n}(t-\tau_1,\ldots,t-\tau_n) E_{\alpha_1}(\mathbf{r},\tau_1)\ldots E_{\alpha_n}(\mathbf{r},\tau_n) d\tau_1\cdots d\tau_n \\ &= \varepsilon_0 \int_{-\infty}^{\infty} \cdots \int_{-\infty}^{\infty} R^{(n)}_{\mu\alpha\cdots\alpha_n}(\tau_1',\ldots,\tau_n') E_{\alpha_1}(\mathbf{r},t-\tau_1')\ldots E_{\alpha_n}(\mathbf{r},t-\tau_n') d\tau_1'\cdots d\tau_n'. \end{aligned} \tag{2.20}$$

Example 2.2. *Third-order response function*
Assume that a third-order response function equals

$$T^{(3)}_{\mu\alpha\beta}(t;t_1,t_2,t_3) = A\delta(t-t_1)f(t_1-t_2)\delta(t_2-t_3). \tag{2.21}$$

In this case the induced polarization equals

$$\begin{aligned} P_i(t) &= \varepsilon_0 \int_{t_1}\int_{t_2}\int_{t_3} T^{(3)}_{ijkl}(t;t_1,t_2,t_3)E_j(t_1)E_k(t_2)E_l(t_3)dt_1dt_2dt_3 \\ &= \varepsilon_0 E_j(t) A \int_{t_2} f(t-t_2)E_k(t_2)E_l(t_2)dt_2. \end{aligned} \tag{2.22}$$

—— ■ ——

2.2 The Born-Oppenheimer approximation*

Within the Born-Oppenheimer approximation it is possible to derive general expressions for the nonlinear susceptibility to various orders. This requires a quantum mechanical approach. However, the reader only interested in the results of the Born-Oppenheimer approximation may read the introduction to Figure 2.2 and jump to the results in Eqn. (2.62), page 56.

It is desirable to set up a model that describes the induced polarization in various materials. Different approaches have been pursued to do this. In the following the so-called Born-Oppenheimer approach is outlined (Not to be confused with the Born approximation, see for example [24]). The description that follows is based on the work of Hellwarth [17]. It is noted that the intention is not to give a full derivation, since this is considered to be outside the scope of this text. The purpose is merely to illustrate and discuss the results of this model, its assumptions, validity of approximations, and its applications. Regarding applications of the model, we focus on the third-order induced polarization.

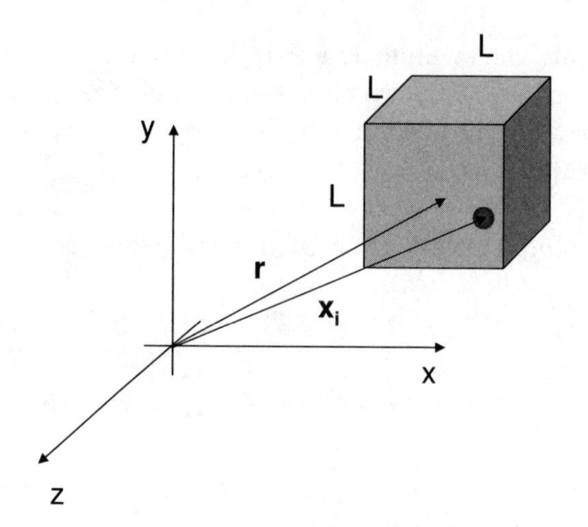

Figure 2.2: Sub-volume $V = L^3$ centered at position $\mathbf{r}$. The sphere indicates a charge at position x_i within the box.

In some descriptions of the nonlinear induced polarization, for example in gasses, atoms or molecules are assumed to be independent from each other. This is not the case in dense media [17] like liquids, glasses and crystalline solids; not even when corrections are made such as local field corrections. For such dense media, atoms or molecules are not independent from each other. As a consequence the Born-Oppenheimer (BO) approximation is very useful. Using the BO approximation, useful relations of the material response is obtained. These relations are nearly exact when all electromagnetic frequencies involved are much lower than any electronic frequencies, around 10^{15} Hz. Then the electronic motions can follow the changing fields and the moving nuclei adiabatically.

In the BO approximation the Hamiltonian, i.e. the operator corresponding to the energy of the system, is separated into a sum of kinetic and potential energy of all particles within the system. In the case described here the Hamiltonian is separated into a sum of the following contributions

- Kinetic energy of the nuclei
- Kinetic energy of the electrons
- Coulomb interactions between nuclei
- Coulomb interactions between nuclei and electrons
- Coulomb forces among electrons
- Interaction between applied field and electrons
- Interaction between applied field and nuclei

The origin of the BO approximation is to assumed that the nuclei in the molecules are virtually standing still relative to the electrons. This approximation is very important and is referred to as the BO approximation. As a consequence of this, the energy of a molecule may be calculated as a sum of electronic energy, vibrational energy, and rotational energy. The validity of this approximation is based on the small electron-to- nuclear mass ratio, which is 1/1800 even for the case of the smallest ratio, i.e. in a hydrogen atom.

The BO approximation is valid in media where the difference between energy levels of electronic states in the media is much higher than the energy corresponding to the energy of the optical frequencies present, i.e. again we restrict our analysis to transparent media where all relevant optical frequencies are well below any electronic resonances.

The expectation value of the induced polarization density $\mathbf{P}(t)$ in units [C/m^2] in a small volume V at position r, see Figure 2.2, may be calculated using the dipole moment density operator. By denoting the μth-component of the dipole moment density operator in the Schrödinger picture, $\hat{p}_\mu^S$ the μth-component of induced polarization density is

$$P_\mu(t) = \langle \hat{U}^{-1}(t)\hat{p}_\mu^S \hat{U}(t)\rangle, \tag{2.23}$$

where $\hat{U}(t)$ is the unitary time evolution operator defined from the BO Hamiltonian. Note, in the Schrödinger picture the dipole moment density operator is time independent. The brackets $\langle\rangle$ indicate expectation value.

By operating from left and right by $\hat{U}^{-1}$ and $\hat{U}$ respectively, the dipole moment density operator is transferred into the Heisenberg picture, in the following denoted $\hat{p}^H$, where the operator is time-dependent.

The evolution operator $\hat{U}(t)$ is found from

$$i\hbar\frac{d\hat{U}}{dt} = [\hat{H}_0 + \hat{H}_E^S]\hat{U}, \tag{2.24}$$

where $\hat{H}_0$ is the unperturbed Hamiltonian, i.e. no applied electric field ($E = 0$), and independent upon time, and where $\hat{H}_E^S$ is the interaction Hamiltonian in the Schrödinger picture describing the interaction between the material, i.e. electrons and nuclei and the macroscopic electric field $E(t)$, which may be time dependent. It is noted that to bring an operator into the interaction picture, $\hat{H}_E^S = 0$ and one is left with $\hat{U}_0$. If $\hat{\mathcal{O}}^S$ is an operator in the Schrödinger picture, then $\hat{\mathcal{O}}^I = \hat{U}_0^\dagger \hat{\mathcal{O}}^S \hat{U}_0$ is the operator in the interaction picture and $\hat{\mathcal{O}}^H = \hat{U}^\dagger \hat{\mathcal{O}}^S \hat{U}$ the operator in the Heisenberg picture. In Eqn. (2.24) the interaction Hamiltonian is in the Schrödinger picture. For linear dipolar interactions it takes the form

$$\hat{H}_E^S = -\hat{\mathbf{Q}}_L^S \cdot \mathbf{E}(t), \tag{2.25}$$

where $\mathbf{E}(t)$ is the applied macroscopic electric field, not an operator, and $\hat{\mathbf{Q}}_L^S$ is the linear dipole moment operator.

In general, a quantum mechanical system is completely described by an operator $\hat{O}$ acting on any state, denoted $|\psi\rangle$, of the system which then produces or changes to a new state. An observable may be associated with an operator. Here, for example, the induced polarization $P(t)$ is associated with the dipole moment density operator $\hat{p}$, either in the Schrödinger, the Heisenberg or the Interaction picture. The mean value, or the expectation value of the observable is found from measurements on an ensemble of identically prepared systems.

If we do not have enough information to specify the state vector of a system but know the probabilities $\wp_n$ that the system is in a normalized state $|\psi_n\rangle$ then the mean value of an observable A with associated operator $\hat{A}$ is

$$A = \sum_n \wp_n \langle\psi_n|\hat{A}|\psi_n\rangle. \tag{2.26}$$

In such cases, it is useful to introduce the density matrix operator $\hat{\rho}$, which is defined through

$$\hat{\rho} = \sum_n \wp_n |\psi_n\rangle\langle\psi_n|. \tag{2.27}$$

The mean value of A is then given by

$$A = \langle\hat{A}\rangle = \mathrm{Tr}(\hat{\rho}\hat{A}). \tag{2.28}$$

In the Heisenberg picture $\hat{\rho}$ is independent upon time [25], and may as well be replaced by ρ_0. It is noted that it does not matter which picture one is in i.e. $\langle\hat{\rho}^H\hat{O}^H\rangle = \langle\hat{\rho}^S\hat{O}^S\rangle$. In the following, we work in the Heisenberg picture where the density operator is independent of time. From this, the expectation value of the dipole moment is

$$\langle P(t)\rangle = \mathrm{Tr}[\hat{\rho}_0\hat{p}^H], \tag{2.29}$$

where $\hat{p}^H$ is the dipole moment density operator in the Heisenberg picture, i.e. timedependent $(\hat{p}^H(t))$, which may be expressed from the dipole moment density operator in the Schrödinger picture $(\hat{p}^S)$ through $\hat{p}^H = \hat{U}^{-1}(t)\hat{p}^S\hat{U}(t)$. Inserting this in Eqn. (2.29) we get

$$\langle P(t)\rangle = \mathrm{Tr}[\hat{\rho}_0\hat{U}^{-1}(t)\hat{p}^S\hat{U}(t)]. \tag{2.30}$$

To continue we note that the solution to Eqn. (2.24) is

$$\hat{U}(t) = \hat{U}(t=0)\exp\left[\hat{H}_o t/i\hbar\right]\exp\left[\int \hat{H}_E^S dt'/i\hbar\right]. \tag{2.31}$$

If $\hat{U}(t=0) = 1$ then this solution may be written as: $\hat{U} = \hat{U}_0\hat{U}_E$ where $\hat{U}_0 = \exp\left[\hat{H}_o t/i\hbar\right]$ and $\hat{U}_E = \exp\left[\int \hat{H}_E^S dt'/i\hbar\right]$, that is $\hat{U}_E = \hat{U}_0^\dagger\hat{U}$.

We may now rewrite the Heisenberg dipole moment operator $\hat{p}^H = \hat{U}^{-1}(t)\hat{p}^S\hat{U}(t)$ in terms of the dipole moment density operator in the interaction picture. This is done by utilizing that $\hat{U}_0$ is unitary i.e., $\hat{U}_0\hat{U}_0^\dagger = \hat{U}_0^\dagger\hat{U}_0 = 1$. Inserting this into the expression for $\hat{p}^H$ we get

$$\hat{U}^{-1}(t)\hat{p}^S\hat{U}(t) = \hat{U}^{-1}(t)\hat{U}_0\hat{U}_0^\dagger\hat{p}^S\hat{U}_0\hat{U}_0^\dagger\hat{U}(t), \tag{2.32}$$

since $\hat{U}_E = \hat{U}_0^\dagger\hat{U}$ then $\hat{U}^{-1}(t)\hat{U}_0\hat{U}_0^\dagger\hat{p}^S\hat{U}_0\hat{U}_0^\dagger\hat{U}(t)$ may be written as $\hat{U}_E^\dagger\hat{U}_0^\dagger\hat{p}^S\hat{U}_0\hat{U}_E = \hat{U}_E^\dagger\hat{p}^I\hat{U}_E$, where $\hat{p}^I = \hat{U}_0^\dagger\hat{p}^S\hat{U}_0$ is the dipole moment density operator in the interaction picture.

We now need to find an expression for $\hat{U}_E(t)$. Since $\hat{U}_E = \hat{U}_0^\dagger\hat{U}$ then by using Eqn. (2.24) we find

$$\begin{aligned} i\hbar\frac{d}{dt}\hat{U}_E &= -(i\hbar\frac{d}{dt}\hat{U}_0)^\dagger\hat{U} + \hat{U}_0^\dagger(i\hbar\frac{d}{dt}\hat{U}) \\ &= -\hat{U}_0^\dagger\hat{H}_0\hat{U} + \hat{U}_0^\dagger(\hat{H}_0 + \hat{H}_E^S)\hat{U} = \hat{U}_0^\dagger\hat{H}_E^S\hat{U}, \end{aligned} \tag{2.33}$$

where we have used $(H_0U_0)^\dagger = U_0^\dagger H_0^\dagger$ and since the Hamiltonian is Hermitian then $U_0^\dagger H_0^\dagger = U_0^\dagger H_0$. Applying as before $\hat{U}_0\hat{U}_0^\dagger$ we may rewrite $\hat{U}_0^\dagger\hat{H}_E^S\hat{U}$ as

$$\hat{U}_0^\dagger\hat{H}_E^S\hat{U}_0\hat{U}_0^\dagger\hat{U} = \hat{H}_E^I\hat{U}_0^\dagger\hat{U} = \hat{H}_E^I\hat{U}_E. \tag{2.34}$$

When $\hat{U}_E(0) = 1$, the solution to this may be written as an expansion

$$\begin{aligned} \hat{U}_E(t) &= 1 + \frac{-i}{\hbar}\int_0^t \hat{H}_E^I(t_1)\hat{U}_E(t_1)dt_1 \\ &= 1 + \frac{-i}{\hbar}\int_0^t \hat{H}_E^I(t_1)\left(1 + \frac{-i}{\hbar}\int_0^{t_1}\hat{H}_E^I(t_2)\hat{U}_E(t_2)dt_2\right)dt_1 \\ &= 1 + \frac{-i}{\hbar}\int_0^t \hat{H}_E^I(t_1)dt_1 + \left(\frac{-i}{\hbar}\right)^2\int_0^{t_1}\int_0^t \hat{H}_E^I(t_1)\hat{H}_E^I(t_2)dt_2dt_1 \\ &+ \left(\frac{-i}{\hbar}\right)^3\int_0^t\int_0^{t_1}\int_0^{t_2}\hat{H}_E^I(t_1)\hat{H}_E^I(t_2)\hat{H}_E^I(t_3)\hat{U}_E(t_3)dt_3dt_2dt_1. \end{aligned} \tag{2.35}$$

It is noted that the integration may start at $-\infty$, since first of all it is assumed that the electric field is non-existing prior to $t = 0$. Consequently $\hat{H}_E^I(t)$ is zero for $t < 0$. We can now let $\hat{U}_E^\dagger$ and $\hat{U}_E$ operate on $\hat{p}^I$ and finally we get

$$\begin{aligned} \hat{p}_i^H(t) = \hat{U}_E^{-1}\hat{p}_i^I\hat{U}_E &= \hat{p}_i^I(t) + \frac{1}{i\hbar}\int_0^t\left[\hat{p}^I(t), \hat{H}_E^I(t_1)\right]dt_1 \\ &+ \left(\frac{-i}{\hbar}\right)^2\int_0^{t_1}\int_0^t\left[\left[\hat{p}_i^I(t), \hat{H}_E^I(t_1)\right], \hat{H}_E^I(t_2)\right]dt_2dt_1 + \cdots. \end{aligned} \tag{2.36}$$

To evaluate the expansion of the dipole moment density operator in the Heisenberg picture i.e. Eqn. (2.36) we adopt the Hamiltonian from Appendix B

$$\begin{aligned}\hat{H}_E^I(t) &= \left[-\hat{m}_i^I E_i - \frac{1}{2}\hat{\alpha}_{ij}^I E_i E_j - \frac{1}{3}\hat{\beta}_{ijk}^I E_i E_j E_k + \cdots\right] V \\ &= \hat{v}_1^I + \hat{v}_2^I + \hat{v}_3^I + \cdots\end{aligned} \tag{2.37}$$

and also from Appendix B the dipole moment density operator

$$\begin{aligned}\hat{p}_i^I(t) &= \hat{m}_i^I + \hat{\alpha}_{ij}^I E_j(t) + \hat{\beta}_{ijk}^I E_j(t) E_k(t) + \hat{\gamma}_{ijkl}^I E_j(t) E_k(t) E_l(t) + \cdots \\ &= \hat{p}_{i0}^I + \hat{p}_{i1}^I + \hat{p}_{i2}^I + \hat{p}_{i3}^I + \cdots .\end{aligned} \tag{2.38}$$

It is noted that the Hamiltonian is a scalar whereas the dipole moment density operator is a vector, of which one component is given by Eqn. (2.38).

We may now insert the expansions of the dipole moment density operator and the interaction Hamiltonian into Eqn. (2.36) to evaluate the dipole moment density operator in the Heisenberg picture. This, and hence also the induced polarization, is an expansion to different orders in the applied electric field.

The **zero-order** or static induced polarization is the corresponding static dipole moment density operator. By inspection of Eqn. (2.36) with the application of Eqn. (2.37) and (2.38), it is seen that only $\hat{p}_{i0}^I$ contribute to this.

The **first-order** dipole moment density operator has two contributions. One from $\hat{p}_{i1}^I$ and by evaluation the term $[\hat{p}_i^I(t), \hat{H}_i^I(t_1)]$, which is calculated through the commutator $[\hat{p}_{i0}^I + \hat{p}_{i1}^I + \hat{p}_{i2}^I + \hat{p}_{i3}^I, \hat{v}_1^I + \hat{v}_2^I + \hat{v}_3^I]$, one identifies the contribution $[\hat{p}_{i0}^I, \hat{v}_1^I]$ to the first-order induced polarization. Thus in summary

$$\hat{p}_i^H\Big|^{(1)}(t) = \hat{p}_{i1}^I + \frac{1}{i\hbar}\int_0^t [\hat{p}_{i0}^I, \hat{v}_1^I]\, dt_1. \tag{2.39}$$

It is noted that all other terms in Eqn. (2.36) are of higher-order in the electric field. Note, the notation $\hat{p}_i^H\Big|^{(1)}(t)$ shows that we evaluate the ith component of the dipole moment density operator in the Heisenberg picture, i.e. the operator is time dependent and that we, for now, just focus on the term that depends linearly on the electric field.

Considering the **third-order** dipole moment density operator, we find that this has a total of eight contributions. One from $\hat{p}_{i3}^I$, three from the commutator $[\hat{p}_i^I(t), \hat{H}_E^I(t_1)]$, more specifically, these are $[\hat{p}_{i0}^I(t), \hat{v}_3^I(t_1)] + [\hat{p}_{i1}^I(t), \hat{v}_2^I(t_1)] + [\hat{p}_{i2}^I(t), \hat{v}_1^I(t_1)]$, three contributions from $[[\hat{p}_i^I(t), \hat{H}_E^I(t_1)], \hat{H}_E^I(t_2)]$ i.e. $[[\hat{p}_{i0}^I(t), \hat{v}_1^I(t_1)], \hat{v}_2^I(t_2)] + [[\hat{p}_{i0}^I, \hat{v}_2^I(t_1)], \hat{v}_1^I(t_2)] + [[\hat{p}_{i1}^I(t), \hat{v}_1^I(t_1)], \hat{v}_1^I(t_1)]$, and finally one from $[[[\hat{p}_i^I(t), \hat{H}_E^I(t_1)], \hat{H}_E^I(t_2)], \hat{H}_E^I(t_3)]$, which equal $[[[\hat{p}_{i0}^I, \hat{v}_1^I(t_1)], \hat{v}_1^I(t_2)], \hat{v}_1^I(t_3)]$.

In summary, the third-order dipole moment density operator has eight contributions which are obtained by inserting Eqn. (2.37) and Eqn. (2.38) into Eqn. (2.36).

$$\begin{aligned}\hat{p}_i^H\Big|^{(3)}(t) &= \hat{p}_{i3}^I(t) + \frac{1}{i\hbar}\int_0^t [\hat{p}_{i0}^I, \hat{v}_3^I(t_1)] + [\hat{p}_{i1}^I(t), \hat{v}_2^I(t_1)] + [\hat{p}_{i2}^I(t), \hat{v}_1^I(t_1)]\, dt_1 \\ &+ \left(\frac{1}{i\hbar}\right)^2 \int_0^{t_1}\int_0^t \left[\left[\hat{p}_{i0}^I, \hat{v}_1^I(t_1)\right], \hat{v}_2^I(t_2)\right] \\ &\quad + \left[\left[\hat{p}_{i0}^I, \hat{v}_2^I(t_1)\right], \hat{v}_1^I(t_2)\right] + \left[\left[\hat{p}_{i1}^I(t), \hat{v}_1^I(t_1)\right], \hat{v}_1^I(t_2)\right] dt_1 dt_2 \\ &+ \left(\frac{1}{i\hbar}\right)^3 \int_0^{t_2}\int_0^{t_1}\int_0^t \left[\left[\left[\hat{p}_{i0}^I, \hat{v}_1^I(t_1)\right], \hat{v}_1^I(t_2)\right], \hat{v}_1^I(t_3)\right] dt_1 dt_2 dt_3.\end{aligned} \tag{2.40}$$

In all of the above, we have evaluated the dipole moment operator in the Heisenberg picture to different orders in the applied electric field. In the following we show an example of prediction of the first-order induced polarization as well as a more detailed discussion of the third-order dipole moment density operator.

2.2.1 First-order polarization

We would like to evaluate the expectation value of $P^{(1)}(t)$ starting from the dipole moment operator $\hat{p}_i^H\Big|^{(1)}(t) = \hat{p}_{i1}^I + \frac{1}{i\hbar}\int_0^t [\hat{p}_{i0}^I, \hat{v}_1^I] dt_1$. To do so we start from Eqn. (2.29) that enables us to calculate the expectation value of an observable O with corresponding operator $\hat{O}(t)$ in an ensemble as

$$\begin{aligned}\langle\hat{O}(t)\rangle = \mathrm{Tr}[\hat{\rho}_0\hat{O}^H] &= \sum_n \langle\psi_n|\hat{\rho}_0\hat{O}^H|\psi_n\rangle \\ &= \sum_{n,m}\langle\psi_n|\hat{\rho}_0|\psi_m\rangle\langle\psi_m|\hat{O}^H|\psi_n\rangle,\end{aligned} \tag{2.41}$$

where $\hat{O}^H$ is the operator in the Heisenberg picture. Using that $\langle\psi_n|\hat{\rho}_0|\psi_m\rangle$ is non-vanishing only for $n = m$ [2], we may write the expectation value of an operator in an ensemble as $\langle\hat{O}(t)\rangle = \sum_n \langle\psi_n|\hat{\rho}_0|\psi_n\rangle\langle\psi_n|\hat{O}^H|\psi_n\rangle$, where we note that $\langle\psi_n|\hat{\rho}_0|\psi_n\rangle$ is the statistical weight factor of states $|\psi_n\rangle$ and is given by $\vartheta_n = e^{-E_n/kT}/\sum_m e^{-E_m/kT}$ [25].

In the considered case, the first contribution originates from Eqn. (2.38) $\hat{p}_{i1}^I = \hat{\alpha}_{ij}E_j$.

The second contribution to the expectation value of the induced polarization originates from the integral of the commutator $[\hat{p}_0^I, \hat{v}_1^I]$. For simplicity we ignore the integral for now, but return to the integral in Eqn. (2.46), and we denote the contribution $\langle[\hat{p}_0^I, \hat{v}_1^I]\rangle$. We then get

$$\langle[\hat{p}_0^I, \hat{v}_1^I]\rangle = \sum_n \vartheta_n\langle\psi_n|[\hat{p}_0^I, \hat{v}_1^I]|\psi_n\rangle. \tag{2.42}$$

Inserting $\hat{p}_0^I = \hat{m}^I(t)$ and $\hat{v}_1^I = \hat{m}^I(s)E(s)V$ we get

$$\begin{aligned}\langle[\hat{p}_0^I, \hat{v}_1^I]\rangle =&V\sum_n \vartheta_n\langle\psi_n|\hat{m}^I(t)\hat{m}^I(s)E(s) - \hat{m}^I(s)E(s)\hat{m}^I(t)|\psi_n\rangle \\ =&V\sum_{n,l}\vartheta_n\langle\psi_n|\hat{m}^I(t)|\psi_l\rangle\langle\psi_l|\hat{m}^I(s)E(s)|\psi_n\rangle - c.c.\end{aligned} \tag{2.43}$$

We have now expressed the contribution by using operators in the interaction picture where the operator $\hat{m}$ is timedependent. To deal with this, we may express the contribution in the Schrödinger picture where the operator is time-independent but where the timedependence is described though time evolution operator $\hat{U}_0$ i.e.

$$\langle[\hat{p}_0^I, \hat{v}_1^I]\rangle = V\sum_{n,l}\vartheta_n\langle\psi_n|\hat{U}_0^\dagger(t)\hat{m}^S\hat{U}_0(t)|\psi_l\rangle\langle\psi_l|\hat{U}_0^\dagger(s)\hat{m}^S E(s)\hat{U}_0(s)|\psi_n\rangle - c.c., \tag{2.44}$$

where we have used that $\sum_n |\psi_n\rangle\langle\psi_n| = 1$. Using that $|\psi_n\rangle$ is an eigenstate to $\hat{H}_0$, i.e. $H_0|\psi_n\rangle = E_n|\psi_n\rangle$ we get $\langle\psi_n|U_0^\dagger(t)\hat{m}^S U_0(t)|\psi_l\rangle = e^{iE_nt/\hbar}e^{-iE_lt/\hbar}\langle\psi_n|\hat{m}^S|\psi_l\rangle$. Remembering that the energy $E_n = \omega_n\hbar$ and by introducing $\omega_{ln} = \omega_l - \omega_n$, then $\langle\psi_n|U_0^\dagger(t)\hat{m}^S U_0(t)|\psi_l\rangle$ may be written as $e^{-i\omega_{ln}t}\langle\psi_n|\hat{m}^S|\psi_l\rangle$, where $\hat{m}^S$ is in the Schrödinger picture. Likewise $\langle\psi_l|U_0^\dagger(s)\hat{m}^S E(s)U_0(s)|\psi_n\rangle = e^{i\omega_{ln}s}\langle\psi_l|\hat{m}^S|\psi_n\rangle E(s)$, and we may rewrite the commutator $\langle[\hat{p}_0^I, \hat{v}_1^I]\rangle$ as

$$\langle[\hat{p}_0^I, \hat{v}_1^I]\rangle = V\sum_{n,l}\vartheta_n e^{i\omega_{ln}(s-t)}\langle\psi_n|\hat{m}^S|\psi_l\rangle\langle\psi_l|\hat{m}^S|\psi_n\rangle E(s) - c.c., \tag{2.45}$$

and inserting this into Eqn. (2.39) we get

$$\begin{aligned}\frac{1}{i\hbar}\int_0^t \left[\hat{p}^I(t), \hat{H}_E^I(t_1)\right] dt_1 =& \frac{1}{i\hbar}\int_0^t \left[\hat{p}_0^I(t), \hat{v}_E^I(t_1)\right] dt_1 \\ =& V\frac{1}{i\hbar}\int_0^t \left[\sum_{n,l}\vartheta_n e^{i\omega_{ln}(s-t)}\langle\psi_n|\hat{m}^S|\psi_l\rangle\langle\psi_l|\hat{m}^S|\psi_n\rangle E(s) - c.c.\right] ds.\end{aligned} \tag{2.46}$$

If the integrand is multiplied by the stepfunction $\Theta(t-s)$ then the upper limit in the integral may be changed to ∞. Note, the step function $\Theta(t)$ is 0 for $t < 0$ and 1 for $t \geq 0$. Remembering that $E(s) = 0$ for $s < 0$, then we may rewrite the integral as

$$\frac{V}{i\hbar}\int_{-\infty}^{\infty}\left[\sum_{n,l}\vartheta_n e^{i\omega_{ln}(s-t)}\langle\psi_n|\hat{m}^S|\psi_l\rangle\langle\psi_l|\hat{m}^S|\psi_n\rangle\Theta(t-s)E(s) - c.c.\right] ds. \tag{2.47}$$

Introducing $R(\tau) = \sum_{n,l}\vartheta_n e^{i\omega_{ln}\tau}\langle\psi_n|\hat{m}^S|\psi_l\rangle\langle\psi_l|\hat{m}^S|\psi_n\rangle\Theta(\tau)$, the integral is as expected a convolution between the response R and the electric field E. Consequently, we may evaluate the induced polarization in the frequency domain as the product of the fourier transform of $R(t)$ and $E(t)$. The Fourier transform of $R(t)$ is

$$R(\omega) = V\sum_{n,l}\vartheta_n\langle\psi_n|\hat{m}^S|\psi_l\rangle\langle\psi_l|\hat{m}^S|\psi_n\rangle\int_{-\infty}^{\infty}(e^{i\omega_{ln}\tau} - c.c)e^{i\omega\tau}d\tau. \tag{2.48}$$

We note that the integral does not converge as τ approaches infinity. To ensure convergence we multiply the integrand by a factor $e^{-\epsilon\tau}$ where ϵ is an infinitesimal small number. With this we get

$$V\sum_{n,l}\vartheta_n\langle\psi_n|\hat{m}^S|\psi_l\rangle\langle\psi_l|\hat{m}^S|\psi_n\rangle\int_{-\infty}^{\infty}(e^{i\omega_{ln}\tau}e^{-\epsilon\tau}-c.c)e^{i\omega\tau}d\tau \quad (2.49)$$

$$=V\sum_{n,l}\vartheta_n\langle\psi_n|\hat{m}^S|\psi_l\rangle\langle\psi_l|\hat{m}^S|\psi_n\rangle\left(\frac{-2i\omega_{ln}}{\omega_{ln}^2-(i\epsilon+\omega)^2}\right).$$

To evaluate the induced polarization we need to multiply Eqn. (2.49) by $(i\hbar)^{-1}$. If the two matrix elements in Eqn. (2.49) are real, then the induced polarization is the sum of narrow band filters around the frequencies ω_{ln}.

The induced polarization calculated here resembles the induced polarization calculated using the Lorentz model in Chapter 1.

2.2.2 Third-order polarization

From Eqn. (2.40), the third-order term of the dipole moment density operator, and hence also the third-order induced polarization, has eight contributions. In the following, the different contributions are discussed under the assumption that the applied electric field is purely optical for example a simple plane wave at frequency ω. The simplest contribution to the third-order dipole density operator in the Heisenberg picture, $\hat{p}_i^H|^{(3)}(t)$ equals $\hat{p}_{30}^I$ is

$$\hat{\gamma}_{ijkl}^I E_j(t)E_k(t)E_l(t). \quad (2.50)$$

Moving on to the contributions originating from the simple commutators in Eqn.(2.40), the first one we discuss is from the commutator $[\hat{p}_{i0}^I(t),\hat{v}_3^I(t_1)]$, inserting expressions for $\hat{p}_{i0}^I(t)$ and $\hat{v}_3^I(t_1)$ we get

$$\frac{V}{i\hbar}\int_{-\infty}^{\infty}\left[\hat{m}_i^I(t),\frac{1}{3}\hat{\beta}_{jkl}^I E_j(t_1)E_k(t_1)E_l(t_1)\right]\Theta(t-t_1)dt_1$$

$$=\frac{V}{i\hbar}\int_{-\infty}^{\infty}\left[\hat{m}_i^I(t),\frac{1}{3}\hat{\beta}_{jkl}^I\right]E_j(t_1)E_k(t_1)E_l(t_1)\Theta(t-t_1)dt_1. \quad (2.51)$$

Since it has been assumed that $E(t)$ is a monochromatic wave of frequency ω, then the cubic product of $E(t)$ results in an electric field consisting of a contribution at three times ω in addition to a frequency component at ω. However, since we have assumed that we are well away from resonances then there is no contribution to the final induced polarization from the term at ω nor the term at three times ω.

Considering now the contribution from the commutator $[\hat{p}_{i1}^I(t),\hat{v}_2^I(t_1)]$, we find by inserting $\hat{p}_{i1}^I(t)$ and $\hat{v}_2^I(t_1)$ that this contribution equals

$$\frac{V}{i\hbar}\int_{-\infty}^{\infty}E_j(t)\left[\hat{\alpha}_{ij}^I(t),\frac{1}{2}\hat{\alpha}_{kl}^I\right]E_k(t_1)E_l(t_1)\Theta(t-t_1)dt_1. \quad (2.52)$$

Again under the assumption that the electric field consists of the frequency ω then the squared dependence $E_k(t_1)E_l(t_1)$ results in a DC component that is within the response of the material. Consequently, this term does contribute to the final induced polarization.

Next we find the last contribution from the simple commutators in Eqn. (2.40), i.e. from the commutator $[\hat{p}^I_{i2}(t), \hat{v}^I_1(t_1)]$, which equals

$$\frac{V}{i\hbar}\int_{-\infty}^{\infty} E_j(t)E_k(t)\left[\hat{\beta}^I_{ijk}, \hat{m}^I_l\right] E_l(t_1)\Theta(t-t_1)dt_1. \tag{2.53}$$

This term does not contribute to the final induced polarization since the integrand, more specifically $E_l(t_1)$ is at optical frequency and the material is not able to respond to this.

From the double commutators that is $[[\hat{p}^I_{i0}(t), \hat{v}^I_1(t_1)], \hat{v}^I_2(t_2)]$, $[[\hat{p}^I_{i0}(t), \hat{v}^I_2(t_1)], \hat{v}^I_1(t_2)]$ and $[[\hat{p}^I_1(t), \hat{v}^I_1(t_1)], \hat{v}^I_1(t_2)]$ we find

$$\left(\frac{1}{i\hbar}\right)^2 \int_0^{t_1}\int_0^t \left[\left[\hat{p}^I_{i0}, \hat{v}^I_1(t_1)\right], \hat{v}^I_2(t_2)\right] dt_1 dt_2 \tag{2.54}$$
$$= \left(\frac{V}{i\hbar}\right)^2 \int_0^{t_1}\int_0^t \left[\left[\hat{m}^I_i(t), \hat{m}^I_j\right], \frac{1}{2}\hat{\alpha}^I_{kl}\right] E_j(t_1)E_k(t_2)E_l(t_2)\Theta(t-t_1)\Theta(t_1-t_2)dt_2 dt_1,$$

which does not contribute to the induced polarization for the same reason as for the simple commutator Eqn. (2.53). The contribution from $\left[\left[\hat{p}^I_{i0}, \hat{v}^I_2(t_1)\right], \hat{v}^I_1(t_2)\right]$ is

$$\left(\frac{1}{i\hbar}\right)^2 \int_0^{t_1}\int_0^t \left[\left[\hat{p}^I_{i0}, \hat{v}^I_2(t_1)\right], \hat{v}^I_1(t_2)\right] dt_1 dt_2 \tag{2.55}$$
$$= \left(\frac{V}{i\hbar}\right)^2 \int_0^{t_1}\int_0^t \left[\left[\hat{m}^I_i(t), \frac{1}{2}\hat{\alpha}^I_{jk}\right]\hat{m}^I_l\right] E_j(t_1)E_k(t_1)E_l(t_2)\Theta(t-t_1)\Theta(t_1-t_2)dt_2 dt_1$$

which does not contribute either, and

$$\left(\frac{1}{i\hbar}\right)^2 \int_0^{t_1}\int_0^t \left[\left[\hat{p}^I_{i0}, \hat{v}^I_1(t_1)\right], \hat{v}^I_2(t_2)\right] dt_1 dt_2 \tag{2.56}$$
$$= \left(\frac{V}{i\hbar}\right)^2 \int_0^{t_1}\int_0^t E_j(t)\left[\left[\hat{\alpha}^I_{ij}(t), \hat{m}^I_k\right], \hat{m}^I_l\right] E_k(t_1)E_l(t_2)\Theta(t-t_1)\Theta(t_1-t_2)dt_2 dt_1$$

and finally from the triple commutator $[[\hat{p}^I_0(t), \hat{v}^I_1(t_1)], \hat{v}^I_1(t_2)], \hat{v}^I_1(t_3)]$, which equals

$$\left(\frac{V}{i\hbar}\right)^3 \int_0^{t_2}\int_0^{t_1}\int_0^t \langle[[[\hat{m}^I_i(t), \hat{m}^I_j]\,\hat{m}^I_k]\,\hat{m}^I_l]\rangle \tag{2.57}$$
$$\times E_j(t_1)E_k(t_2)E_l(t_3)\Theta(t-t_1)\Theta(t_1-t_2)\Theta(t_2-t_3)dt_3 dt_2 dt_1,$$

which also vanishes and hence does not contribute to $\hat{p}^H_i\big|^{(3)}(t)$.

In summary, the induced third-order dipole moment density operator and thus the third-order induced polarization only consist of two contributions

$$\hat{\gamma}_{ijkl}E_j(t)E_k(t)E_l(t)+\frac{V}{i\hbar}\int_{-\infty}^{\infty}E_j(t)\left[\hat{\alpha}^I_{ij}(t),\frac{1}{2}\hat{\alpha}^I_{kl}\right]E_k(t_1)E_l(t_1)\Theta(t-t_1)dt_1. \quad (2.58)$$

Unfortunately, it is impossible to make any further general statements. However, the commutators in the integrals represent material response functions, and are all functions that depend on the difference between their arguments only. If the response is rather slow compared to optical frequencies, then the response acts like a low pass filter. Consequently, only low frequency components of the relevant products of the applied field give a contribution in the integrals. Thus in such situations the only relevant terms are

$$P_i^{(3)}(t)=\langle\gamma_{ijkl}\rangle E_jE_kE_l+i\hbar^{-1}VE_j(t)\int\langle[\alpha_{ij}(t),\frac{1}{2}\alpha_{kl}(s)]\rangle E_k(s)E_l(s)ds, \quad (2.59)$$

where the first contribution is included, since all frequencies of the applied electrical field are below electronic transition frequencies, yet optical.

Fortunately, in many nonlinear effects all frequencies of the applied electric field as well as of the induced polarization, are optical, that is, well above any of the frequencies of the commutator, i.e the nuclear response. Thus the response reduces to two contributions in Eqn. (2.59), which may be rewritten as

$$P_i^{(3)}(t)=\varepsilon_0\sigma_{ijkl}E_j(t)E_k(t)E_l(t)+E_j(t)\varepsilon_0\int_{-\infty}^{\infty}d_{ijkl}(t-s)E_k(s)E_l(s)ds, \quad (2.60)$$

where

$$d_{ijkl}(t)=\frac{1}{2}i\hbar^{-1}\langle[\tilde{\alpha}_{ij}(t),\tilde{\alpha}_{kl}(0)]\rangle\Theta(t)V, \quad (2.61)$$

where $\Theta(t)$ is the unit step function which equals 0 for $t<0$ and 1 for $t\geq 0$. It is interesting to note that the induced polarization in Eqn. (2.60) cannot last any longer than the duration of the electric field.

In accordance with the above, the corresponding material response in the time domain may then be written as

$$T^{(3)}_{ijkl}(t;t_1,t_2,t_3)=\sigma_{ijkl}\delta(t-t_1)\delta(t_1-t_2)\delta(t_2-t_3)+\delta(t-t_1)d_{ijkl}(t_1-t_2)\delta(t_2-t_3). \quad (2.62)$$

Here the first term on the right-hand side accounts for the electronic response, i.e the interaction between the electric field and the electros, and characterized by a constant tensor which must be invariant with respect to any interchange of its indices. The second term accounts for the nuclear response, more specifically the interaction between the electric field and the nucleus.

Eqn. (2.62) is the main result of the BO approximation. It is interesting to note that the nuclear contribution to the induced polarization may never extent beyond $E(t)$, i.e. no delayed response is possible within the BO approximation. Another very important consequence of the BO-approximation, is that all electronic contributions to the induced polarization are instantaneous and totally symmetric under the interchange of their space coordinates. This symmetry is often referred to as the **Kleinman symmetry**. We return to this symmetry in Chapter 4.

Under the BO-approximation, d_{ijkl} is invariant to the interchange of its first or last two indices, and under the exchange of its first pair of indices with its last [17] i.e.

$$\begin{aligned} d_{ijkl} &= d_{jikl} = d_{ijlk} = d_{jilk}, \\ d_{ijkl} &= d_{klij}. \end{aligned} \tag{2.63}$$

As mentioned in the introduction to the BO approximation, a necessary condition for the approximation to be valid is that all frequencies in the electric field and the induced polarization are optical, i.e. well above the frequencies in the nuclear response, 10^{12} Hz, while still below electronic frequencies, 10^{16} Hz. Examples where this holds include some of the most studied kinds of third-order nonlinear effects such as Raman scattering, four-wave mixing, self-focusing, intensity-induced refractive index changes and third-harmonic generation [17].

Other terms are required for effects depending on low-frequency components such as the DC Kerr effect and field induced second-harmonic generation.

Example 2.3. *Third-order response function*
An example of an instantaneous third-order response is

$$T^{(3)}_{ijkl}(t;t_1,t_2,t_3) = \sigma_{ijkl}\delta(t-t_1)\delta(t_1-t_2)\delta(t_2-t_3), \tag{2.64}$$

which results in the induced polarization

$$P^{(3)}_i(t) = \varepsilon_0 \int_{t_1}\int_{t_2}\int_{t_3} \sigma_{ijkl}\delta(t-t_1)\delta(t_1-t_2)\delta(t_2-t_3)E_j(t_1)E_k(t_2)E_l(t_3)dt_1dt_2dt_3. \tag{2.65}$$

By carrying out the integration over t_3 the induced polarization may be rewritten as

$$P^{(3)}_i(t) = \varepsilon_0 \int_{t_1}\int_{t_2} \sigma_{ijkl}\delta(t-t_1)\delta(t_1-t_2)E_j(t_1)E_k(t_2)E_l(t_2)dt_1dt_2. \tag{2.66}$$

The integration over t_2 leads to

$$P^{(3)}_i(t) = \varepsilon_0 \int_{t_1} \sigma_{ijkl}\delta(t-t_1)E_j(t_1)E_k(t_1)E_l(t_1)dt_1, \tag{2.67}$$

and finally by integrating over t_1 we arrive at the result

$$P_i^{(3)}(t) = \varepsilon_0 \sigma_{ijkl} E_j(t) E_k(t) E_l(t). \tag{2.68}$$

The same induced polarization may be obtained from the third-order response

$$T_{ijkl}^{(3)}(t; t_1, t_2, t_3) = \sigma_{ijkl} \delta(t - t_1) \delta(t - t_2) \delta(t - t_3). \tag{2.69}$$

Both of the above examples of response functions result in the same induced polarization. However, when the response is modified slightly to have a finite response time, then there is a significant difference. Let us compare the induced polarization in Eqn. (2.60), more specifically the delayed part of that induced polarization resulting from the response function in Eqn. (2.62), against the induced polarization from the the third-order response

$$T_{ijkl}^{(3)}(t - t_1, t - t_2, t - t_3) = R_{ijkl}(t - t_1) \delta(t - t_2) \delta(t - t_3). \tag{2.70}$$

The induced polarization from this response function is

$$P_i^{(3)}(t) = \varepsilon_0 \int_{t_1} \int_{t_2} \int_{t_3} R_{ijkl}(t - t_1) \delta(t - t_2) \delta(t - t_3) E_j(t_1) E_k(t_2) E_l(t_3) dt_1 dt_2 dt_3. \tag{2.71}$$

The integration over t_3 leads to

$$P_i^{(3)}(t) = \varepsilon_0 E_l(t) \int_{t_1} \int_{t_2} R_{ijkl}(t - t_1) \delta(t - t_2) E_j(t_1) E_k(t_2) dt_1 dt_2, \tag{2.72}$$

and finally, the integration over t_2 results in

$$P_i^{(3)}(t) = \varepsilon_0 E_k(t) E_l(t) \int_{t_1} R_{ijkl}(t - t_1) E_j(t_1) dt_1. \tag{2.73}$$

This induced polarization is significantly different from the induced polarization expressed by the second term in Eqn. (2.60).

— ■ —

2.2.3 Time-domain response for cubic and isotropic materials

Cubic and isotropic materials are characterized by having linear optical properties that are directional-invariant. Such materials include glasses and fluids.

In cubic and isotropic materials the electronic part of the induced polarization may for all practical purposes be considered instantaneous since it responds to field changes within a few electronic cycles (10^{-16} sec) [26],[27],[28]. In such materials σ_e is a scalar parameter, consequently, the induced polarization may be written as

$$\mathbf{P}(t) = \frac{1}{2} \varepsilon_0 \sigma_e \left(\mathbf{E}(t) \cdot \mathbf{E}(t) \right) \mathbf{E}(t). \tag{2.74}$$

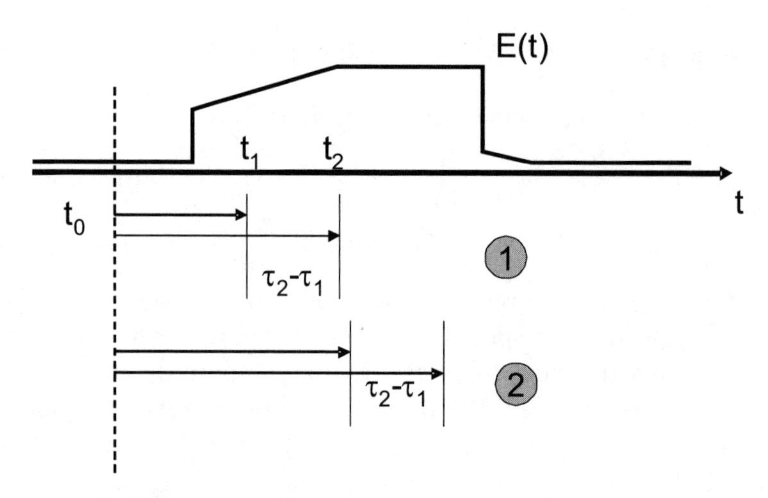

Figure 2.3: $E(t)$ is the applied electric field, t_0 is the time at which the induced polarization is to be determined. For a second-order nonlinear process, the induced polarization depends on the electric field at two different times t_1 and t_2. As discussed in the text, it is only the difference between the time of the applied electric field and the observation time that is important i.e. $\tau_1 = t_1 - t_0$ and $\tau_2 = t_2 - t_0$, respectively. The figure illustrates two different situations, where τ_1 and τ_2 are different, while the difference between τ_2 and τ_1 is the same. A given material may have a response function that depends on the difference between τ_1 and τ_2 as in Eqn. (2.64), and hence gives the same contribution in both cases, or a response function that depends on the absolute values of τ_1 and τ_2 as in Eqn. (2.69), and hence gives different contributions in the two cases.

Applying the BO-approximation to cubic and isotropic materials, the electronic response tensor σ_{ijkl} is symmetric among its indices, and for cubic materials it may be written in the form [17]

$$\sigma_{ijkl} = \frac{1}{6}\sigma_e \left(\delta_{ij}\delta_{kl} + \delta_{ik}\delta_{jl} + \delta_{il}\delta_{jk}\right) + \frac{1}{2}\tau\delta_{ijkl}, \tag{2.75}$$

where σ_e and τ are constants, δ_{ij} is the Kronecker delta, i.e. $\delta_{ij} = 1$ when $i = j$, and $\delta_{ij} = 0$ otherwise. Similarly, $\delta_{ijkl} = 1$ when $i = j = k = l$ and 0 otherwise. For isotropic materials (glasses and fluids) the response simplifies since $\tau = 0$ for these materials.

As opposed to the electronic susceptibility, the nuclear susceptibility has a non-instantaneous time dependence. For cubic materials the susceptibility has the form [17]

$$d_{ijkl}(t) = a(t)\delta_{ij}\delta_{kl} + \frac{1}{2}b(t)\left(\delta_{il}\delta_{jk} + \delta_{ik}\delta_{jl}\right) + c(t)\delta_{ijkl}. \tag{2.76}$$

For isotropic materials $c(t) = 0$, leading to an even simpler response. With this, the third-order induced polarization in isotropic materials is

$$\mathbf{P}^{(3)}_{\text{nuclear}}(t) = \varepsilon_0 \mathbf{E}(t) \int_s a(t-s)\left(\mathbf{E}(s)\cdot\mathbf{E}(s)\right)ds + \varepsilon_0 \int_s \mathbf{E}(s) b(t-s) \mathbf{E}(t)\cdot\mathbf{E}(s) ds. \tag{2.77}$$

2.3 Raman scattering response function of silica

When light propagates through a material, it scatters off the rotating and vibrating molecules in the material. In the Raman process, energy is typically transferred from the incoming light to the material, and the scattered light has a lower frequency. The frequency shift is characteristic for a given material and has been used to characterize materials for decades and, more recently, to optically amplify signals in optical communication systems, as well as to act as a gain medium in Raman gain based laser systems.

The Raman effect may be described as a third-order nonlinearity, i.e. through the third-order induced polarization. Thus, in this case the starting point is

$$P^{(3)}_{\mu}(\mathbf{r},t) = \varepsilon_0 \int_{-\infty}^{\infty}\int_{-\infty}^{\infty}\int_{-\infty}^{\infty} T^{(3)}_{\mu\alpha\beta\gamma}(t;\tau_1,\tau_2,\tau_3) E_\alpha(\mathbf{r},\tau_1) E_\beta(\mathbf{r},\tau_2) E_\gamma(\mathbf{r},\tau_3) d\tau_1 d\tau_2 . d\tau_3. \tag{2.78}$$

Considerable simplifications occur if the nonlinear response is assumed to be instantaneous so that the time dependence in $T^{(3)}_{\mu\alpha\beta}(t;\tau_1,\tau_2,\tau_3)$ is given by three delta functions in the form $\delta(t-t_i)$. However, this simplification is not always valid. More realistically, the electronic response time may be assumed instantaneous but the nuclei response time is inherently slower.

In the example of Raman scattering in silica glass, the induced polarization from the nuclear part is expressed through Eqn. (2.77), and $a(t)+b(t)$ is related to the coupling between parallel polarizations in the excitation light and the Stokes shifted light, whereas $b(t)$ is related to the coupling between orthogonal polarizations in the excitation light and the Stokes shifted light. However, the coupling between orthogonal polarizations is weak for most frequency shifts, leading to the approximation $b(t)=0$. As a consequence, the nonlinear polarization reduces to a scalar equation

$$P^{(3)}(\mathbf{r},t) = \varepsilon_0 \mathcal{A} E(\mathbf{r},t) \int_{-\infty}^{\infty} R^{(3)}_R(\tau_1) \left|E(\mathbf{r},t-\tau_1)\right|^2 d\tau_1, \tag{2.79}$$

where $\mathcal{A}$ is an amplitude, $R^{(3)}_R(t-t_1)$ is the nonlinear response function normalized according to $\int_{-\infty}^{\infty} \mathcal{A} R^{(3)}_R(t)dt = \mathcal{A}$, and $R^{(3)}_R(\tau)$ must be zero for $\tau < 0$.

The time scale of the Raman response function $R^{(3)}_R(t)$ in silica is on the order of hundreds of femtoseconds. The longer time scale is largely given by the phonon lifetime. In addition, the Raman response consists of a nearly instantaneous response

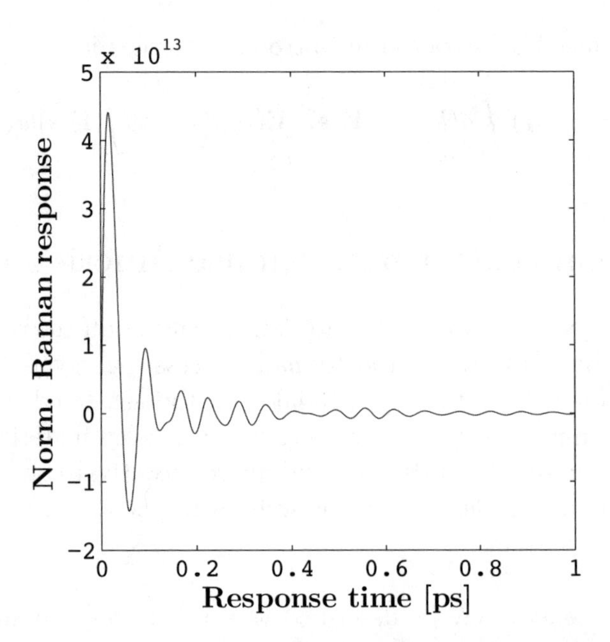

Figure 2.4: Raman response function in silica fiber [29], normalized such that the response integrated from time equal zero to infinity equals one.

similar to a non-response vertual electronic transition, which may be modelled by a delta function. Thus

$$R_R^{(3)}(t) = (1 - f_R)\delta(t) + f_R h_R(t). \tag{2.80}$$

Three models of the Raman response function $h_R(t)$ are presented in the literature. The simplest is an instantaneous model where the response is simply a delta function. A slightly more complicated model is by assuming a single damped harmonic oscillator as in [30]

$$h_R(t) = \exp(-t/\tau_2)\sin(t/\tau_1), \tag{2.81}$$

where τ_1 is related to the phonon frequency and τ_2 is related to the attenuation of the network of vibrating atoms. The best fit to the actual Raman response is obtained using $\tau_1 = 12.2$ fs and $\tau_2 = 32.2$ fs, see Figure 2.5.

A more accurate model of the Raman response of silica is much more complicated. An example is obtained using an ensemble of damped harmonic oscillators where the resonant frequencies have a Gaussian distribution. In 1982 Walrafen and coworkers [31] found that 13 oscillators are needed to obtain an accurate fit to the Raman gain spectrum. The response function reflects this and shows good agreement with Figure 2.4.

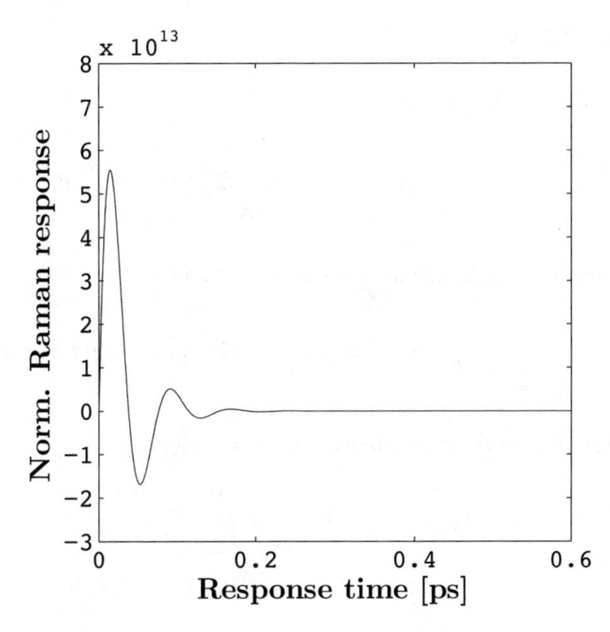

Figure 2.5: Simple response function in silica fiber based on single damped oscillator model.

2.4 Summary

- Linear induced polarization

$$P_\mu^{(1)}(\mathbf{r},t) = \varepsilon_0 \int_{-\infty}^{\infty} T_{\mu\alpha}^{(1)}(t;\tau)E_\alpha(\mathbf{r},\tau)d\tau, \tag{2.2}$$

- Linear induced polarization by convolution

$$P_\mu^{(1)}(\mathbf{r},t) = \varepsilon_0 \int_{-\infty}^{\infty} R_{\mu\alpha}^{(1)}(\tau)E_\alpha(\mathbf{r},t-\tau)d\tau. \tag{2.4}$$

- Nonlinear induced polarization to nth-order

$$\begin{aligned} P_\mu^{(n)}(\mathbf{r},t) = \varepsilon_0 \int_{-\infty}^{\infty} \cdots \int_{-\infty}^{\infty} & T_{\mu\alpha_1\cdots\alpha_n}^{(n)}(t;\tau_1,\ldots,\tau_n) \\ & \times E_{\alpha_1}(\mathbf{r},\tau_1)\cdots E_{\alpha_n}(\mathbf{r},\tau_n)d\tau_1\cdots d\tau_n, \end{aligned} \tag{2.18}$$

  and

$$\begin{aligned} P_\mu^{(n)}(\mathbf{r},t) = \varepsilon_0 \int_{-\infty}^{\infty} \cdots \int_{-\infty}^{\infty} & R_{\mu\alpha\cdots\alpha_n}^{(n)}(\tau_1',\ldots,\tau_n') \\ & \times E_{\alpha_1}(\mathbf{r},t-\tau_1')\ldots E_{\alpha_n}(\mathbf{r},t-\tau_n')d\tau_1'\cdots d\tau_n'. \end{aligned} \tag{2.20}$$

- The Born-Oppenheimer approximation enables a general expression for the third-order induced polarization

$$P_i^{(3)}(t) = \varepsilon_0\sigma_{ijkl}E_j(t)E_k(t)E_l(t) + E_j(t)\varepsilon_0 \int_{-\infty}^{\infty} d_{ijkl}(t-s)E_k(s)E_l(s)ds. \tag{2.60}$$

Chapter 3

Material response in the frequency domain, susceptibility tensors

Contents

Until this point we have been focusing on the material response in the time domain. However, in many cases it is more convenient to work in the frequency domain since we often describe monochromatic waves, or a sum of monochromatic waves, or reduce problems such that we can apply a description using monochromatic waves, or a sum of monochromatic waves.

In the time domain we worked on the material response function, or material response tensors, whereas in the frequency domain this translates into susceptibility tensors. The frequency domain description is the focus of this chapter.

The susceptibility tensors in the frequency domain arise when the electric field $E(t)$ of the light is expressed in terms of its Fourier transform $E(\omega)$, by means of the Fourier transformation

$$E(\omega) = \frac{1}{2\pi} \int_{-\infty}^{\infty} E(t) \exp(i\omega t) dt, \tag{3.1}$$

with the inverse relation

$$E(t) = \int_{-\infty}^{\infty} E(\omega) \exp(-i\omega t) d\omega. \tag{3.2}$$

The convention of the factor of $1/(2\pi)$, as well as the $\pm i\omega t$ in the exponential function, follows [2]. However, it is emphasized that this convention may be different in other textbooks. The sign convention leads to wave solutions of the form $f(\beta z - \omega t)$ for monochromatic waves propagating in the positive z-direction.

3.1 The susceptibility tensor

In the previous chapter the induced polarization was described in the time domain. This was done using the material response tensor in the time domain. In the following we describe how the time response function is translated into a frequency response, i.e. the susceptibility tensor.

3.1.1 First-order susceptibility tensor

From the discussion of the material response in the time domain, the linear polarization equals

$$P_{\mu}^{(1)}(\mathbf{r}, t) = \varepsilon_0 \int_{-\infty}^{\infty} R_{\mu\alpha}^{(1)}(\tau) E_{\alpha}(\mathbf{r}, t - \tau) d\tau, \tag{3.3}$$

where a summation over α is implied, as introduced in Chapter 2. By using the Fourier transformation set, the induced polarization may be written as

$$\begin{aligned} P_{\mu}^{(1)}(\mathbf{r}, t) &= \varepsilon_0 \int_{-\infty}^{\infty} R_{\mu\alpha}^{(1)}(\tau) \left[\int_{-\infty}^{\infty} E_{\alpha}(\mathbf{r}, \omega) e^{i\omega\tau} e^{-i\omega t} d\omega \right] d\tau \\ &= \varepsilon_0 \int_{-\infty}^{\infty} \chi_{\mu\alpha}^{(1)}(-\omega; \omega) E_{\alpha}(\mathbf{r}, \omega) e^{-i\omega t} d\omega, \end{aligned} \tag{3.4}$$

where the linear susceptibility is introduced through

$$\chi_{\mu\alpha}^{(1)}(-\omega; \omega) = \int_{-\infty}^{\infty} R_{\mu\alpha}^{(1)}(\tau) e^{i\omega\tau} d\tau, \tag{3.5}$$

which defines the linear susceptibility tensor. Note that the susceptibility in Eqn. (3.5) is not a simple Fourier transformation of the material response function in the time domain since the constant $1/(2\pi)$ is missing.

Regarding the linear susceptibility, the arguments are $(-\omega;\omega)$, where the first argument is the output frequency whereas the second argument in the linear case the last argument, is the input frequency. There are mainly two reasons for this sign convention, one is that the sum of the frequencies equals zero, this will always be the case using this notation. The other is that when using symmetry, more specifically overall permutation symmetry, that is discussed in Chapter 4 and used for example in Chapter 5, the sign is carried over to the input frequencies, see the discussions on the Manley-Rowe relations in Chapter 5.

In most applications of Fourier transforms the frequency ω may be considered as real valued and certainly for all practical purposes only real frequencies are meaningful. Nevertheless in the treatment of susceptibilities we shall allow the frequency ω to be **complex**, but constrained to the upper half of the complex plane, such that the integrals in Eqn. (3.5) are convergent. The reason for this is a matter of mathematical convenience, since in practical physical systems the response function $R_{\mu\alpha}(\tau) \to 0$ as $\tau \to 0$.

Any realistic physical medium is subject to relaxation as $\tau \to \infty$ so that the frequency ω can be taken real - however, it simplifies our analysis if relaxation processes are neglected for the time being. If we do this then: $R(\tau)$ remains finite in the remote future, i.e $\tau \to \infty$ and ω must be taken to lie in the upper half plane to ensure convergence of the integral

$$\chi^{(1)}_{\mu\alpha}(-\omega;\omega) = \int_{-\infty}^{\infty} R^{(1)}_{\mu\alpha}(\tau)e^{i\omega\tau}d\tau. \tag{3.6}$$

Inserting ω as a complex frequency with real part ω_r and imaginary part ω_i, i.e. $\omega = \omega_r + i\omega_i$, $\chi^{(1)}_{\mu\alpha}(-\omega,\omega)$ is rewritten as

$$\chi^{(1)}_{\mu\alpha}(-\omega;\omega) = \int_{-\infty}^{\infty} R^{(1)}_{\mu\alpha}(\tau)e^{i\omega_r\tau}e^{-\omega_i\tau}d\tau, \tag{3.7}$$

from this it is immediately clear that since we are in the upper half of the complex plane then the convergence of this integral is secured.

Since $\mathbf{E}(t)$ and $\mathbf{P}(t)$ are real, it follows that the susceptibility tensor must satisfy

$$\left[\chi^{(1)}_{\mu\alpha}(-\omega;\omega)\right]^* = \chi^{(1)}_{\mu\alpha}(\omega^*;-\omega^*). \tag{3.8}$$

Example 3.1. *Spectrum of relaxation oscillation*
We consider two examples. One where the material response is a simple exponential decay

$$R_1(\tau) = \exp(-\tau/\tau_0), \quad \tau > 0, \tag{3.9}$$

where τ_0 is a constant, and one where the response is a decay multiplied by a harmonic time varying function, corresponding to a damped harmonic oscillator

$$R_2(\tau) = \sin(\omega_m t)\exp(-\tau/\tau_2), \quad \tau > 0, \tag{3.10}$$

where ω_m is the resonance frequency of the material and τ_2 the damping of the oscillator.

Using Eqn. (3.5) we find

$$\chi_1(\omega) = \frac{\tau}{1 - i\omega\tau}, \tag{3.11}$$

and

$$\chi_2(\omega) = \frac{\omega_m}{(i\omega - \gamma)^2 + \omega_m^2}, \tag{3.12}$$

where $(\gamma = 1/\tau_2)$. Note that the subscripts 1 and 2 in $\chi_1(\omega)$ and $\chi_2(\omega)$, respectively, do not refer to anything else than the two different response functions considered, i.e. it should not be confused with any subindex of the material response tensor. The susceptibility in Eqn. (3.12) may also be rewritten as

$$\chi_2(\omega) = \frac{\omega_m}{\omega_m^2 - \omega^2 - 2i\gamma\omega + \gamma^2}. \tag{3.13}$$

The peak frequency of the frequency response is determined by ω_m and the width is determined by γ. That is, if the frequency response has a large frequency shift then it corresponds to a fast oscillation and if the peak is wide, it corresponds to a slowly damped oscillation of the material. As it will be shown in Section 3.6, absorption or amplification is always associated by a change in the refractive index.

— ■ —

3.1.2 Second-order susceptibility tensor

Similar to the linear response, the second-order polarization is

$$P_\mu^{(2)}(\mathbf{r},t) = \varepsilon_0 \int_{-\infty}^{\infty}\int_{-\infty}^{\infty} R_{\mu\alpha\beta}^{(2)}(\tau_1,\tau_2) E_\alpha(\mathbf{r}, t-\tau_1) E_\beta(\mathbf{r}, t-\tau_2) d\tau_2 d\tau_1. \tag{3.14}$$

By applying a Fourier transformation this may be written as

$$\begin{aligned} P_\mu^{(2)}(\mathbf{r},t) &= \varepsilon_0 \int_{-\infty}^{\infty}\int_{-\infty}^{\infty} R_{\mu\alpha\beta}^{(2)}(\tau_1,\tau_2) \\ &\quad \times \left[\int_{-\infty}^{\infty}\int_{-\infty}^{\infty} E_\alpha(\mathbf{r},\omega_1)E_\beta(\mathbf{r},\omega_2)e^{i(\omega_1\tau_1+\omega_2\tau_2)}e^{-i(\omega_1+\omega_2)t}d\omega_1 d\omega_2\right] d\tau_1 d\tau_2 \\ &= \varepsilon_0 \int_{-\infty}^{\infty}\int_{-\infty}^{\infty} \chi_{\mu\alpha\beta}^{(2)}(-\omega_\sigma;\omega_1,\omega_2)E_\alpha(\mathbf{r},\omega_1)E_\beta(\mathbf{r},\omega_2)e^{-i\omega_\sigma t}d\omega_1 d\omega_2, \end{aligned} \tag{3.15}$$

where ω_σ represents a sum of two frequencies from the quadratic interaction of the electric field, i.e. $\omega_\sigma = \omega_1 + \omega_2$ and the quadratic susceptibility is introduced through

$$\chi^{(2)}_{\mu\alpha\beta}(-\omega_\sigma; \omega_1, \omega_2) = \int_{-\infty}^{\infty}\int_{-\infty}^{\infty} R^{(2)}_{\mu\alpha\beta}(\tau_1, \tau_2) e^{i(\omega_1\tau_1 + \omega_2\tau_2)} d\tau_1 d\tau_2. \tag{3.16}$$

As noted in the discussion of the linear susceptibility, our notation follows the convention of [2], that is $(-\omega_\sigma; \omega_1, \omega_2)$ is the argument to the susceptibility. The output frequency i.e. the frequency of the induced polarization is ω_σ, which is created by the mutual interactions of the electric field at the frequencies ω_1 and ω_2. That is, the input frequencies are ω_1 and ω_2 and the output frequency is ω_σ. The sum of the frequencies in the argument equals zero: $0 = -\omega_\sigma + \omega_1 + \omega_2$. Similar to the linear case we have

$$\left[\chi^{(2)}_{\mu\alpha\beta}(-\omega_\sigma; \omega_1, \omega_2)\right]^* = \chi^{(2)}_{\mu\alpha\beta}(\omega_\sigma^*; -\omega_1^*, -\omega_2^*). \tag{3.17}$$

The intrinsic permutation symmetry of the second-order polarization response function $R^{(2)}$ carries over to the second-order susceptibility tensor i.e.

$$\chi^{(2)}_{\mu\alpha\beta}(-\omega_\sigma; \omega_1, \omega_2) = \chi^{(2)}_{\mu\beta\alpha}(-\omega_\sigma; \omega_2, \omega_1). \tag{3.18}$$

This illustrates that the second-order susceptibility is invariant under any of the $2! = 2$ pairwise permutations of (α, ω_1) and (β, ω_2).

3.1.3 Higher-order susceptibility tensor

In analogy to the first- and second-order polarization, the μth-component of the nth-order polarization is

$$\begin{aligned}
P^{(n)}_\mu(\mathbf{r}, t) &= \varepsilon_0 \int_{-\infty}^{\infty} \cdots \int_{-\infty}^{\infty} R^{(n)}_{\mu\alpha_1,\cdots,\alpha_n}(\tau_1, \ldots, \tau_n) \qquad (3.19)\\
&\times \Bigg[\int_{-\infty}^{\infty} \cdots \int_{-\infty}^{\infty} E_{\alpha_1}(\mathbf{r}, \omega_1) \cdots E_{\alpha_n}(\mathbf{r}, \omega_n) \\
&\quad e^{i(\omega_1\tau_1 + \cdots + \omega_n\tau_n)} e^{-i(\omega_1 + \cdots + \omega_n)t} d\omega_1 \cdots d\omega_n\Bigg] d\tau_1 \cdots d\tau_n \\
&= \varepsilon_0 \int_{-\infty}^{\infty} \cdots \int_{-\infty}^{\infty} \chi^{(n)}_{\mu\alpha_1,\ldots,\alpha_n}(-\omega_\sigma; \omega_1, \ldots, \omega_n) E_{\alpha_1}(\mathbf{r}, \omega_1) \cdots E_{\alpha_n}(\mathbf{r}, \omega_n) e^{-i\omega_\sigma t} d\omega_1 \cdots d\omega_n,
\end{aligned}$$

where

$$\begin{aligned}
&\chi^{(n)}_{\mu\alpha_1\cdots\alpha_n}(-\omega_\sigma; \omega_1, \ldots, \omega_n) \qquad (3.20)\\
&= \int_{-\infty}^{\infty} \cdots \int_{-\infty}^{\infty} R^{(n)}_{\mu\alpha_1\cdots\alpha_n}(\tau_1, \ldots, \tau_n) e^{i(\omega_1\tau_1 + \cdots + \omega_n\tau_n)} d\tau_1 \cdots d\tau_n,
\end{aligned}$$

and where the output frequency ω_σ is related to the input frequencies, $\omega_1, \omega_2, \cdots, \omega_n$ through: $\omega_\sigma = \omega_1 + \omega_2 + \ldots + \omega_n$. The fact that the response function $R^{(n)}_{\mu\alpha_1\cdots\alpha_n}(\tau_1, \ldots, \tau_n)$ is real leads to the relation

$$\left[\chi^{(n)}_{\mu\alpha_1\cdots\alpha_n}(-\omega_\sigma; \omega_1, \ldots, \omega_n)\right]^* = \left[\chi^{(n)}_{\mu\alpha_1\cdots\alpha_n}(\omega_\sigma^*; -\omega_1^*, \ldots, -\omega_n^*)\right]. \tag{3.21}$$

The intrinsic permutation symmetry, discussed for the time response in Chapter 2, also applies in the frequency domain. Thus, for example

$$\chi^{(n)}_{\mu\alpha_1\alpha_2\cdots\alpha_n}(-\omega_\sigma;\omega_1,\omega_2,\ldots,\omega_n) = \chi^{(n)}_{\mu\alpha_2\alpha_1\cdots\alpha_n}(-\omega_\sigma;\omega_2,\omega_1,\ldots,\omega_n). \tag{3.22}$$

3.2 The induced polarization in the frequency domain

Now that we have demonstrated how the susceptibility tensor and the material response in the time domain are related, we can evaluate the induced polarization in the frequency domain. In general, this is done by taking the Fourier transform of the induced polarization in the time domain. That is, we can find the induced polarization in the frequency domain to order 'n' by taking the Fourier transform of the induced polarization in the time domain of order 'n'. More specifically,

$$\mathbf{P}^{(n)}(\omega) = \frac{1}{2\pi}\int_{-\infty}^{\infty}\mathbf{P}^{(n)}(\mathbf{r},t)e^{i\omega t}dt. \tag{3.23}$$

Inserting $\mathbf{P}^{(n)}(\mathbf{r},t)$ from Eqn. (3.19) and using Eqn. (3.23) the μth-component of the induced polarization is

$$P^{(n)}_\mu(\omega) = \frac{\varepsilon_0}{2\pi}\int_{-\infty}^{\infty}\int_{-\infty}^{\infty}\cdots\int_{-\infty}^{\infty}\chi^{(n)}_{\mu\alpha_1,\ldots,\alpha_n}(-\omega_\sigma;\omega_1,\ldots,\omega_n) \\ \times E_{\alpha_1}(\mathbf{r},\omega_1)\cdots E_{\alpha_n}(\mathbf{r},\omega_n)e^{-i\omega_\sigma t}e^{i\omega t}dtd\omega_1\cdots d\omega_n. \tag{3.24}$$

Using the relation $(2\pi)^{-1}\int_{-\infty}^{\infty}\exp[ix(y-y_1)]dx = \delta(y-y_1)$, we may replace the integration over t by a delta function and $P^{(n)}_\mu(\omega)$ may be rewritten as

$$P^{(n)}_\mu(\omega) = \varepsilon_0\int_{-\infty}^{\infty}\cdots\int_{-\infty}^{\infty}\chi^{(n)}_{\mu\alpha_1,\ldots,\alpha_n}(-\omega_\sigma;\omega_1,\ldots,\omega_n) \\ \times E_{\alpha_1}(\mathbf{r},\omega_1)\cdots E_{\alpha_n}(\mathbf{r},\omega_n)\delta(\omega-\omega_\sigma)d\omega_1\cdots d\omega_n. \tag{3.25}$$

It is important to note that the delta function $\delta(\omega-\omega_\sigma)$ ensures that the sum of the frequencies that contribute to the integration always equals ω_σ.

3.3 Sum of monochromatic fields

The analysis of nonlinear optical phenomena is simplified if the electrical fields are mono-chromatic or at least treated as monochromatic fields. However, it is emphasized that even though the electric field via the Fourier integral can be seen as a superposition of infinitely many monochromatic components, the superposition principle known from linear optics, which states that the wave equations may be independently solved for each component, in general does not hold in nonlinear optics. Often nonlinear optics deals with phenomena related to the coupling of power between different frequencies.

For a sum of monochromatic fields the electric field in the time-domain may be written as

$$\mathbf{E}(r,t) = \frac{1}{2}\sum_{\omega_k\geq 0}\left[\mathbf{E}^0_{\omega_k}e^{-i\omega_k t} + \mathbf{E}^0_{-\omega_k}e^{i\omega_k t}\right], \tag{3.26}$$

which in the frequency domain is expressed through the sum of delta functions as

$$\mathbf{E}(r,\omega) = \frac{1}{2}\sum_{\omega_k\geq 0}\left[\mathbf{E}^0_{\omega_k}\delta(\omega-\omega_k) + \mathbf{E}^0_{-\omega_k}\delta(\omega+\omega_k)\right]. \tag{3.27}$$

Inserting this form into the polarization density, one obtains

$$\mathbf{P}^{(n)}(r,t) = \frac{1}{2}\sum_{\omega_\sigma\geq 0}\left[(\mathbf{P}^0_{\omega_\sigma})^{(n)}e^{-i\omega_\sigma t} + (\mathbf{P}^0_{-\omega_\sigma})^{(n)}e^{i\omega_\sigma t}\right]. \tag{3.28}$$

Note once again $\mathbf{P}(t)$ is real thus $(\mathbf{P}^0_{-\omega})^{(n)} = ((\mathbf{P}^0_{\omega})^{(n)})^*$. In cartesian coordinates a single component of the amplitude of the induced polarization is

$$\begin{aligned}((P^0_{\omega_\sigma})^{(n)})_\mu = 2\varepsilon_0\sum_{\alpha_1\cdots\alpha_n}\Big[&\chi^{(n)}_{\mu\alpha_1\cdots\alpha_n}(-\omega_\sigma;\omega_1,\omega_2,\ldots,\omega_n)\frac{1}{2}(E^0_{\omega_1})_{\alpha_1}\frac{1}{2}(E^0_{\omega_2})_{\alpha_2}\cdots\frac{1}{2}(E^0_{\omega_n})_{\alpha_n}\\ &+\chi^{(n)}_{\mu\alpha_1\cdots\alpha_n}(-\omega_\sigma;\omega_2,\omega_1,\ldots,\omega_n)\frac{1}{2}(E^0_{\omega_2})_{\alpha_1}\frac{1}{2}(E^0_{\omega_1})_{\alpha_2}\cdots\frac{1}{2}(E^0_{\omega_n})_{\alpha_n}+\cdots\Big],\end{aligned} \tag{3.29}$$

where, as previously $\omega_\sigma = \omega_1 + \omega_2 + \ldots + \omega_n$. The factor of two on the right-hand side of Eqn. (3.29) appears since we wish to express the induced polarization with the prefactor of one half e.g. Eqn. (3.28). On the right-hand side of Eqn. (3.29), the summation is performed over all distinguishable terms, i.e. over all possible combinations of $\omega_1, \omega_2, \ldots, \omega_n$ that give rise to the particular frequency ω_σ. With respect to this, a certain frequency and its negative counterpart are to be considered as distinct frequencies when appearing in the set. Keeping the intrinsic permutation symmetry in mind, it is clear that only one term needs to be written, and the number of times this term appears in the expression for the polarization density should consequently be equal to the number of distinguishable combinations of $\omega_1, \omega_2, \ldots, \omega_n$ resulting in the frequency ω_s.

As for example, if in Eqn. (3.29) $\omega_1 = \omega_2$ then it is seen that the second term is indistinguishable from the first term and does therefore **not** occur. Consider on the other hand the situation $\omega_1 \neq \omega_2$; the first term and the second term are then distinguishable and both must be included in the expansion in Eqn. (3.29). However, in this case by

- relabeling the subscripts (α_1, α_2) in the second term as (α_2, α_1)

and

- recognizing factors in the term using intrinsic permutation symmetry, i.e. interchanging (α_1, ω_1) and (α_2, ω_2),

it is seen that the two terms are in fact the same.

This example demonstrates that although two or more terms occur in the summation in Eqn. (3.29) may be distinguishable as defined above and therefore must be included, they nevertheless each have the same value because of intrinsic permutation symmetry.

Example 3.2. *Raman Scattering*
We consider again Raman scattering (the Stokes case) where a pump at frequency ω_1 is generating a Stokes wave at frequency ω_2, that is the electrical field is

$$\mathbf{E} = \frac{1}{2}\left[\begin{pmatrix} (E^0_{\omega_1})_x \\ (E^0_{\omega_1})_y \\ (E^0_{\omega_1})_z \end{pmatrix} e^{-i\omega_1 t} + \begin{pmatrix} (E^0_{\omega_2})_x \\ (E^0_{\omega_2})_y \\ (E^0_{\omega_2})_z \end{pmatrix} e^{-i\omega_2 t} + c.c.\right] . \tag{3.30}$$

If we now consider the Raman scattering described through the third-order susceptibility, and furthermore restrict ourselves to a case of a transverse electromagnetic wave, where $E^0_z = 0$, which is valid in many cases, including waveguiding structures, then we are considering terms in the susceptibility that generate the Stoke wave ω_2. These can only originate from $\omega_1 - \omega_1 + \omega_2$ or from $\omega_2 - \omega_2 + \omega_2$. The first combination is responsible for Raman scattering whereas the latter is responsible for the so-called optical Kerr effect or the intensity dependent refractive index. In addition, the nonlinear polarization can only have contributions from the x- and/or y-component respectively. All possible terms can be tabulated as shown in Table 3.1.

From this the induced third-order polarization at frequency ω_2 can be written as a sum of $2^3 = 8$ contributions. $\chi^{(3)}_{\mu xxx}$, $\chi^{(3)}_{\mu xyy}$, $\chi^{(3)}_{\mu yyx}$, $\chi^{(3)}_{\mu yxy}$, $\chi^{(3)}_{\mu xxy}$, $\chi^{(3)}_{\mu xyx}$, $\chi^{(3)}_{\mu yxx}$, $\chi^{(3)}_{\mu yyy}$. However, because of other symmetry rules that we return to later, not all terms are present in the final expression for the induced third-order polarization.

—— ■ ——

3.4 The prefactor to the induced polarization

To ease the evaluation of the induced polarization at any order, [2] provides a 'recipe' for evaluating the induced polarization at a given order

$$\left((P^0_{\omega_\sigma})^{(n)}\right)_\mu = \varepsilon_0 \sum_{\alpha_1\cdots\alpha_n} \sum_{\omega} K(-\omega_\sigma; \omega_1, \ldots, \omega_n) \tag{3.31}$$
$$\chi^{(n)}_{\mu\alpha_1\cdots\alpha_n}(-\omega_\sigma; \omega_1, \ldots, \omega_n)(E^0_{\omega_1})_{\alpha_1} \cdots (E^0_{\omega_n})_{\alpha_n}.$$

Table 3.1: The table illustrates the permutation symmetry in the Raman example. The top row indicates the possible contributions from $-\omega_1+\omega_1+\omega_2$ in the first column to $-\omega_1+\omega_2+\omega_1$ in the second column etc. In each column all the options for Cartesian components are listed. The linked contributions are symmetric and represent permutation symmetric terms.

-1 1 2	-1 2 1	2 -1 1	1 -1 2	1 2 -1	2 1 -1
x x x	x x x	x x x	x x x	x x x	x x x
x y x	x y x	x y x	x y x	x y x	x y x
y x x	y x x	y x x	y x x	y x x	y x x
x x y	x x y	x x y	x x y	x x y	x x y
x y y	x y y	x y y	x y y	x y y	x y y
y x y	y x y	y x y	y x y	y x y	y x y
y y x	y y x	y y x	y y x	y y x	y y x
y y y	y y y	y y y	y y y	y y y	y y y

Comparing to Eqn. (3.29) and Eqn. (3.31) it is seen that all the factors of 1/2 is now included in the K-factor.

The first summation in Eqn. (3.31) over $\alpha_1, \ldots, \alpha_n$ is simply a summation over the coordinates of the relevant vectors of the electric field, for example in Cartesian coordinates a summation over x, y, z. The summation sign $\sum_\omega$, however, serves as a reminder that the expression that follows is to be summed over all distinct sets of $\omega_1, \ldots, \omega_n$. By all distinct sets of $\omega_1 \ldots, \omega_n$ is meant that the induced polarization at a particular frequency may arise due to different combinations of the frequency components.

For example, if a scalar initial field has two frequencies ω_1 and ω_2, i.e.

$$E = \frac{1}{2}\left(E^0_{\omega_1} e^{-i\omega_1 t} + E^0_{\omega_2} e^{-i\omega_2 t} + c.c.\right), \tag{3.32}$$

then the induced polarization at ω_1 consists of two terms, one solely due to the frequency component at ω_1

$$(P^0_{\omega_1})^{(3)} = \varepsilon_0 K(-\omega_1; \omega_1, -\omega_1, \omega_1)\chi^{(3)}(-\omega_1; \omega_1, -\omega_1, \omega_1)|E^0_{\omega_1}|^2 E^0_{\omega_1}, \tag{3.33}$$

and one due to the frequency component at ω_2 together with the frequency component at ω_1

$$(P^0_{\omega_1})^{(3)} = \varepsilon_0 K(-\omega_1; \omega_2, -\omega_2, \omega_1)\chi^{(3)}(-\omega_1; \omega_2, -\omega_2, \omega_1)|E^0_{\omega_2}|^2 E^0_{\omega_1}. \tag{3.34}$$

The first contribution corresponds to self phase modulation whereas the second contribution corresponds to for example a cross-phase modulation term or a Raman scattering term.

Thus the summation sign $\sum_\omega$ in Eqn. (3.31) represents a sum over the above two terms.

It is noted that because of the intrinsic permutation symmetry the frequency arguments appearing in Eqn. (3.31) may be written in arbitrary order.

In Eqn. (3.31), the prefactor K is formally described as

$$K(-\omega_\sigma;\omega_1,\ldots,\omega_n) = 2^{l+m-n}p, \tag{3.35}$$

where

$$\begin{aligned} p &= \text{the number of distinct permutations of } \omega_1,\omega_2,\ldots,\omega_n, \quad (3.36)\\ n &= \text{the order of nonlinearity,}\\ m &= \text{the number of angular frequencies } \omega_k \text{ that are zero,}\\ l &= \begin{cases} 1, & \text{if } \omega_\sigma \neq 0,\\ 0, & \text{otherwise.}\end{cases} \end{aligned}$$

In other words, m is the number of DC electric fields present, and $l = 0$ if the nonlinearity, that we consider, is static. Table 3.2 provides the prefactor for various typical nonlinear phenomena.

Table 3.2: Prefactor, K, of typical nonlinear optical phenomena, see [2].

Process	Order	Frequency	K
Linear absorption/emission and refractive index	1	$(-\omega;\omega)$	1
Optical rectification (optically induced DC fields)	2	$(0;\omega,-\omega)$	1/2
Pockels effect	2	$(-\omega;0,\omega)$	2
Second-harmonic generation	2	$(-2\omega;\omega,\omega)$	1/2
Sum and difference frequency mixing Parametric amplification	2	$(-\omega_3;\omega_1,\pm\omega_2)$	1
DC Kerr effect	3	$(-\omega;0,0,\omega)$	3
Third-harmonic generation	3	$(-3\omega;\omega,\omega,\omega)$	1/4
Cross phase modulation Stimulated Raman scattering, Stimulated Brillouin scattering	3	$(-\omega_s;\omega_p,-\omega_p,\omega_s)$	3/2
Optical Kerr effect, intensity dep. refractive index self-focussing, self-phase modulation	3	$(-\omega;\omega,-\omega,\omega)$	3/4
Degenerate four wave mixing	3	$(-\omega_s;\omega_p,-\omega_i,\omega_p)$	3/4

3.5 Third-order polarization in the Born-Oppenheimer approximation in the frequency domain

From Chapter 2, the induced polarization consists of two contributions, one from interactions between the applied electric field and electrons, considered instantaneous, and described through the parameter σ, and one contribution from the interaction between the applied electric field and the nuclei.

The first one is the interaction between the applied electric field and the electron cloud described through $P = \sigma/2|E|^2 E$ while the second one is the interaction between the electric field and the molecular structure of the material the light propagates through. Where the first contribution is very simple, the second one deserves more consideration.

Recall the expression for the third-order polarization in the time domain from Chapter 2, Eqn. (2.77)

$$\begin{aligned}\left(P^{(3)}_{\text{nuclear}}(t)\right)_\mu &= E_\mu(t)\varepsilon_0 \int_{-\infty}^{\infty} a(t-s)\,[\mathbf{E}(s)\cdot\mathbf{E}(s)]ds \\ &+ \varepsilon_0 \int_{-\infty}^{\infty} E_\mu(s) b(t-s)\,[\mathbf{E}(t)\cdot\mathbf{E}(s)]ds.\end{aligned} \tag{3.37}$$

Note that this is the molecular contribution only. In the frequency domain this corresponds to

$$\left(P^{(3)}_{\text{nuclear}}(\omega)\right)_\mu = \varepsilon_0 \left[E_\mu(\omega) * \left[A(\omega)\mathcal{F}\left(\sum_i E_i^2\right)\right] + \sum_i E_i(\omega) * [B(\omega)\mathcal{F}(E_\mu E_i)]\right], \tag{3.38}$$

where $*$ represents a convolution, $\mathcal{F}$ is the Fourier transform, and the coefficients $A(\omega)$ and $B(\omega)$ are given by $A(\omega) = \int_{-\infty}^{\infty} a(t)e^{i\omega t}dt$ and $B(\omega) = \int_{-\infty}^{\infty} b(t)e^{i\omega t}dt$. Eqn. (3.38) is rather complicated however, but with the assumption that the input electric field is a monochromatic wave (or a sum of discrete frequencies), Eqn. (3.38) simplifies significantly.

Example 3.3. *Optical Kerr effect*
As an example we evaluate the induced polarization for the case of self phase modulation, i.e. the induced polarization that a monochromatic electric field generates at the frequency of the electric field itself. Thus the input field is

$$\mathbf{E}(t) = \frac{1}{2}\left(\mathbf{E}^0 e^{-i\omega_0 t} + c.c.\right), \tag{3.39}$$

where $\mathbf{E}^0$ is a constant amplitude vector. Correspondingly in the frequency domain

$$\mathbf{E}(\omega) = \frac{1}{2}\left[\mathbf{E}^0\delta(\omega-\omega_0) + (\mathbf{E}^0)^*\delta(\omega+\omega_0)\right]. \tag{3.40}$$

We start with the term in Eqn. (3.37) described through $a(t)$, and for simplicity let us evaluate the x-component of the induced polarization due to $a(t)$, which is

$$\left(P^{(3)}_{\text{nuclear}}(t)\right)_x/\varepsilon_0 = E_x(t)\sum_i \int_{-\infty}^{\infty} a(t-s)E_i(s)E_i(s)ds. \tag{3.41}$$

That is, we need to evaluate the convolution of $a(t)$ and terms of the form $E_i(t)E_i(t)$, where in the latter $i \in (x, y, z)$. In the frequency domain this is a simple product, i.e.

$$A(\omega)\mathcal{F}[E_i(t)E_i(t)] = \frac{1}{4}\Big[A(2\omega_0)(E_i^0)^2\delta(\omega-2\omega_0) + 2A(0)|E_i^0|^2 \\ + A(-2\omega_0)\left[(E_i^0)^*\right]^2\delta(\omega+2\omega_0)Big], \tag{3.42}$$

where $A(\omega)$ is defined as shown below Eqn. (3.38). When including the factor $E_x(t)$ in Eqn. (3.41) we need to find the product of $E_x(t)$ and the convolution $\int_{-\infty}^{\infty} a(t-s)E_i(s)E_i(s)ds$, which in the frequency domain is the convolution between $E_x(\omega) = \frac{1}{2}\left[E_x{}^0\delta(\omega-\omega_0) + (E_x{}^0)^*\delta(\omega+\omega_0)\right]$ and Eqn. (3.42). From this convolution we get the following contributions to the induced polarization at the positive frequency ω_0, corresponding to the three terms in Eqn. (3.41),

$$\frac{1}{8}\left[(E_x^0)^*A(2\omega_0)\sum_i(E_i^0E_i^0) + E_x^0 2A(0)\sum_i|E_i^0|^2\right], \tag{3.43}$$

where $A(0)$ and $A(2\omega_0)$ are the amplitudes of the material response at frequencies 0 and $2\omega_0$ respectively, and where $i \in (x, y, z)$. As pointed out in Chapter 2, in the discussion of the Born-Oppenheimer approximation, we assume that a material is not able to respond to optical frequencies, hence $A(2\omega_0) = 0$ in agreement with [26] and [27]. Consequently Eqn. (3.43) may be written as

$$\frac{1}{8}E_x^0 2A(0)\left(\mathbf{E}^0\cdot(\mathbf{E}^0)^*\right). \tag{3.44}$$

Finally Eqn. (3.44) is the amplitude of the x-component of the induced polarization at the positive frequency ω_0 due to $a(t)$. Thus the induced polarization due to $a(t)$ may be written as

$$(P(\omega))_x = \varepsilon_0\frac{1}{8}\left[E_x^0 2A(0)\left(\mathbf{E}^0\cdot(\mathbf{E}^0)^*\right)\delta(\omega-\omega_0) \\ + (E_x^0)^*2\left[A(0)\right]^*\left(\mathbf{E}^0\cdot(\mathbf{E}^0)^*\right)^*\delta(\omega+\omega_0)\right]. \tag{3.45}$$

We now evaluate the contribution to the induced polarization from $b(t)$, and for simplicity we evaluate again the x-component. Separating the s and t dependence we get

$$\left(P^{(3)}_{\text{nuclear}}(t)\right)_x/\varepsilon_0 = \sum_i E_i(t)\int_{-\infty}^{\infty} b(t-s)E_x(s)E_i(s)ds. \tag{3.46}$$

In the Fourier domain the convolution becomes a product and since the electric field is a monochromatic wave we find in analogy to Eqn. (3.42)

$$B(\omega)\mathcal{F}\left[E_x(s)E_i(s)\right] = \frac{1}{4}\left[B(2\omega_0)E_x^0E_i^0\delta(\omega-2\omega_0) + B(0)E_x^0(E_i^0)^* \right. \\ \left. + B(0)(E_x^0)^*(E_i^0) + B(-2\omega_0)(E_x^0E_i^0)^*\delta(\omega+2\omega_0)\right], \tag{3.47}$$

where $i \in (x, y, z)$. In the frequency domain the induced polarization is the convolution of the electric field in the frequency domain, i.e. Eqn. (3.40), and Eqn. (3.47). This results in the amplitude of the induced polarization due to $b(t)$ at the positive frequency ω_0,

$$\frac{1}{8}\left[E_i^0B(0)E_x^0(E_i^0)^* + E_i^0B(0)(E_x^0)^*E_i^0\right], \tag{3.48}$$

where we have assumed that $B(2\omega_0) = 0$ as we did for $A(2\omega_0)$. Now we get the following contribution to the x-component of the induced polarization at the positive frequency ω_0

$$\frac{1}{8}\left[E_x^0B(0)\left(E_x^0(E_x^0)^* + (E_x^0)^*E_x^0\right) + E_y^0B(0)\left(E_x^0(E_y^0)^* + (E_x^0)^*E_y^0\right) \right. \\ \left. + E_z^0B(0)\left(E_x^0(E_z^0)^* + (E_x^0)^*E_z^0\right)\right]. \tag{3.49}$$

By collecting terms proportional to $(E_x^0)^*$ and terms proportional to E_x^0, we arrive at

$$\frac{1}{8}\left[(E_x^0)^*B(0)\left(\mathbf{E}^0\cdot\mathbf{E}^0\right) + E_x^0B(0)\left((\mathbf{E}^0)^*\cdot\mathbf{E}^0\right)\right]. \tag{3.50}$$

Combining Eqn. (3.44) and Eqn. (3.50) we get the amplitude to the positive frequency ω_0

$$\left(P_{\omega_0}^0\right)_x = 2\varepsilon_0\frac{1}{8}\left[(E_x^0)^*B(0)\left(\mathbf{E}^0\cdot\mathbf{E}^0\right) + E_x^0\left[2A(0)+B(0)\right]\left((\mathbf{E}^0)^*\cdot\mathbf{E}^0\right)\right], \tag{3.51}$$

where it is noted that the amplitude includes a factor of 2 since we would like to write the induced polarization in the form $\mathbf{P} = \frac{1}{2}\left(\mathbf{P}_{\omega_0}^0 e^{-i\omega t} + \mathbf{P}_{-\omega_0}^0 e^{i\omega t}\right)$ that is used throughout this book.

As pointed out in [1], the first term corresponds to a change in the handedness of the field while the second term conserves the handedness of the field. In addition, the induced polarization is written in terms of two parameters $A(0)$ and $B(0)$ as opposed to tensor elements $\chi_{ijkl}^{(3)}$.

We now complete the description using the BO-approximation by including the induced polarization due to the interaction between the applied electric field and the electron. From [26] the amplitude of the induced polarization at the positive frequency ω_0 due to this interaction is

$$\mathbf{P}_{\omega_0}^0 = \frac{\sigma_e}{2}\varepsilon_0\mathbf{E}^0\sum_i |E_i^0|^2 = \frac{\sigma_e}{2}\varepsilon_0\mathbf{E}^0\left((\mathbf{E}^0)^*\cdot\mathbf{E}^0\right), \tag{3.52}$$

where $i \in (x, y, z)$ and where the factor 1/2 is included since the material only responds to the time average of the applied electric field, and σ_e defined in Eqn. (2.74).

Combining this with Eqn. (3.51), i.e. the molecular contribution, we write the induced polarization as

$$P_x^0/\varepsilon_0 = \frac{\sigma_e}{2}\left((\mathbf{E}^0)^* \cdot \mathbf{E}^0\right) E_x^0 + \frac{2}{8}\left[(E_x^0)^* B(0)\left(\mathbf{E}^0 \cdot \mathbf{E}^0\right) + E_x^0\left(2A(0) + B(0)\right)\left((\mathbf{E}^0)^* \cdot \mathbf{E}^0\right)\right] \tag{3.53}$$

Following the notation of Maker and Terhune [32] and Boyd [33] this may be rewritten in terms of two new parameters $\tilde{A}$ and $\tilde{B}$

$$P_x^0/\varepsilon_0 = E_x^0 \tilde{A}\left(\mathbf{E}^0 \cdot (\mathbf{E}^0)^*\right) + (E_x^0)^* \frac{1}{2}\tilde{B}\left(\mathbf{E}^0 \cdot \mathbf{E}^0\right), \tag{3.54}$$

where $\tilde{A} = \frac{2}{8}\left[2A(0) + B(0)\right] + \frac{\sigma_e}{2}$ and $\frac{1}{2}\tilde{B} = \frac{2}{8}B(0)$.

We would now like to compare the above with the results we obtain using the susceptibility tensor. Since we consider self induced refractive index changes in an isotropic material (for example glass) we are concerned with contributions to the induced polarization obtained from propagation of a monochromatic beam. This is characterized through $\chi^{(3)}(-\omega; \omega, -\omega, \omega)$. To evaluate terms in the induced polarization of the form

$$\mathbf{P}^{(n)} = \frac{1}{2}\left[(\mathbf{P}_\omega^0)^{(n)} e^{-i\omega t} + c.c\right], \tag{3.55}$$

we need to identify products that oscillate at the frequency ω. These can only originate from terms of the type $E_\omega^0 E_{-\omega}^0 E_\omega^0$ and thus the amplitude of the third-order induced polarization at frequency ω is written as

$$\left((P_\omega^0)^{(3)}\right)_i = \frac{3}{4}\varepsilon_0 \chi_{ijkl}^{(3)} E_j^0 (E_k^0)^* E_l^0, \tag{3.56}$$

where the factor 3/4 is the prefactor that was introduced in Section 3.4.

Applying intrinsic permutation symmetry, $\chi_{ijkl}^{(3)}(-\omega; \omega, -\omega, \omega) = \chi_{ilkj}^{(3)}(-\omega; \omega, -\omega, \omega)$, i.e. exchanging input no 1 and no 3 yields $\chi_{xxyy}^{(3)}(-\omega, \omega, -\omega, \omega)) = \chi_{xyyx}^{(3)}(-\omega, \omega, -\omega, \omega)$. With this, the susceptibility of isotropic materials reduces to

$$\chi_{ijkl}^{(3)} = \chi_{xxyy}^{(3)}(\delta_{ij}\delta_{kl} + \delta_{il}\delta_{jk}) + \chi_{xyxy}^{(3)}\delta_{ik}\delta_{jl}. \tag{3.57}$$

We return to a more detailed discussion of the susceptibility tensor for isotropic materials in Chapter 4.

For simplicity we consider only the x-component of the induced polarization. When writing this in a form similar to Eqn. (3.51)

$$\begin{aligned}\frac{4}{3\varepsilon_0}P_x^0 &= \chi_{xjkl}^{(3)}(-\omega;\omega,-\omega,\omega)E_j^0(E_k^0)^*E_l^0 \\ &= E_x^0(\chi_{xxyy}^{(3)}(E_y^0)^*E_y^0 + \chi_{xyyx}^{(3)}E_y^0(E_y^0)^* + \chi_{xxzz}^{(3)}(E_z^0)^*E_z^0 + \chi_{xzzx}^{(3)}E_z^0(E_z^0)^*) \\ &\quad + (E_x^0)^*(\chi_{xyxy}^{(3)}E_y^0E_y^0 + \chi_{xzxz}^{(3)}E_z^0E_z^0) + \chi_{xxxx}^{(3)}E_x^0(E_x^0)^*E_x^0.\end{aligned} \tag{3.58}$$

For isotropic materials $\chi_{iijj}^{(3)} = \chi_{iikk}^{(3)}$ and $\chi_{iiii}^{(3)} = \chi_{iijj}^{(3)} + \chi_{ijij}^{(3)} + \chi_{ijji}^{(3)}$ and thus we arrive at

$$\frac{4}{3\varepsilon_0}P_i^0 = E_x^0(\chi_{iijj}^{(3)} + \chi_{ijji}^{(3)})((\mathbf{E}^0)^* \cdot \mathbf{E}^0) + (E_x^0)^*\chi_{ijij}^{(3)}(\mathbf{E}^0 \cdot \mathbf{E}^0), \tag{3.59}$$

that is, by comparing against Eqn. (3.51) we find $\frac{3}{4}\chi_{ijij}^{(3)} = \frac{2}{8}B(0)$ and $\frac{3}{4}\left(\chi_{iijj}^{(3)} + \chi_{ijji}^{(3)}\right) = \frac{2}{8}\left(2A(0) + B(0)\right) + \sigma/2$.

We return to the Kerr effect for example in Chapter 5 where we discuss the impact on the intensity dependent refractive index due to the state of polarization of the electric field.

Because of intrinsic permutation symmetry we have been free to choose either $E_\omega^0 E_{-\omega}^0 E_\omega^0$ or $E_\omega^0 E_\omega^0 E_{-\omega}^0$, since we get the same from either starting point.

— ■ —

Example 3.4. *Raman scattering*
Returning to Raman scattering Example 3.2, we consider the case where the electric field has two frequencies; one at the excitation frequency ω_p and one at the scattered frequency ω_s, that is

$$\mathbf{E}(t) = \frac{1}{2}\left(\mathbf{E}_p^0 e^{-i\omega_p t} + c.c.\right) + \frac{1}{2}\left(\mathbf{E}_s^0 e^{-i\omega_s t} + c.c.\right). \tag{3.60}$$

Applying the BO-approximation, we know from Chapter 2 that the induced polarization is expressed as

$$P_x = \varepsilon_0 E_x(t)\int_{-\infty}^{\infty} a(t-s)\left[\mathbf{E}(s)\cdot\mathbf{E}(s)\right]ds + \varepsilon_0\int_{-\infty}^{\infty} E_x(s)b(t-s)\left[\mathbf{E}(t)\cdot\mathbf{E}(s)\right]ds. \tag{3.61}$$

Similar to the Kerr effect we evaluate the two terms corresponding to $a(t)$ and $b(t)$ individually. Starting with the contribution from $a(t)$, we evaluate the quadratic electric field from Eqn. (3.60). For simplicity we consider one component, here the x-component, the y- and z-component have similar forms,

$$E_x^2 = \frac{1}{4}\left(E_p^0 e^{-i\omega_p t} + c.c\right)^2 + \frac{1}{4}\left(E_s^0 e^{-i\omega_s t} + c.c\right)^2 + \frac{1}{2}\left(E_p^0 e^{-i\omega_p t} + c.c\right)\left(E_s^0 e^{-i\omega_s t} + c.c\right). \tag{3.62}$$

As noted earlier, a selects only the DC components or near DC frequency components of the mutual interaction among frequency components of the optical field due to the response time of the material, thus we find

$$A\,(0)\,\frac{1}{2}E_p^0(E_p^0)^* + A_\Delta\frac{1}{2}(E_p^0)^*E_s^0 e^{-i(\omega_s-\omega_p)t}, \tag{3.63}$$

where $A_\Delta = A\,(\omega_s - \omega_p)$. The amplitude of the induced polarization at the positive frequency ω_s is

$$P_{\omega_s}^0 = 2\varepsilon_0\left((E_s^0)_x A\,(0)\,\frac{1}{2}\left(\mathbf{E}_p^0\cdot(\mathbf{E}_p^0)^*\right) + A_\Delta\frac{1}{2}(E_p^0)_x\left((\mathbf{E}_p^0)^*\cdot\mathbf{E}_s^0\right)\right). \tag{3.64}$$

It is noted that the imaginary part of A (and B) gives rise to Raman scattering whereas the real part gives rise to an induced change in the refractive index. Hence, $A\,(0)$ and $B\,(0)$ which are purely real give no Raman scattering.

Now, we consider the contribution due to the term $b(t)$

$$\int_{-\infty}^{\infty} E_x(s)b(t-s)\left[\mathbf{E}(t)\cdot\mathbf{E}(s)\right]ds, \tag{3.65}$$

i.e. the x-component

$$\begin{aligned}&\int_{-\infty}^{\infty} E_x(s)b(t-s)\left[E_x(t)E_x(s) + E_y(t)E_y(s) + E_z(t)E_z(s)\right]ds\\ &\quad= E_x(t)\int_{-\infty}^{\infty} b(t-s)E_x(s)E_x(s)ds + E_y(t)\int_{-\infty}^{\infty} b(t-s)E_x(s)E_y(s)ds\\ &\quad\quad+ E_z(t)\int_{-\infty}^{\infty} b(t-s)E_x(s)E_z(s)ds.\end{aligned} \tag{3.66}$$

Let us look at the second term and insert $E_x = \frac{1}{2}\left((E_p^0)_x e^{-i\omega_p t} + c.c\right) + \frac{1}{2}\left((E_s^0)_x e^{-i\omega_s t} + c.c\right)$ and $E_y = \frac{1}{2}\left((E_p^0)_y e^{-i\omega_p t} + c.c\right) + \frac{1}{2}\left((E_s^0)_y e^{-i\omega_s t} + c.c\right)$. With this we get

$$\begin{aligned}4E_xE_y = &\left((E_p^0)_x e^{-i\omega_p t} + c.c\right)\left((E_p^0)_y e^{-i\omega_p t} + c.c\right)\\ &+\left((E_p^0)_x e^{-i\omega_p t} + c.c\right)\left((E_s^0)_y e^{-i\omega_s t} + c.c\right)\\ &+\left((E_s^0)_x e^{-i\omega_s t} + c.c\right)\left((E_p^0)_y e^{-i\omega_p t} + c.c\right)\\ &+\left((E_s^0)_x e^{-i\omega_s t} + c.c\right)\left((E_s^0)_y e^{-i\omega_s t} + c.c\right).\end{aligned} \tag{3.67}$$

Selecting DC or near DC terms we get

$$\begin{aligned}4E_xE_y \approx\ &(E_p^0)_x(E_p^0)_y^* + (E_p^0)_x^*(E_p^0)_y + (E_p^0)_x e^{-i\omega_p t}(E_s^0)_y^* e^{i\omega_s t}\\ &+ (E_p^0)_x^* e^{i\omega_p t}(E_s^0)_y e^{-i\omega_s t} + (E_s^0)_x e^{-i\omega_s t}(E_p^0)_y^* e^{i\omega_p t} + (E_s^0)_x^* e^{i\omega_s t}(E_p^0)_y e^{-i\omega_p t}\\ &+ (E_s^0)_x e^{-i\omega_s t}(E_s^0)_y^* e^{i\omega_s t} + (E_s^0)_x^* e^{i\omega_s t}(E_s^0)_y e^{-i\omega_s t}.\end{aligned} \tag{3.68}$$

When multiplying with the electric field to get the final cubic term $E_y^0 E_x^0 E_y^0$ and looking for terms that oscillate at the positive frequency ω_s, we see that the terms $(E_p^0)_x e^{-i\omega_p t}(E_s^0)_y^* e^{i\omega_s t}$, and $(E_s^0)_x^* e^{i\omega_s t}(E_p^0)_y e^{-i\omega_p t}$ do not contribute. Having in mind that we eventually are interested in Raman scattering then $B(0)$ does not contribute, since it is purely real. Including contributions from $E_z^0 E_x^0 E_z^0$ and $E_x^0 E_x^0 E_x^0$ we then get

$$B_\Delta \left\{ (E_p^0)_x \left[(\mathbf{E}_p^0)^* \cdot \mathbf{E}_s^0 \right] + (E_s^0)_x \left[\mathbf{E}_p^0 \cdot (\mathbf{E}_p^0)^* \right] \right\}. \tag{3.69}$$

Adding the two contributions, from A_Δ and B_Δ we get

$$(E_p^0)_x A_\Delta \frac{1}{2} \left((\mathbf{E}_p^0)^* \cdot \mathbf{E}_s^0 \right) e^{-i\omega_s t} + B_\Delta e^{-i\omega_s t} \left[(E_p^0)_x^* (\mathbf{E}_p^0 \cdot \mathbf{E}_s^0) + (E_s^0)_x (\mathbf{E}_p^0 \cdot (\mathbf{E}_p^0)^*) \right]. \tag{3.70}$$

This is a very important result that enables us to compare two cases: One where $\mathbf{E}_p^0$ is parallel to $\mathbf{E}_s^0$ and one where $\mathbf{E}_p^0$ is orthogonal to $\mathbf{E}_s^0$.

If E_p^0 and E_s^0 are orthogonal $\mathbf{E}_p^0 \cdot \mathbf{E}_s^0 = \left(\mathbf{E}_p^0\right)^* \cdot \mathbf{E}_s^0 = 0$,

$$P_x^0(\omega_s)/\varepsilon_0 = B_\Delta \left((E_s^0)_x (\mathbf{E}_p \cdot (\mathbf{E}_p^0)^*) \right). \tag{3.71}$$

This is often referred to as the orthogonal case, see Figure 3.1.

If E_p^0 and E_s^0 are parallel $\mathbf{E}_p^0 \cdot \mathbf{E}_s^0 \neq 0$, the induced polarization equals

$$P_x^0(\omega_s)/\varepsilon_0 = A_\Delta \frac{1}{2} \left((\mathbf{E}_p^0)^* \cdot \mathbf{E}_s^0 \right) + B_\Delta \left[(E_p^0)_x^* (\mathbf{E}_p^0 \cdot \mathbf{E}_s^0) + (E_s^0)_x (\mathbf{E}_p^0 \cdot (\mathbf{E}_p^0)^*) \right]. \tag{3.72}$$

This is often referred to as the parallel case see, Figure 3.1.

For comparison against the description using susceptibility, it is noted that Raman scattering is a third-order nonlinear process i.e. in relation to Table 3.2, $n = 3$ resulting in the coefficient $(1/2)^3$, the number of distinct perturbations equals $6(= n!)$, i.e $p = 6$, and finally, the frequency of the scattering is different from a DC that is when using the notation $\mathbf{P} = 1/2(\mathbf{P}^0 e^{-i\omega_s t} + \left(\mathbf{P}^0\right)^* e^{i\omega_s t})$, that is $l = 1$. Finally, $m = 0$ since none of the generating frequencies equal zero. Putting all this into the equation for the K-factor one arrives at

$$K = 2^{l+m-n} p = 2^{-2} 6 = 3/2, \tag{3.73}$$

which is the number that appears in Table 3.2.

When describing the induced polarization corresponding to Raman scattering, the amplitude of the induced polarization equals

$$P_i^0 = \frac{3}{2} \varepsilon_0 \chi_{ijkl}^{(3)}(\omega_s; \omega_p, -\omega_p, \omega_s)(E_p^0)_j (E_p^0)_k^* (E_s^0)_l. \tag{3.74}$$

For simplicity, but without loss of generality, we consider the induced polarization along the x-coordinate which has 7 contributions since the indices $ijkl$ have to

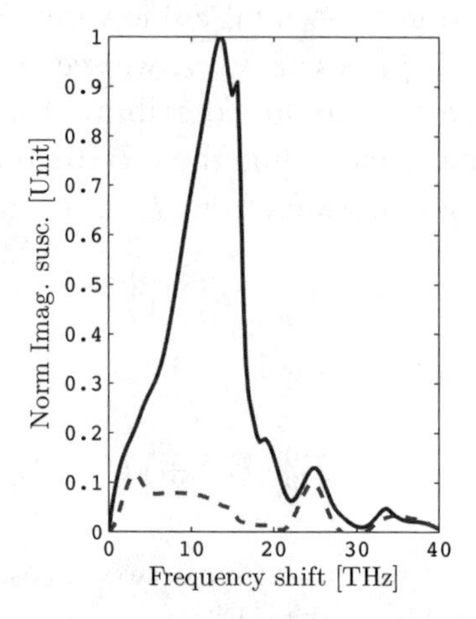

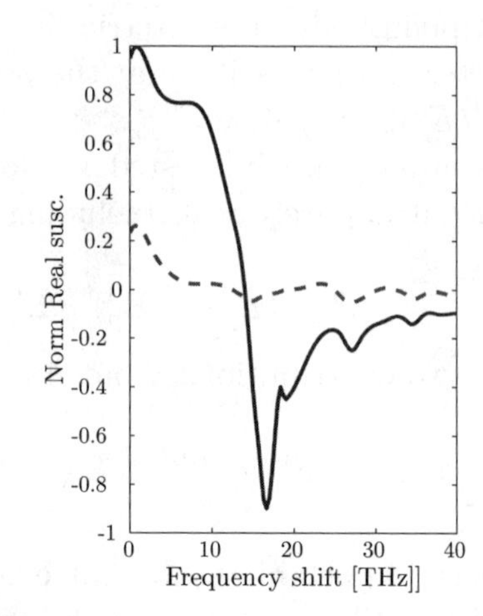

Figure 3.1: Left Imaginary part of Raman susceptibility, proportional to Raman gain versus frequency shift in (THz) for silica glass. The solid curve is the parallel case ($A_\Delta + B_\Delta$), whereas the dashed curve is the orthogonal case (B_Δ). Both curves are normalized to the peak of the parallel case. **Right** The real part of the Raman susceptibility proportional to change in refractive index versus frequency shift in (THz) for silica glass. The solid curve is the parallel case ($A_\Delta + B_\Delta$), whereas the dashed curve is the orthogonal case (B_Δ). Both curves are normalized to the peak of the parallel case. Data adapted from [17].

appear in pairs or all be the same. Let us again consider an isotropic material (for example silica glass) thus $\chi^{(3)}_{iiii} = \chi^{(3)}_{iijj} + \chi^{(3)}_{ijij} + \chi^{(3)}_{ijji}$, such that the induced polarization equals

$$P^0_x = \frac{3}{2}\varepsilon_0 \left[\chi^{(3)}_{iijj}(E^0_p)_x \left[(\mathbf{E}^0_p)^* \cdot \mathbf{E}^0_s\right] + \chi^{(3)}_{ijij}(E^0_p)^*_x \left[(\mathbf{E}^0_p) \cdot \mathbf{E}^0_s\right] + \chi^{(3)}_{ijji}(E^0_s)_x \left[(\mathbf{E}^0_p)^* \cdot \mathbf{E}^0_p\right]\right]. \tag{3.75}$$

This may be compared against Eqn. (3.72), showing how A_Δ and B_Δ relate to the nonlinear susceptibility.

—— ■ ——

3.6 Kramers-Kronig relations

We have already discussed that the susceptibility is complex valued. To exemplify this, the real part may be associated with a change in the refractive index whereas the imaginary part is associated with a loss or amplification depending on the sign of the imaginary part. Thus, often either the real part or the imaginary part of the susceptibility is known from measurements, and it is desirable to obtain knowledge of the other part. This may be obtained by using the Kramers-Kronig relations.

These are known from linear optics where the real and imaginary parts of the susceptibility are related through

$$\mathrm{Re}[\chi^{(1)}(\omega)] = \frac{1}{\pi}\mathrm{p.v.}\int_{-\infty}^{\infty} \frac{\mathrm{Im}[\chi^{(1)}(\omega')]d\omega'}{\omega' - \omega}, \tag{3.76a}$$

$$\mathrm{Im}[\chi^{(1)}(\omega)] = -\frac{1}{\pi}\mathrm{p.v.}\int_{-\infty}^{\infty} \frac{\mathrm{Re}[\chi^{(1)}(\omega')]d\omega'}{\omega' - \omega}, \tag{3.76b}$$

where p.v. is the principal value. The principal value of an integral of a function f with poles in $x_1, x_2, \cdots, x_n$ is defined as the limit of

$$\mathrm{limit}_{\epsilon\to 0}\left[\int_{-\infty}^{x_1-\epsilon} f(x)dx + \int_{x_1+\epsilon}^{x_2-\epsilon} f(x)dx + \cdots + \int_{x_{n-1}+\epsilon}^{x_n-\epsilon} f(x)dx + \int_{x_n+\epsilon}^{\infty} f(x)dx\right], \tag{3.77}$$

assuming the limit exists and is finite. This is often referred to as the Cauchy principal value.

3.6.1 Kramers-Kronig relations in linear optics

Using that the real part of the susceptibility is symmetric, that is $\mathrm{Re}\left[\chi^{(1)}(-\omega)\right] = \mathrm{Re}\left[\chi^{(1)}(\omega)\right]$, whereas the imaginary part is anti-symmetric, that is $\mathrm{Im}\left[\chi^{(1)}(-\omega)\right] = -\mathrm{Im}\left[\chi^{(1)}(\omega)\right]$, one may rewrite the Kramers-Kronig relations as

$$\mathrm{Re}\left[\chi^{(1)}\right] = \frac{2}{\pi}\mathrm{p.v.}\int_{0}^{\infty} \frac{\omega'\mathrm{Im}\left[\chi^{(1)}(\omega')\right]}{\omega'^2 - \omega^2}d\omega' \tag{3.78a}$$

$$\mathrm{Im}\left[\chi^{(1)}\right] = -\frac{2\omega}{\pi}\mathrm{p.v.}\int_{0}^{\infty} \frac{\mathrm{Re}\left[\chi^{(1)}(\omega')\right]}{\omega'^2 - \omega^2}d\omega'. \tag{3.78b}$$

In linear optics the real and imaginary part of the susceptibility are often replaced by the refractive index and an absorption coefficient. More specifically the squared refractive index is identified with the susceptibility through the relation $n^2 = 1 + \chi^{(1)}$. Since $\chi^{(1)}$ is a complex number the refractive index is allowed to be complex and the imaginary part of the refractive index describes absorption. Propagation of an electric field E from z to $z + \delta z$ is then described through

$$E(z + \delta z) = E(z)\exp[ik_0 n\delta z - (\alpha/2)\delta z], \tag{3.79}$$

where $k_0 = \omega/c$ is the wave number, n is the conventional real refractive index, and α is the absorption coefficient. Using the Kramers-Kronig relations it may be shown that the refractive index is related to the absorption coefficient through

$$n - 1 = \frac{c}{\pi}\mathrm{p.v.}\int_{0}^{\infty} \frac{\alpha(\omega')}{\omega'^2 - \omega^2}d\omega'. \tag{3.80}$$

3.6.2 Kramers-Kronig relations in nonlinear optics

Relations similar to the Kramers-Kronig relations for the linear response can be deduced for some but not all nonlinear optical interactions [33]. As a rule of thumb, when the nonlinear effect may be described as an effective first-order susceptibility, the real and imaginary parts of the nonlinear susceptibilities are related using the Kramers-Kronig relations. Examples include: Cross-phase modulation, Raman and Brillouin scattering. All of these are described by a susceptibility of the form. $\chi^{(3)}_{ijkl}(-\omega_k; -\omega_1, \omega_1, \omega_k)$. In addition to these, the Kramers-Kronig relations are also valid when describing second- and third-harmonic generation.

The Kramers-Kronig relations are on the other hand **not** valid when describing effects where the applied field is causing an interaction with itself as for example in self induced changes in the refractive index, i.e. the optical Kerr effect.

Raman scattering continued

An accurate mathematical fit to the Raman response function of silica glass in the time domain is provided in [34]

$$R_R^{(3)}(\tau) = h_R(\tau) = \sum_{i=1}^{13} \frac{A_i'}{\omega_{\nu,i}} \exp(-\gamma_i\tau)\exp(-\Gamma_i^2\tau^2/4)\sin(\omega_{\nu,i}\tau), \quad \tau > 0, \tag{3.81}$$

where the sum is over i vibrational modes, each with amplitude A_i' and vibrational frequency $\omega_{\nu,i}$, and γ_i is the Lorentzian linewidth for mode i and Γ_i the Gaussian linewidth of mode i. The Fourier transform of the Raman response function $h_R(\tau)$ is the third-order susceptibility. The Raman gain is related to the imaginary part of the susceptibility, thus the imaginary part of the Fourier transform of the response function should match the measurable gain spectrum.

The relation between the Raman gain and the imaginary part of the susceptibility is given by

$$g_R(\Omega) = \frac{\omega_0}{cn_0}\chi^{(3)}\mathrm{Im}\left[h_R(\Omega)\right], \tag{3.82}$$

where ω_0 is the frequency of the optical wave being Raman scattered and n_0 the refractive index at the optical frequency. The imaginary part of the Fourier transform of the response function equals

$$\mathrm{Im}\left[h_R(\Omega)\right] = \sum_{i=1}^{13} \frac{A_i'}{\omega_{\nu,i}} \int_{\tau=0}^{\infty} \exp(-\gamma_i\tau)\exp(-\Gamma_i^2\tau^2/4)\sin(\omega_{\nu,i}\tau)\sin(\omega t)d\tau. \tag{3.83}$$

The real part, which is responsible for a change in the refractive index, is

$$\mathrm{Re}[h_R(\Omega)] = \sum_{i=1}^{13} \frac{A_i'}{\omega_{\nu,i}} \int_{\tau=0}^{\infty} \exp(-\gamma_i\tau)\exp(-\Gamma_i^2\tau^2/4)\sin(\omega_{\nu,i}\tau)\cos(\omega\tau)d\tau. \tag{3.84}$$

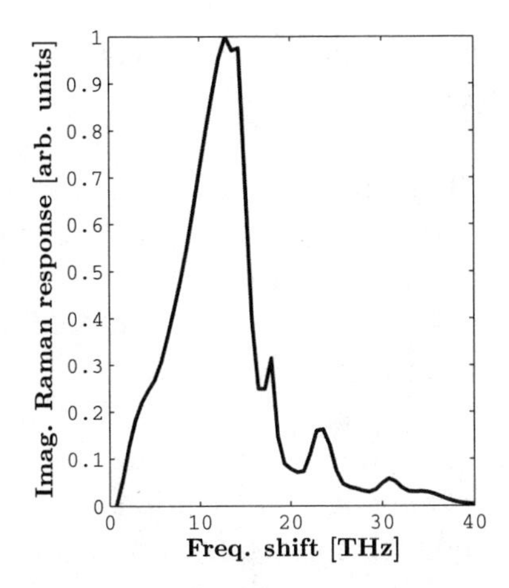

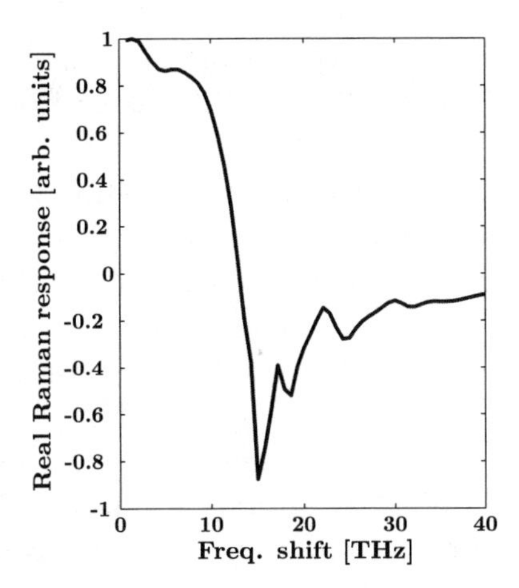

Figure 3.2: Normalized spectra of the Raman susceptibility. **Left** proportional to the imaginary part i.e. gain, **Right** proportional to the real part i.e. refractive index. Data adapted from [17].

The real part of $h_R(\Omega)$ is related to the imaginary part of $h_R(\Omega)$ by the Kramers-Kronig relations. This is described in the following section.

Because the Raman susceptibility may be considered as an effective first-order nonlinearity $\chi^{(1)}$, the real part of $\chi^{(3)}$ is related to the imaginary part of $\chi^{(3)}$ by the Kramers-Kronig relations, i.e.

$$\mathrm{Re}[\chi^{(3)}(\omega')] = \frac{1}{\pi}\mathrm{p.v.}\int_{-\infty}^{\infty}\frac{\mathrm{Im}[\chi^{(3)}]}{\omega' - \omega}d\omega', \tag{3.85}$$

and

$$\mathrm{Im}[\chi^{(3)}(\omega')] = -\frac{1}{\pi}\mathrm{p.v.}\int_{-\infty}^{\infty}\frac{\mathrm{Re}[\chi^{(3)}]}{\omega' - \omega}d\omega'. \tag{3.86}$$

Assuming in the simplest case that the response function is a decaying exponential, $h_R(t) = \frac{1}{\tau}\exp(-t/\tau)\Theta(t)$, as a Debye model described in Appendix C, one finds that the real and imaginary part equals

$$\mathrm{Re}[H_R(\Omega)] = \frac{N_R}{1+(\Omega\tau)^2}, \tag{3.87a}$$

$$\mathrm{Im}[H_R(\Omega)] = \frac{N_R\Omega\tau}{1+(\Omega\tau)^2}, \tag{3.87b}$$

where N_R is the Raman contribution to the intensity dependent refractive index [29], Ω is the frequency separation between the pump and the Stokes shifted frequency, and τ is the damping of the exponential decay. For comparison, consider the more accurate description in Eqn. (3.81).

The value of the real part of $H_R(\Omega)$ at $\Omega = 0$ gives a contribution to the nonlinear refractive index, the self induced refractive index. This is simply the Raman response when the pump and signal have identical frequencies, i.e. the contribution to the nonlinear refractive index from molecular vibrations.

Denoting the nonlinear refractive index, n_{20}. This may be written as a sum of an instantaneous, i.e. time scale of few fs, contribution from electronic transitions n_∞ and a contribution from the Raman scattering ($n_R = \mathrm{Re}[H_R(\Omega = 0)]$), i.e. $n_{20} = n_\infty + n_R$. Some times n is written in terms of a fraction to the nonlinear refractive index from Raman scattering f_R as $n_\infty = n_{10}(1 - f_R)$

From measurements of the Raman gain, the imaginary part of the 'Raman'-susceptibility is directly evaluated. Using the Kramers-Kronig relations, the real part of the 'Raman'-susceptibility is then determined. From this, the fraction of the nonlinearity originating from the Raman process and from the electronic transitions is evaluated. It turns out that the contribution from the Raman scattering, f_R approximates 20% of the refractive index change in silica glass.

3.7 Summary

- The nth-order induced polarization in the time domain is written as

$$P_{\mu}^{(n)}(\mathbf{r},t) = \varepsilon_0 \int_{-\infty}^{\infty} \cdots \int_{-\infty}^{\infty} \chi_{\mu\alpha_1,\ldots,\alpha_n}^{(n)}(-\omega_\sigma;\omega_1,\ldots,\omega_n) E_{\alpha_1}(\mathbf{r},\omega_1)\cdots E_{\alpha_n}(\mathbf{r},\omega_n) e^{-i\omega_\sigma t} d\omega_1 \cdots d\omega_n, \tag{3.19}$$

where the nth-order susceptibility is given as

$$\chi_{\mu\alpha_1\cdots\alpha_n}^{(n)}(-\omega_\sigma;\omega_1,\ldots,\omega_n) = \int_{-\infty}^{\infty} \cdots \int_{-\infty}^{\infty} R_{\mu\alpha_1\cdots\alpha_n}^{(n)}(\tau_1,\ldots,\tau_n) e^{i(\omega_1\tau_1+\cdots+\omega_n\tau_n)} d\tau_1 \cdots d\tau_n, \tag{3.20}$$

and

$$\omega_\sigma = \omega_1 + \omega_2 + \ldots + \omega_n,$$

and

$$\left[\chi_{\mu\alpha_1\cdots\alpha_n}^{(n)}(-\omega_\sigma;\omega_1,\ldots,\omega_n)\right]^* = \left[\chi_{\mu\alpha_1\cdots\alpha_n}^{(n)}(\omega_\sigma^*;-\omega_1^*,\ldots,-\omega_n^*)\right]. \tag{3.21}$$

- The nth-order induced polarization in the frequency domain is written as

$$P^{(n)}(\omega) = \varepsilon_0 \int_{-\infty}^{\infty} \cdots \int_{-\infty}^{\infty} \chi_{\mu\alpha_1,\ldots,\alpha_n}^{(n)}(-\omega_\sigma;\omega_1,\ldots,\omega_n) E_{\alpha_1}(\mathbf{r},\omega_1)\cdots E_{\alpha_n}(\mathbf{r},\omega_n) \delta(\omega-\omega_\sigma) d\omega_1 \cdots d\omega_n. \tag{3.25}$$

- In Cartesian coordinates a single component of the induced polarization of a monochromatic field is

$$\left((P_{\omega_\sigma}^0)^{(n)}\right)_\mu = \varepsilon_0 \sum_{\alpha_1\cdots\alpha_n} \sum_{\omega} K(-\omega_\sigma;\omega_1,\ldots,\omega_n) \chi_{\mu\alpha_1\cdots\alpha_n}^{(n)}(-\omega_\sigma;\omega_1,\ldots,\omega_n)(E_{\omega_1}^0)_{\alpha_1}\cdots(E_{\omega_n}^0)_{\alpha_n}, \tag{3.31}$$

where

$$K(-\omega_\sigma;\omega_1,\ldots,\omega_n) = 2^{l+m-n} p, \tag{3.35}$$

and

$$\begin{aligned} p &= \text{the number of distinct permutations of } \omega_1,\omega_2,\ldots,\omega_n, \\ n &= \text{the order of nonlinearity}, \\ m &= \text{the number of angular frequencies } \omega_k \text{ that are zero}, \\ l &= \begin{cases} 1, & \text{if } \omega_\sigma \neq 0, \\ 0, & \text{otherwise.} \end{cases} \end{aligned}$$

- Kramers-Kronig relations known from linear optics also apply to some, but not all, susceptibilities. In general, if a nonlinear susceptibility may be written, as an effective first-order susceptibility, the Kramers-Kronig relations apply.

Chapter 4

Symmetries in nonlinear optics

Contents

The induced polarization in the frequency domain may be mathematically described through the use of the susceptibility tensors. The tensors corresponding to the nth-order nonlinear induced polarization contain 3^{n+1} elements i.e. 81 elements for the susceptibility tensor describing the third-order induced polarization. Fortunately, they typically reduce because of symmetry relations. In the following we focus on these symmetry relations.

There are essentially four classes of symmetries that we encounter.

Intrinsic permutation symmetry: Applied field frequencies can be permutated if the Cartesian indices are simultaneously permutated, i.e. in $\chi^{(n)}_{\mu\alpha_1\cdots\alpha_n}(-\omega_\sigma;\omega_1,\cdots,\omega_n)$ the pairs (α_1,ω_1), through (α_n,ω_n) may be permuted, resulting in $n!$ permutations.

For example in a process described by a $\chi^{(3)}$ tensor intrinsic permutation symmetry has the consequence that

$$\begin{aligned}\chi^{(3)}_{\mu\alpha\beta\gamma}(-\omega_\sigma;\omega_1,\omega_2,\omega_3) &= \chi^{(3)}_{\mu\beta\alpha\gamma}(-\omega_\sigma;\omega_2,\omega_1,\omega_3) \\ &= \chi^{(3)}_{\mu\beta\gamma\alpha}(-\omega_\sigma;\omega_2,\omega_3,\omega_1) \\ &= \chi^{(3)}_{\mu\gamma\beta\alpha}(-\omega_\sigma;\omega_3,\omega_2,\omega_1),\end{aligned} \tag{4.1}$$

that is, the susceptibility is invariant under the 3! possible permutations of the pairs $(\alpha,\omega_1),(\beta,\omega_2)\cdots(\gamma,\omega_3)$. As discussed in Chapter 2 this symmetry applies generally, and is a direct consequence of the constraint to the material response that enables one to write the material response as a symmetric tensor.

Overall permutation symmetry: All frequencies can be permutated if the Cartesian indices are simultaneously permutated, i.e. in $\chi^{(n)}_{\mu\alpha_1\cdots\alpha_n}(-\omega_\sigma;\omega_1,\cdots,\omega_n)$ the pairs (α_1,ω_1) through (α_n,ω_n) in addition to $(\mu,-\omega_\sigma)$ may be permuted, resulting in $(n+1)!$ permutations.

For example in a process described by a $\chi^{(3)}$ tensor overall permutation symmetry has the consequence that

$$\begin{aligned}\chi^{(3)}_{\mu\alpha\beta\gamma}(-\omega_\sigma;\omega_1,\omega_2,\omega_3) &= \chi^{(3)}_{\alpha\mu\beta\gamma}(\omega_1;-\omega_\sigma,\omega_2,\omega_3) \\ &= \chi^{(3)}_{\alpha\beta\mu\gamma}(\omega_1;\omega_2,-\omega_\sigma,\omega_3,) \\ &= \chi^{(3)}_{\alpha\beta\gamma\mu}(\omega_1;\omega_2,\omega_3,-\omega_\sigma).\end{aligned} \tag{4.2}$$

This symmetry applies when all optical frequencies appearing in the susceptibility are far from transition frequencies of the medium - i.e. the medium is transparent at all relevant frequencies.

An important consequence of overall permutation symmetry is the Manley-Rowe relations that we return to in Chapter 5.

Kleinman symmetry: All Cartesian indices can be permutated independent of the frequencies. That is in the susceptibility $\chi^{(n)}_{\mu\alpha_1\cdots\alpha_n}(-\omega_\sigma;\omega_1,\cdots,\omega_n)$ all indexes $\mu\alpha_1\cdots\alpha_n$ may be freely permuted completely independent of the frequencies. This results in $(n+1)!$ permutations.

For example in a process described by a $\chi^{(3)}$ tensor intrinsic permutation symmetry has the consequence that

$$\chi^{(3)}_{\mu\alpha\beta\gamma}(-\omega_\sigma;\omega_1,\omega_2,\omega_3) = \chi^{(3)}_{\alpha\mu\beta\gamma}(-\omega_\sigma;\omega_1,\omega_2,\omega_3) = \chi^{(3)}_{\alpha\beta\mu\gamma}(-\omega_\sigma;\omega_1,\omega_2,\omega_3) = \chi^{(3)}_{\alpha\beta\gamma\mu}(-\omega_\sigma;\omega_1,\omega_2,\omega_3). \quad (4.3)$$

This symmetry applies as a consequence of the overall permutation symmetry, and applies in the low-frequency limit of non-resonant media [2].

Spatial symmetries: All material possess some degree of symmetry. One special case is where the light material interaction is completely independent of the propagation direction of light through the crystal. However, typically crystals possess a well defined symmetry and are consequently categorized into crystal classes in accordance with their symmetry.

In the following we discuss these crystal classes, their symmetry and the implications on the propagation of light.

4.1 Spatial symmetries

In the previous section we have been concerned with 'internal symmetry', i.e. intrinsic and overall permutation symmetry but not spatial symmetry, that is the symmetry among indices of a susceptibility tensor element and its arguments - which is relevant for example to study what happens when the launched electric field does not match with the orientation of the material, for example the crystal orientation.

Some materials are naturally independent in their orientation; these are referred to as isotropic materials. It is important here to stress the fact that a nonlinear process, for example Raman scattering, may be polarization dependent even in an isotropic material like fused silicate glass. Glass is an important material mainly because of its use as host material in optical fibers and optical integrated circuits. Consequently glass is often used as an example in the following.

First of all, it is noted that the mathematical description of wave propagation in a nonlinear material may be carried out in different frames. However, the final result needs to be independent on what particular frame that is used. To ease the mathematical task it may be easier to use different coordinate frames.

All tensors that we are concerned with are so-called polar tensors, see Appendix A. Polar tensors may be transferred from one to another coordinate system using the transformation rule [35],

$$T'_{\mu\alpha_1\cdots\alpha_n} = R_{\mu u}R_{\alpha_1 a_1}\cdots R_{\alpha_n a_n}T_{ua_1\cdots a_n}, \quad (4.4)$$

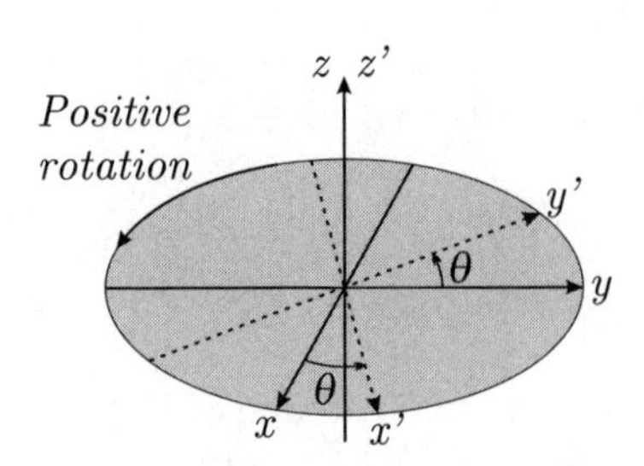

Figure 4.1: A positive rotation around the z-axis means counterclockwise rotation of the xy-plane by an angle θ.

where $\underline{\underline{R}}$ is a rotation matrix in which each element is defined through $R_{ij} = \cos(\theta_{ij})$, where θ_{ij} is the angle between the ith new axis and the jth original axis. Summation over repeated indices i.e. $ua_1 \cdots a_n$, as implicitly assumed was introduced in the previous chapters.

To emphasize the relation between a new coordinate frame, indicated by a prime (x', y', z'), and an old frame with unprimed coordinates (x, y, z), the transformation matrix is denoted

$$\underline{\underline{R}} = \begin{pmatrix} R_{x'x} & R_{x'y} & R_{x'z} \\ R_{y'x} & R_{y'y} & R_{y'z} \\ R_{z'x} & R_{z'y} & R_{z'z} \end{pmatrix}. \tag{4.5}$$

From this a positive rotation of the coordinate system by an angle θ around the z-axis as in Figure 4.1 is then

$$\underline{\underline{R}} = \begin{pmatrix} \cos(\theta) & \sin(\theta) & 0 \\ -\sin(\theta) & \cos(\theta) & 0 \\ 0 & 0 & 1 \end{pmatrix}. \tag{4.6}$$

The tensor elements in the primed coordinate system are then evaluated by using Eqn. (4.6) together with Eqn. (4.4).

4.1.1 Crystal classes/Nomenclature

Crystals are commonly classified in seven **crystal classes** depending on their degree of symmetry: triclinic (the least symmetrical), monoclinic, orthombic, tetragonal, hexagonal, trigonal and isometric (the most symmetric). The seven classes are in turn divided into **point group's** according to their symmetry with respect to a point. There are 32 such point groups.

In crystallography the properties of a crystal are described in terms of the natural coordinate system provided by the crystal itself. The axes of this natural system are the edges of a unit cell. That is the basis vectors that define the natural coordinate system may not be of equal lengths nor be perpendicular to each other. The

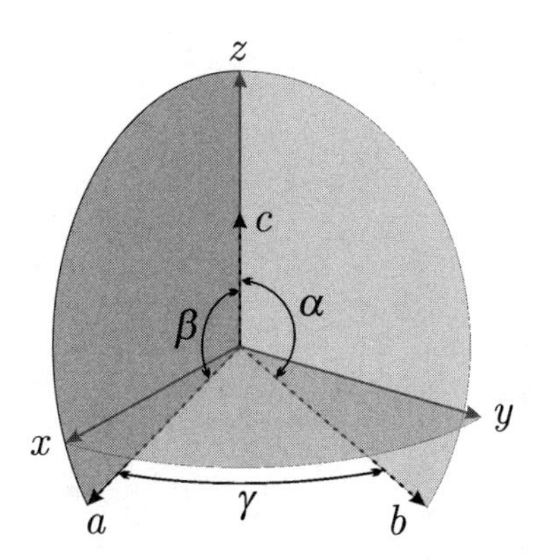

Figure 4.2: Coordinate definitions from ITU standard committee.

natural axes are referred to as a, b, c and α, β, γ the angles between b and c, c and a and a and b respectively as shown in Figure 4.2.

In relation to a rectangular coordinate system x, y, z which is used in the theoretical treatment of electromagnetic problems, as described in the context of this book, a standard which is described in [36] is used:

1. the z-axis is chosen parallel to the c-axis
2. the x-axis is chosen perpendicular to the c-axis in the ac-plane, pointing in the positive a-direction
3. the y-axis is normal to the ac-plane pointing in the positive b-direction and forming a right-handed coordinate system with z and x.

The designation for any class of symmetry is made of one, two, or three symbols.

The first symbol refers to the principal axis of crystal if any, indicating the type of symmetry of that axis and the existence of reflection planes perpendicular to that axis. The notation for an axis rotation is 1,2,3,4,6, where the number indicates through its reciprocal the part of full rotations about the axis which is required to bring the crystal back to an equivalent position in regard to its structural properties. **The second symbol** refers to the secondary axis on the crystal. **The third symbol** names the tertiary axis if such exists. However, since the aim of this book is not to go into an in-depth description regarding crystal classes etc., we only adopt the necessary results for the description of nonlinear optics. For a more detailed description of crystallography and tensors related to crystals the reader is referred to more comprehensive texts on this topic, for example [37]. In the following we refer to crystals in accordance with their so-called 'point-group' and 'space group'. This notation was introduced by Hermann-Mauguin and sometimes referred to as the Hermann-Mauguin notation [38].

The 32 crystallographic point groups are:

- $1, \bar{1}$
- $2, m, 2/m$
- $222, mm2, mmm$
- $4, \bar{4}, 4/m, 422, 4mm, \bar{4}2m, 4/mmm$
- $3, \bar{3}, 32, 3m, \bar{3}m$
- $6, \bar{6}, 6/m, 622, 6mm, \bar{6}2m, 6/mmm$
- $23, m\bar{3}, 432, \bar{4}3m, m\bar{3}m$

4.1.2 Neumann principle

When describing propagation of light through a material, or crystal, it is essential to know the properties of the tensor that describes a given material. These properties are defined on the basis of the symmetry of the material. To evaluate the tensor corresponding to a given material the so-called Neumann principle is essential. According to this, the elements of a tensor representing a material property must remain invariant under a transformation of coordinates governed by a symmetry operation valid for the point group of the material.

Applying Neumann principle to all elements of the tensor corresponding to a distinct physical property, each element of the susceptibility tensor must remain invariant under relevant transformations, i.e.

$$T'_{\mu\alpha_1\cdots\alpha_n} = T_{\mu\alpha_1\cdots\alpha_n}, \tag{4.7}$$

where the prime indicate a new frame while the unprimed tensor is in the original frame.

This is a very important, yet intuitive, relation, that we use intensively in the following which illustrates what elements of a tensor are non-zero in specific crystals, and how others are mutually related.

For every tensor component an equation of the type Eqn. (4.7) should be valid so that the tensor component must satisfy a system of equations. Since this holds for every symmetry operation of a given crystal class, the number of the systems of equations between the tensor components equals the number of the symmetry operations which may be performed in the given crystal class. However, to obtain every relationship among the components of a tensor representing any physical property in the case of a given crystal class, it is not necessary to write down for every symmetry operation the system of equations of the type Eqn. (4.7). It is known from group theory that for various crystal classes every symmetry operation may be deduced from a few basic symmetry operations. The application of the matrices corresponding to these basic operations (the generating matrices) are sufficient to obtain the effect due to the symmetry of the crystal class on the given tensor in question. The table in Appendix D.1 provides a series of the generating matrices for every conventional crystal class. For further details see [37].

4.2 Second-order materials

Several crystals are used to take advantage of second-order nonlinear effects. Some of the most applied crystals include KTP (Potassium Titanium Oxide Phosphate ($KTiOPO_4$)), $LiNbO_3$ Lithium Niobate, and BBO beta-BaB_2O_4 Beta Barium Borate. Table 4.1 lists their respective crystal classes. Applications of second-order nonlinear effects include frequency doubling, frequency mixing, parametric amplification, and electro optic modulation.

In Chapter 6 we return to examples of applications of second-order nonlinear materials. However, in the following we show a few examples of transformation of tensors between different frames and applications of Neumann's principle. To make it easier to read these examples the reader is reminded of the generic notation of a general second-order susceptibility tensor, which is

$$\chi^{(2)} = \begin{pmatrix} xxx & xyy & xzz & xyz & xzy & xzx & xxz & xxy & xyx \\ yxx & yyy & yzz & yyz & yzy & yzx & yxz & yxy & yyx \\ zxx & zyy & zzz & zyz & zzy & zzx & zxz & zxy & zyx \end{pmatrix}. \tag{4.8}$$

It is noted that this is the second-order tensor for a triclinic material, and each element is for simplicity just shown by corresponding index i.e. $\chi^{(2)}_{xxx}$ is simply written xxx.

Example 4.1. *Potassium Titanium Oxide Phosphate ($KTiOPO_4$) KTP*
KTP belongs to point symmetry group mm2 and the symmetry is described by the generating matrices

$$M_5 = \begin{pmatrix} -1 & 0 & 0 \\ 0 & 1 & 0 \\ 0 & 0 & 1 \end{pmatrix}, \tag{4.9}$$

corresponding to a reflection in the $yz-$plane [37] and

$$M_2 = \begin{pmatrix} -1 & 0 & 0 \\ 0 & -1 & 0 \\ 0 & 0 & 1 \end{pmatrix}, \tag{4.10}$$

corresponding to a two-fold rotation about the z-axis.

Table 4.1: Second-order nonlinear crystals.

Crystal	Structure	Spacegroup (Point group)	Generating Matrices
KTP	Orthorhombic	mm2	M_5 , M_2
$LiNbO_3$	Trigonal	3m	M_9 , M_5
BBO	Trigonal	3m	M_9 , M_5

Considering first the point symmetry imposed by the M_5 matrix we find that the second-order susceptibility in the frame described by the reflection in the yz-plane is

$$\chi^{(2)\,\prime}_{111} = R_{1\mu}R_{1\alpha}R_{1\beta}\chi^{(2)}_{\mu\alpha\beta}, \tag{4.11}$$

where a summation over $\mu\alpha\beta$ is implicitly assumed, and R_{ij} is element (ij) in the generating matrix. Here and in the following we use numbers as indices in the framed coordinate system and letters as indices in the original frame. It is noted that since the generating matrices are by standard notation referred to by an index, for example index 5 for M_5, R_{ij} is used for elements in the matrix M. Inserting the transformation matrix M_5 we find

$$\chi^{(2)\,\prime}_{111} = R_{1x}R_{1x}R_{1x}\chi^{(2)}_{xxx} = (-1)(-1)(-1)\chi^{(2)}_{xxx} = -\chi^{(2)}_{xxx}, \tag{4.12}$$

where we have used that all terms other than those with $\mu = \alpha = \beta = 1$ vanish. That is, $\chi^{(2)}_{xxx}$ in the new frame is identical to $\chi^{(2)}_{xxx}$ multiplied by minus one. According to Neumann's principle, $\chi^{(2)}_{xxx}$ has to be the same in the two frames, and thus $\chi^{(2)}_{xxx}$ must equal zero, i.e.: $\chi^{(2)}_{xxx} = 0$.

It is noted that since M_5 only contains elements in the main-diagonal, then all elements $\chi^{(2)\prime}_{ijk}$ are determined from single elements $\chi^{(2)}_{ijk}$ with similar indices and only the sign may change. Evaluating the remaining 26 elements of the second-order susceptibility tensor as shown above leads to the intermediate result as a consequence of the point symmetry related to M_5

$$\chi^{(2)} = \begin{pmatrix} 0 & 0 & 0 & 0 & 0 & xzx & xxz & xxy & xyx \\ yxx & yyy & yzz & yyz & yzy & 0 & 0 & 0 & 0 \\ zxx & zyy & zzz & zyz & zzy & 0 & 0 & 0 & 0 \end{pmatrix}. \tag{4.13}$$

The tensor elements that have not been identified as zero may now be re-evaluated using the point symmetry imposed by the M_2 matrix, i.e. the frame described by a two-fold rotation about the z-axis. Considering the element $\chi^{(2)\,\prime}_{131}$ we find

$$\begin{aligned}\chi^{(2)\,\prime}_{131} &= R_{1\mu}R_{3\alpha}R_{1\beta}\chi^{(2)}_{\mu\alpha\beta} = R_{1x}R_{3z}R_{1x}\chi^{(2)}_{xzx} \\ &= (-1)(1)(-1)\chi^{(2)}_{xzx} = \chi^{(2)}_{xzx}\end{aligned} \tag{4.14}$$

i.e. from the two generating matrices we cannot conclude other than $\chi^{(2)}_{131}$ is a nonzero element in the material KTP. Evaluating the remaining 13 elements we find after some straightforward calculations the result

$$\chi^{(2)} = \begin{pmatrix} 0 & 0 & 0 & 0 & 0 & xzx & xxz & 0 & 0 \\ 0 & 0 & 0 & yyz & yzy & 0 & 0 & 0 & 0 \\ zxx & zyy & zzz & 0 & 0 & 0 & 0 & 0 & 0 \end{pmatrix}. \tag{4.15}$$

—— ■ ——

Example 4.2. *Lithium Niobate* $LiNbO_3$
Lithium Niobate is an often used second-order nonlinear material for example in optical modulators. Lithium Niobate has a refractive index close to 2, a trigonal crystal structure and belongs to space group 3m. Its susceptibility tensor has 11 non-zero elements of which 4 are independent.

$LiNbO_3$ is slightly more complicated than KTP since its point symmetry involves a threefold rotation around the z-axis in addition to the reflection in the yz-plane, similar to the mm2 point symmetry for KTP. The generating matrices in the case of $LiNbO_3$ are

$$M_9 = \frac{1}{2}\begin{pmatrix} -1 & -\sqrt{3} & 0 \\ \sqrt{3} & -1 & 0 \\ 0 & 0 & 2 \end{pmatrix}, \tag{4.16}$$

and

$$M_5 = \begin{pmatrix} -1 & 0 & 0 \\ 0 & 1 & 0 \\ 0 & 0 & 1 \end{pmatrix}, \tag{4.17}$$

respectively.

Since the reflection in the yz-plane is the simplest operation, we start with evaluation of all elements using M_5. We see immediately that since M_5 only has diagonal elements then $\chi^{(2)'}_{ijk}$ is related to a single element $\chi^{(2)}_{ijk}$ and all elements which have x appearing one or three times have to be identical zero. Consequently, the full tensor (triclinic) is reduced to

$$\chi^{(2)} = \begin{pmatrix} 0 & 0 & 0 & 0 & 0 & xzx & xxz & xxy & xyx \\ yxx & yyy & yzz & yyz & yzy & 0 & 0 & 0 & 0 \\ zxx & zyy & zzz & zyz & zzy & 0 & 0 & 0 & 0 \end{pmatrix}. \tag{4.18}$$

We now turn to the three fold rotation around the $z-$axis i.e. using M_9 and starting with the simplest terms i.e.

$$\chi^{(2)\,\prime}_{333} = R_{3\mu}R_{3\alpha}R_{3\beta}\chi^{(2)}_{\mu\alpha\beta} = R_{3z}R_{3z}R_{3z}\chi^{(2)}_{zzz} = \chi^{(2)}_{zzz}, \tag{4.19}$$

which simply states that $\chi^{(2)\,\prime}_{333}$ is identical to $\chi^{(2)}_{zzz}$. We then take the elements with two indices equal to z,

$$\begin{aligned} \chi^{(2)\,\prime}_{233} &= R_{2\mu}R_{3\alpha}R_{3\beta}\chi^{(2)}_{\mu\alpha\beta} = R_{2\mu}\chi^{(2)}_{\mu zz} \\ &= -\frac{1}{2}\chi^{(2)}_{xzz} - \frac{\sqrt{3}}{2}\chi^{(2)}_{yzz} \end{aligned} \tag{4.20}$$

since $\chi^{(2)}_{xzz} = 0$ then $\chi^{(2)}_{yzz} = 0$. Through similar calculations we find $\chi^{(2)}_{zyz} = 0$, and finally $\chi^{(2)}_{zzy} = 0$ which is found by using the earlier result, i.e. $\chi^{(2)}_{yzz} = 0$.

To summarize at this point

$$\chi^{(2)} = \begin{pmatrix} 0 & 0 & 0 & 0 & 0 & xzx & xxz & xxy & xyx \\ yxx & yyy & 0 & yyz & yzy & 0 & 0 & 0 & 0 \\ zxx & zyy & zzz & 0 & 0 & 0 & 0 & 0 & 0 \end{pmatrix}. \tag{4.21}$$

We now continue by considering tensor elements that include only one z index, i.e. $\chi^{(2)}_{xzx}$, $\chi^{(2)}_{xxz}$ and $\chi^{(2)}_{zxx}$. We find that

$$\begin{aligned}\chi^{(2)\,\prime}_{131} &= R_{1\mu}R_{3\alpha}R_{1\beta}\chi^{(2)}_{\mu\alpha\beta} \\ &= \frac{1}{4}\left(\chi^{(2)}_{xzx} + 3\chi^{(2)}_{yzy}\right),\end{aligned} \tag{4.22}$$

since $\chi^{(2)\,\prime}_{131}$ has to equal $\chi^{(2)}_{xzx}$ from Neumann's principle, then $\chi^{(2)}_{xzx} = \chi^{(2)}_{yzy}$. By similar approach we find $\chi^{(2)}_{xxz} = \chi^{(2)}_{yyz}$, and $\chi^{(2)}_{zxx} = \chi^{(2)}_{zyy}$. That is

$$\chi^{(2)} = \begin{pmatrix} 0 & 0 & 0 & 0 & 0 & 131 & 113 & xxy & xyx \\ yxx & yyy & 0 & 113 & 131 & 0 & 0 & 0 & 0 \\ 311 & 311 & 333 & 0 & 0 & 0 & 0 & 0 & 0 \end{pmatrix}. \tag{4.23}$$

In Eqn. (4.23) a tensor element given by a number f.ex. 131 for $\chi^{(2)\,\prime}_{131}$ in the primed coordinate system indicate that the element has been evaluated, as in the used example $\chi^{(2)\,\prime}_{131}$ in Eqn. (4.22).

We are now left with four terms that are somewhat more complex. Considering $\chi^{(2)\,\prime}_{112}$ we find

$$\begin{aligned}\chi^{(2)\,\prime}_{112} &= R_{1\mu}R_{1\alpha}R_{2\beta}\chi^{(2)}_{\mu\alpha\beta} \\ &= R_{1\mu}R_{1\alpha}\left\{R_{2x}\chi^{(2)}_{\mu\alpha x} + R_{2y}\chi^{(2)}_{\mu\alpha y}\right\} \\ &= R_{1\mu}\left\{R_{1x}R_{2x}\chi^{(2)}_{\mu xx} + R_{1x}R_{2y}\chi^{(2)}_{\mu xy} + R_{1y}R_{2x}\chi^{(2)}_{\mu yx} + R_{1y}R_{2y}\chi^{(2)}_{\mu yy}\right\} \\ &= R_{1x}R_{1x}R_{2x}\chi^{(2)}_{xxx} + R_{1x}R_{1x}R_{2y}\chi^{(2)}_{xxy} + R_{1x}R_{1y}R_{2x}\chi^{(2)}_{xyx} + R_{1x}R_{1y}R_{2y}\chi^{(2)}_{xyy} \\ &\quad + R_{1y}R_{1x}R_{2x}\chi^{(2)}_{yxx} + R_{1y}R_{1x}R_{2y}\chi^{(2)}_{yxy} + R_{1y}R_{1y}R_{2x}\chi^{(2)}_{yyx} + R_{1y}R_{1y}R_{2y}\chi^{(2)}_{yyy}.\end{aligned} \tag{4.24}$$

Using the already obtained results i.e. $\chi^{(2)}_{xxx} = \chi^{(2)}_{xyy} = \chi^{(2)}_{yxy} = \chi^{(2)}_{yyx} = 0$ and inserting the elements for the matrix M_9, we find by Neumann's principle

$$3\chi^{(2)}_{xxy} = \chi^{(2)}_{xyx} + \chi^{(2)}_{yxx} - \chi^{(2)}_{yyy}. \tag{4.25}$$

Following a similar approach we find for the other three elements

$$3\chi^{(2)}_{xyx} = \chi^{(2)}_{xxy} + \chi^{(2)}_{yxx} - \chi^{(2)}_{yyy} \tag{4.26}$$

$$3\chi^{(2)}_{yxx} = \chi^{(2)}_{xxy} + \chi^{(2)}_{xyx} - \chi^{(2)}_{yyy} \tag{4.27}$$

$$3\chi^{(2)}_{yyy} = -\left(\chi^{(2)}_{xxy} + \chi^{(2)}_{xyx} + \chi^{(2)}_{yxx}\right). \tag{4.28}$$

The set of equations Eqn. (4.25) through Eqn. (4.28) has the nontrivial solution

$$\chi^{(2)}_{xxy} = \chi^{(2)}_{xyx} = \chi^{(2)}_{yxx} = -\chi^{(2)}_{yyy}. \tag{4.29}$$

From this we arrive at the final result

$$\chi^{(2)} = \begin{pmatrix} 0 & 0 & 0 & 0 & 0 & xzx & xxz & \overline{yyy} & \overline{yyy} \\ \overline{yyy} & yyy & 0 & xxz & xzx & 0 & 0 & 0 & 0 \\ zxx & zxx & zzz & 0 & 0 & 0 & 0 & 0 & 0 \end{pmatrix}, \tag{4.30}$$

where a bar is the standard notation used for a negative element [2]. Eqn. (4.30) is the final result for the susceptibility tensor for Lithium Niobate.

—— ■ ——

4.3 Third-order nonlinear materials

The susceptibility tensor for a third-order nonlinear material has in general 81 elements. However, we focus on cubic and isotropic materials, more specifically glass and silicon, which because of symmetry have 21 out of 81 non-zero elements. For both types of materials, the non-zero elements are $\chi^{(3)}_{iijj}$, $\chi^{(3)}_{ijji}$, $\chi^{(3)}_{ijij}$ and $\chi^{(3)}_{iiii}$. From Chapter 2 and 3 we know that applying the Born-Oppenheimer approximation the nonlinear response may be written as

$$d_{ijkl}(t) = a(t)\delta_{ij}\delta_{kl} + \frac{1}{2}b(t)\left(\delta_{il}\delta_{jk} + \delta_{ik}\delta_{jl}\right) + c(t)\delta_{ijkl}, \tag{4.31}$$

and

$$\sigma_{ijkl} = \frac{1}{6}\sigma_e\left(\delta_{ij}\delta_{kl} + \delta_{ik}\delta_{jl} + \delta_{il}\delta_{jk}\right) + \frac{1}{2}\tau\delta_{ijkl}, \tag{4.32}$$

where $c(t) \neq 0$, $\tau \neq 0$ for cubic materials, for example silicon, and $c(t) = \tau = 0$ for isotropic materials, for example glass.

Example 4.3. *Silicon*
Silicon is a very interesting material for applications within photonics. There are several reasons for this, for example the high nonlinearity of silicon and its relatively low loss for wavelengths exceeding one micrometer. In addition, silicon is a well known material from electronics. Until this point in time most work within silicon photonics has been on waveguide circuitry, however there have also been recent demonstrations of silicon fibers.

Silicon is a semiconductor with a band gap near 1.12 eV, and it becomes nearly transparent beyond 1.1 μm. Its linear refractive index is 3.4 at wavelengths exceeding 1500 nm and its nonlinear refractive index is reported about 100 times larger than that of silica and its Raman scattering is about 1000 times larger than that of silica.

This example focuses on the third-order susceptibility, and since silicon belongs to the m3m point group symmetry, the third-order susceptibility has 21 non-zero elements of which four are independent

$$\begin{aligned} xxxx = yyyy = zzzz, \\ yyzz = zzyy = zzxx = xxzz = xxyy = yyxx, \\ yzyz = zyzy = zxzx = xzxz = xyxy = yxyx, \\ yzzy = zyyz = zxxz = xzzx = xyyx = yxxy. \end{aligned}$$

In addition, the third-order susceptibility has mainly two contributions, one from the interaction between the electric field and bound electrons, and one from the interaction between the electric field and phonons, i.e. Raman scattering. From Chapter 2 the susceptibility tensor may be written as

$$\chi^{(3)}_{ijkl} = \chi^{(3)}_{iijj}\delta_{ij}\delta_{kl} + \chi^{(3)}_{ijij}\delta_{ik}\delta_{jl} + \chi^{(3)}_{ijji}\delta_{il}\delta_{jk} + \chi_d\delta_{ijkl}. \tag{4.33}$$

In this example we focus solely on the tensor that describes Raman scattering. Along a basis aligned with the crystallographic axes, the Raman tensor is given by [39] and references herein. In practice, waveguides are fabricated along for example the $[1, \bar{1}, 0]$ direction,on the surface of a wafer. The surface may for example be the (001) surface, see Figure 4.3.

We use the tensor [39]

$$\begin{aligned} \chi_{ijkl}(-\omega_i; \omega_j - \omega_k, \omega_l) &= g'\tilde{H}_R(\omega_l - \omega_k)(\delta_{ik}\delta_{jl} + \delta_{il}\delta_{jk} - 2\delta_{ijkl}) \\ &+ g'\tilde{H}_R(\omega_j - \omega_k)(\delta_{ik}\delta_{jl} + \delta_{ij}\delta_{kl} - 2\delta_{ijkl}). \end{aligned} \tag{4.34}$$

The Raman response function $\tilde{H}_R$ depends only on the difference between the pump and signal frequency

$$\tilde{H}_R(\Omega) = \frac{\Omega_R^2(\Omega_R^2 - \Omega^2)}{(\Omega_R^2 - \Omega^2)^2 + (2\Gamma_R\Omega)^2} + i\frac{\Omega_R^2 2\Gamma_R\Omega}{(\Omega_R^2 - \Omega^2)^2 + (2\Gamma_R\Omega)^2}, \tag{4.35}$$

where Ω_R is the peak frequency of the Raman spectrum (the imaginary part) and Ω the actual difference between the pump and signal frequency and Γ_R is related to the width of the imaginary part of the Raman spectrum.

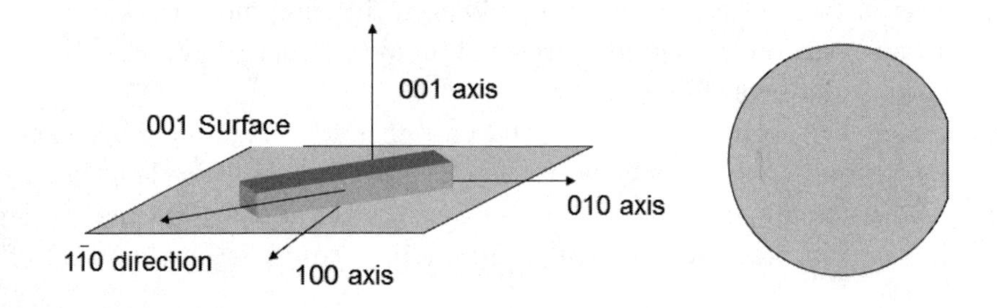

Figure 4.3: Left An example of a waveguide fabricated along the $[1, \bar{1}, 0]$ direction. **Right** On a wafer the orientation of the silicon crystals are described using 'flats'. These are flat segments on the wafer cut relative to the crystal orientation. When necessary, a large and a small flat is used to distinguish different crystal orientations i.e. 111 type from 100.

The real part gives rise to a Raman induced change in the refractive index whereas the imaginary part gives rise to Raman amplification or attenuation. The above tensor Eqn. (4.34) is given in a coordinate frame (x, y, z) oriented along the crystallographic axes [39]. In the remaining of this example we restrict ourselves only to Raman amplification i.e. the imaginary part of $\tilde{H}_R$, and we consider the impact of working with waveguides created on an axis different from the crystallographic axis, or more specifically when the propagating light is not polarized along the directions of the silicon.

To identify the direction of the crystal-planes in a wafer, a footprint is often made for example as a flat cut on one side of the wafer, see the right part of Figure 4.3.

We start with the susceptibility tensor of silicon as given in its reference system, and we consider a coordinate system that is rotated 45° around the x-axis. To do this we use the rotation matrix

$$\underline{\underline{R}} = \begin{pmatrix} \frac{1}{\sqrt{2}} & \frac{1}{\sqrt{2}} & 0 \\ -\frac{1}{\sqrt{2}} & \frac{1}{\sqrt{2}} & 0 \\ 0 & 0 & 1 \end{pmatrix}. \tag{4.36}$$

In this example we wish to evaluate specific elements of the susceptibility tensor in the new coordinate system, the primed coordinate system, i.e.: $\chi^{(3)}_{1111}{}'$, $\chi^{(3)}_{1122}{}'$, $\chi^{(3)}_{1212}{}'$, $\chi^{(3)}_{1221}{}'$. Let us start with the element $\chi^{(3)}_{1111}{}'$ in the primed coordinate system.

$$\chi^{(3)}_{1111}{}' = R_{1q}R_{1r}R_{1s}R_{1t}\chi^{(3)}_{qrst}, \tag{4.37}$$

where a summation over $qrst$ is assumed. Writing $\chi^{(3)}_{1111}{}'$ by all non-zero elements of the transformation matrix R, we get

$$\begin{aligned}
\chi^{(3)}_{1111}{}' &= R_{1q}R_{1r}R_{1s}\left(R_{1x}\chi^{(3)}_{qrsx} + R_{1y}\chi^{(3)}_{qrsy}\right) \qquad (4.38)\\
&= R_{1q}R_{1r}\left(R_{1x}R_{1x}\chi^{(3)}_{qrxx} + R_{1y}R_{1x}\chi^{(3)}_{qryx} + R_{1x}R_{1y}\chi^{(3)}_{qrxy} + R_{1y}R_{1y}\chi^{(3)}_{qryy}\right)\\
&= R_{1q}\Big(R_{1x}R_{1x}R_{1x}\chi^{(3)}_{qxxx} + R_{1x}R_{1y}R_{1x}\chi^{(3)}_{qxyx} + R_{1x}R_{1x}R_{1y}\chi^{(3)}_{qxxy}\\
&\qquad + R_{1x}R_{1y}R_{1y}\chi^{(3)}_{qxyy} + R_{1y}R_{1x}R_{1x}\chi^{(3)}_{qyxx} + R_{1y}R_{1y}R_{1x}\chi^{(3)}_{qyyx}\\
&\qquad + R_{1y}R_{1x}R_{1y}\chi^{(3)}_{qyxy} + R_{1y}R_{1y}R_{1y}\chi^{(3)}_{qyyy}\Big)\\
&= R_{1x}R_{1x}R_{1x}R_{1x}\chi^{(3)}_{xxxx} + R_{1x}R_{1x}R_{1y}R_{1x}\chi^{(3)}_{xxyx} + R_{1x}R_{1x}R_{1x}R_{1y}\chi^{(3)}_{xxxy}\\
&\quad + R_{1x}R_{1x}R_{1y}R_{1y}\chi^{(3)}_{xxyy} + R_{1y}R_{1x}R_{1x}R_{1x}\chi^{(3)}_{yxxx} + R_{1y}R_{1x}R_{1y}R_{1x}\chi^{(3)}_{yxyx}\\
&\quad + R_{1y}R_{1x}R_{1x}R_{1y}\chi^{(3)}_{yxxy} + R_{1y}R_{1x}R_{1y}R_{1y}\chi^{(3)}_{yxyy} + R_{1x}R_{1y}R_{1x}R_{1x}\chi^{(3)}_{xyxx}\\
&\quad + R_{1x}R_{1y}R_{1y}R_{1x}\chi^{(3)}_{xyyx} + R_{1x}R_{1y}R_{1x}R_{1y}\chi^{(3)}_{xyxy} + R_{1x}R_{1y}R_{1y}R_{1y}\chi^{(3)}_{xyyy}\\
&\quad + R_{1y}R_{1y}R_{1x}R_{1x}\chi^{(3)}_{yyxx} + R_{1y}R_{1y}R_{1x}R_{1y}\chi^{(3)}_{yyxy} + R_{1y}R_{1y}R_{1y}R_{1x}\chi^{(3)}_{yyyx}\\
&\quad + R_{1y}R_{1y}R_{1y}R_{1y}\chi^{(3)}_{yyyy}.
\end{aligned}$$

Inserting the matrix elements for R and using the fact that the indices have to appear in pairs, this reduces to

$$\chi^{(3)}_{1111}{}' = \frac{1}{4}\left(\chi^{(3)}_{xxxx} + \chi^{(3)}_{xxyy} + \chi^{(3)}_{yxyx} + \chi^{(3)}_{yxxy} + \chi^{(3)}_{xyyx} + \chi^{(3)}_{xyxy} + \chi^{(3)}_{yyxx} + \chi^{(3)}_{yyyy}\right). \tag{4.39}$$

In analog

$$\chi^{(3)}_{1122}{}' = \frac{1}{4}\left(\chi^{(3)}_{xxxx} + \chi^{(3)}_{xxyy} - \chi^{(3)}_{yxxy} - \chi^{(3)}_{yxyx} - \chi^{(3)}_{xyxy} - \chi^{(3)}_{xyyx} + \chi^{(3)}_{yyxx} + \chi^{(3)}_{yyyy}\right), \tag{4.40}$$

$$\chi^{(3)}_{1212}{}' = \frac{1}{4}\left(\chi^{(3)}_{xxxx} - \chi^{(3)}_{xxyy} - \chi^{(3)}_{yxxy} + \chi^{(3)}_{yxyx} + \chi^{(3)}_{xyxy} - \chi^{(3)}_{xyyx} - \chi^{(3)}_{yyxx} + \chi^{(3)}_{yyyy}\right), \tag{4.41}$$

and

$$\chi^{(3)}_{1221}{}' = \frac{1}{4}\left(\chi^{(3)}_{xxxx} - \chi^{(3)}_{xxyy} + \chi^{(3)}_{yxxy} - \chi^{(3)}_{yxyx} - \chi^{(3)}_{xyxy} + \chi^{(3)}_{xyyx} - \chi^{(3)}_{yyxx} + \chi^{(3)}_{yyyy}\right). \tag{4.42}$$

— ■ —

4.3.1 Isotropic materials

An isotropic material is a material that has the same optical properties regardless of the direction in which light propagates through the material.

For a given experimental setup it is often convenient to introduce a reference frame in which one for example expresses the wave propagation as a linear motion along some Cartesian coordinate axis. This reference frame might be chosen for example with the z-axis coinciding with the direction of propagation of the optical wave.

In some cases it might be that this reference frame coincides with the natural coordinate frame of the nonlinear material, in which case the coordinate indices of the linear as well as the nonlinear susceptibility tensors are identical. However, we cannot generally assume that the coordinate frame of the material coincides with a chosen reference frame, and this implies that we should be prepared to transform the susceptibility tensors to arbitrarily rotated coordinate frames.

In order to translate a tensor from one coordinate frame to another one needs the matrix that rotates between the two different coordinate frames. Once this is defined, the tensor may be translated according to the transformation rules as outlined in Appendix A.

These spatial transformation rules are also directly beneficial when verifying the tensor properties of nonlinear optical interactions or when reducing the susceptibility tensors to a minimal set of non-zero elements. This is typically performed by using the knowledge of material symmetry, or in the case of crystals, the so-called crystallographic point symmetry group of the medium, which is essentially a description of the spatial operations that define the symmetry operations of the medium (rotations, inversions, etc.).

In general, all frequencies of the input field are arbitrarily interchangeable as illustrated in the case of Raman scattering in Chapter 3. Thus, $\chi^{(3)}_{ijkl}(-\omega_\sigma;\omega,\omega_2,\omega_3)$ is often simply written in the reduced form $\chi^{(3)}_{ijkl}$. In addition, because of inversion symmetry in an isotropic material, it can be shown by spatial symmetry that such a material consists of 21 non-zero elements out of 81 and that the non-zero elements are having indices in pairs or all identical. Thus, an isotropic material is characterized with the tensor elements

$$\begin{aligned}
&\chi^{(3)}_{xxxx} = \chi^{(3)}_{yyyy} = \chi^{(3)}_{zzzz} \\
&\chi^{(3)}_{xxyy} = \chi^{(3)}_{xxzz} = \chi^{(3)}_{yyxx} = \chi^{(3)}_{yyzz} = \chi^{(3)}_{zzxx} = \chi^{(3)}_{zzyy} \\
&\chi^{(3)}_{xyxy} = \chi^{(3)}_{xzxz} = \chi^{(3)}_{yzyz} = \chi^{(3)}_{yxyx} = \chi^{(3)}_{zxzx} = \chi^{(3)}_{zyzy} \\
&\chi^{(3)}_{xyyx} = \chi^{(3)}_{xzzx} = \chi^{(3)}_{yxxy} = \chi^{(3)}_{yzzy} = \chi^{(3)}_{zxxz} = \chi^{(3)}_{zyyz}.
\end{aligned} \tag{4.43}$$

Furthermore,

$$\chi^{(3)}_{xxxx} = \chi^{(3)}_{xxyy} + \chi^{(3)}_{xyxy} + \chi^{(3)}_{xyyx}. \tag{4.44}$$

Eqn. (4.43) and Eqn. (4.44) can be combined and written in the more compact form

$$\chi^{(3)}_{ijkl} = \chi^{(3)}_{xxyy}\delta_{ij}\delta_{kl} + \chi^{(3)}_{xyxy}\delta_{ik}\delta_{jl} + \chi^{(3)}_{xyyx}\delta_{il}\delta_{jk}. \tag{4.45}$$

This demonstrates an important property of an isotropic material, i.e. the third-order susceptibility of an isotropic material in general has **three** independent elements.

4.4 Cyclic coordinate-system

In the following we return to the discussion in Section 4.1 related to the transformation of a tensor between various coordinate systems. More specifically the focus of this section is to discuss transformation of a tensor into a cyclic coordinate system, as may be advantageous when describing circularly/elliptically polarized light. As pointed out in Section 4.1 transformation between various coordinate systems is essential when considering which tensor elements that are identical zero for example due to symmetry, but also if it is more practical to work in a coordinate system that is rotated relative to a reference frame or when describing propagation of for example circularly polarized light.

To discuss the transformation we evaluate the induced polarization vector **P** and transform this into different coordinate systems. To translate between two coordinate systems, one needs a transformation matrix $\underline{\underline{R}}$ as described in in Section 4.1.

4.4.1 Linear optics

Let us start by discussing the simplest case, i.e. the linear case. With $\chi^{(1)}$ given in a standard Cartesian (x, y, z) frame, the induced first-order polarization is

$$\begin{pmatrix} (P_x^0)^{(1)} \\ (P_y^0)^{(1)} \\ (P_z^0)^{(1)} \end{pmatrix} = \varepsilon_0 \chi^{(1)} \begin{pmatrix} E_x^0 \\ E_y^0 \\ E_z^0 \end{pmatrix}. \tag{4.46}$$

Expressing the induced polarization in a new coordinate-system we multiply $\mathbf{P}^0$ by the transformation matrix $\underline{\underline{R}}$, i.e.

$$\begin{pmatrix} (P_{c1}^0)^{(1)} \\ (P_{c2}^0)^{(1)} \\ (P_{c3}^0)^{(1)} \end{pmatrix} = \underline{\underline{R}} \begin{pmatrix} (P_x^0)^{(1)} \\ (P_y^0)^{(1)} \\ (P_z^0)^{(1)} \end{pmatrix} = \varepsilon_0 \underline{\underline{R}} \chi^{(1)} \begin{pmatrix} E_x^0 \\ E_y^0 \\ E_z^0 \end{pmatrix}. \tag{4.47}$$

Using the inverse transformation matrix $\underline{\underline{R}}^{-1}$ enables us to express the electric field vector $(E_x^0, E_y^0, E_z^0)^T$ in terms of the new coordinates $(E_{c1}^0, E_{c2}^0, E_{c3}^0)^T$. From this we find

$$\begin{pmatrix} (P_{c1}^0)^{(1)} \\ (P_{c2}^0)^{(1)} \\ (P_{c3}^0)^{(1)} \end{pmatrix} = \varepsilon_0 \underline{\underline{R}} \chi^{(1)} \underline{\underline{R}}^{-1} \begin{pmatrix} E_{c1}^0 \\ E_{c2}^0 \\ E_{c3}^0 \end{pmatrix}. \tag{4.48}$$

After some trivial calculations we identify a tensor, $\mathcal{C}$ in the new frame

$$\mathcal{C}_{\mu\alpha}^{(1)} = \sum_{ij} r_{\mu i} \chi_{ij}^{(1)} r_{j\alpha}^{-1}. \tag{4.49}$$

If the transformation matrix is described by an orthogonal matrix[1], as for example a rotation, then the inverse of the transformation matrix equals its transpose

$$r_{ji}^{-1} = r_{ij}. \tag{4.50}$$

In this case Eqn. (4.49) reduces to

$$\mathcal{C}_{\mu\alpha}^{(1)} = \sum_{ij} r_{\mu i} \chi_{ij}^{(1)} r_{\alpha j}. \tag{4.51}$$

4.4.2 Higher-order tensors

In nonlinear optics the susceptibility tensor is more complicated. However, the method introduced may be extended to tensors of rank higher than two.

[1] An orthogonal matrix is defined as a square matrix with real entries whose columns and rows are orthogonal unit vectors (i.e., orthonormal vectors). Examples include a rotation, reflection.

4.4.2.1 Transformation of third-rank tensor

With $\chi^{(2)}$ given in a standard (x, y, z) frame, the induced second-order polarization is

$$\begin{pmatrix} (P_x^0)^{(2)} \\ (P_y^0)^{(2)} \\ (P_z^0)^{(2)} \end{pmatrix} = \varepsilon_0 \chi^{(2)} \begin{pmatrix} E_x^0 \\ E_y^0 \\ E_z^0 \end{pmatrix} \begin{pmatrix} E_x^0 \\ E_y^0 \\ E_z^0 \end{pmatrix}. \tag{4.52}$$

Expressing the induced polarization in the new coordinate-system using $\underline{\underline{R}}$ we get

$$\begin{pmatrix} (P_{c1}^0)^{(2)} \\ (P_{c2}^0)^{(2)} \\ (P_{c3}^0)^{(2)} \end{pmatrix} = \underline{\underline{R}} \begin{pmatrix} (P_x^0)^{(2)} \\ (P_y^0)^{(2)} \\ (P_z^0)^{(2)} \end{pmatrix} = \varepsilon_0 \underline{\underline{R}} \chi^{(2)} \begin{pmatrix} E_x^0 \\ E_y^0 \\ E_z^0 \end{pmatrix} \begin{pmatrix} E_x^0 \\ E_y^0 \\ E_z^0 \end{pmatrix}. \tag{4.53}$$

Using $\underline{\underline{R}}^{-1}$ to express the Cartesian coordinates of the electric field in terms of the new coordinates we find

$$\begin{pmatrix} (P_{c1}^0)^{(2)} \\ (P_{c2}^0)^{(2)} \\ (P_{c3}^0)^{(2)} \end{pmatrix} = \varepsilon_0 \underline{\underline{R}} \chi^{(2)} \underline{\underline{R}}^{-1} \begin{pmatrix} E_{c1}^0 \\ E_{c2}^0 \\ E_{c3}^0 \end{pmatrix} \underline{\underline{R}}^{-1} \begin{pmatrix} E_{c1}^0 \\ E_{c2}^0 \\ E_{c3}^0 \end{pmatrix}. \tag{4.54}$$

At this point, it is important to note that in nonlinear optics the induced polarization evaluated above may originate from an interaction of positive as well as negative frequency components, i.e. from terms $\mathbf{E}\mathbf{E}^*$. In this case the inverse transformation matrix $\underline{\underline{R}}^{-1}$ corresponding to the complex conjugate component of the electrical field also need to be the complex conjugate $(\underline{\underline{R}}^{-1})^*$.

From Eqn. (4.54) we may now identify the susceptibility tensor in the new coordinate system

$$\mathcal{C}_{\mu\alpha\beta}^{(2)} = \sum_{ijk} r_{\mu i} r_{j\alpha}^{-1} r_{k\beta}^{-1} \chi_{ijk}^{(2)}. \tag{4.55}$$

where $r_{j\alpha}^{-1}$ indicate an element from $\underline{\underline{R}}^{-1}$.

As in the case of the first-order susceptibility this may be rewritten if the transformation matrix is orthogonal as

$$\mathcal{C}_{\mu\alpha\beta}^{(2)} = \sum_{ijk} r_{\mu i} r_{\alpha j} r_{\beta k} \chi_{ijk}^{(2)}. \tag{4.56}$$

4.4.2.2 Transformation of fourth-, and higher-rank tensor

Without any proof we extend the above to any higher-order case, that is; if the susceptibility is $\chi^{(n)}$, then in the new coordinate system the susceptibility equals

$$\mathcal{C}_{\mu\alpha_1\cdots\alpha_n}^{(n)} = \sum_{ij_1\cdots j_n} r_{\mu i} r_{j_1\alpha_1}^{-1} \cdots r_{j_n\alpha_n}^{-1} \chi_{ij_1\cdots j_n}^{(n)}. \tag{4.57}$$

As noted explicitly in the case of the second-order nonlinearity, in the general case $\underline{\underline{R}}^{-1}$ may need to appear in its complex conjugated form when the induced polarization is due to an interaction involving negative frequency components. However, when considering an orthogonal transformation Eqn. (4.57) may be rewritten as

$$C^{(n)}_{\mu\alpha_1\cdots\alpha_n} = \sum_{ij_1\cdots j_n} r_{\mu i} r_{\alpha_1 j_1} \cdots r_{\alpha_n j_n} \chi^{(n)}_{ij_1\cdots j_n}. \tag{4.58}$$

Example 4.4. *Cyclic basis, an isotropic material, Kerr effect*
In accordance with section 1.1.8, the cyclic basis, (σ^+, σ^-, z), σ^+ corresponding to left handed and σ^- to right handed polarized light, is

$$\boldsymbol{\sigma}^+ = \frac{1}{\sqrt{2}}(\mathbf{e}_x + i\mathbf{e}_y) \quad \boldsymbol{\sigma}^- = \frac{1}{\sqrt{2}}(\mathbf{e}_x - i\mathbf{e}_y) \quad \boldsymbol{C_3} = \mathbf{e}_z. \tag{4.59}$$

To transform a vector $\mathbf{V} = V_x\mathbf{e}_x + V_y\mathbf{e}_y + V_z\mathbf{e}_z$ to the vector in cyclic coordinates we multiply $\mathbf{V}$ by the transformation matrix R

$$\underline{\underline{R}} = \frac{1}{\sqrt{2}}\begin{pmatrix} 1 & -i & 0 \\ 1 & i & 0 \\ 0 & 0 & \sqrt{2} \end{pmatrix} \tag{4.60}$$

to go from a vector in the cyclic coordinate system to the standard Cartesian system we need $\underline{\underline{R}}^{-1}$

$$\underline{\underline{R}}^{-1} = \frac{1}{\sqrt{2}}\begin{pmatrix} 1 & 1 & 0 \\ i & -i & 0 \\ 0 & 0 & \sqrt{2} \end{pmatrix}. \tag{4.61}$$

We now consider circularly polarized light, more specifically right-handed $E_y = iE_x$, and we assume $E_z = 0$. In addition, we also consider a monochromatic electrical field and we evaluate the third-order induced polarization at the frequency of the applied electric field. In the original (x, y, z) frame the x-component of the induced third order polarization is

$$P^0_x(4/(3\varepsilon_0)) = \chi^{(3)}_{xxxx}|E^0_x|^2 E^0_x + \chi^{(3)}_{xyyx}|E^0_y|^2 E^0_x + \chi^{(3)}_{xyxy} E^0_y (E^0_x)^* E^0_y + \chi^{(3)}_{xxyy} E^0_x (E^0_y)^* E^0_y. \tag{4.62}$$

And we find

$$\begin{aligned} P^0_x(4/(3\varepsilon_0)) &= \chi^{(3)}_{xxxx}|E^0_x|^2 E^0_x + (\chi^{(3)}_{xxyy} + \chi^{(3)}_{xyyx})|E^0_y|^2 E^0_x + \chi^{(3)}_{xyxy}(E^0_y)^2 (E^0_x)^* \\ P^0_y(4/(3\varepsilon_0)) &= \chi^{(3)}_{yyyy}|E^0_y|^2 E^0_y + (\chi^{(3)}_{yyxx} + \chi^{(3)}_{yxxy})|E^0_x|^2 E^0_y + \chi^{(3)}_{yxyx}(E^0_x)^2 (E^0_y)^* \\ P^0_z &= 0. \end{aligned} \tag{4.63}$$

For completeness, the reader is reminded that

$$\chi^{(3)}_{xxxx} = a, \quad \chi^{(3)}_{xxyy} = a/3, \quad \chi^{(3)}_{xyyx} = a/3, \quad \chi^{(3)}_{xyxy} = a/3. \tag{4.64}$$

Using the transformation matrix from Eqn. (4.60) we can express the electric field in the new coordinate system based on the coordinates in the original coordinate system

$$\mathbf{E}^0_\sigma = \underline{\underline{R}}\mathbf{E}^0 = \frac{1}{\sqrt{2}}\begin{pmatrix} 1 & -i & 0 \\ 1 & i & 0 \\ 0 & 0 & \sqrt{2} \end{pmatrix}\begin{pmatrix} E^0_x \\ E^0_y \\ 0 \end{pmatrix} = \frac{1}{\sqrt{2}}\begin{pmatrix} E^0_x - iE^0_y \\ E^0_x + iE^0_y \\ 0 \end{pmatrix}. \tag{4.65}$$

A similar transformation applies to the induced polarization. That is $P^0_{\sigma+} = (P^0_x - iP^0_y)/\sqrt{2}$. Inserting from Eqn. (4.63) we now find the induced polarization in the (σ^+, σ^-, z) frame

$$P^0_{\sigma+} = (P^0_x + (-i)P^0_y)/\sqrt{2} = \frac{1}{\sqrt{2}}(3\varepsilon_0/4)\Big(\chi^{(3)}_{xxxx}|E^0_x|^2E^0_x + (\chi^{(3)}_{xxyy} + \chi^{(3)}_{xyyx})|E^0_y|^2E^0_x$$
$$+\chi^{(3)}_{xyxy}(E^0_y)^2(E^0_x)^* - i\chi^{(3)}_{yyyy}|E^0_y|^2E^0_y - i(\chi^{(3)}_{yyxx} + \chi^{(3)}_{yxxy})|E^0_x|^2E^0_y - i\chi^{(3)}_{yxyx}(E^0_x)^2(E^0_y)^*\Big). \tag{4.66}$$

To find the induced polarization in (σ^+, σ^-, z) on the basis of the electric field also in the (σ^+, σ^-, z) frame, we express E^0_x, E^0_y, and E^0_z by $E^0_{\sigma+}$, $E^0_{\sigma-}$, and E^0_z i.e.

$$\begin{pmatrix} E^0_x \\ E^0_y \\ E^0_z \end{pmatrix} = \underline{\underline{R}}^{-1}\begin{pmatrix} E^0_{\sigma+} \\ E^0_{\sigma-} \\ E^0_{c3} \end{pmatrix}. \tag{4.67}$$

That is

$$E^0_x = \frac{1}{\sqrt{2}}(E^0_{\sigma+} + E^0_{\sigma-}) \tag{4.68}$$
$$E^0_y = \frac{i}{\sqrt{2}}(E^0_{\sigma+} - E^0_{\sigma-})$$

Inserting this into Eqn. (4.66) we get by using $\chi^{(3)}_{xxxx} - (\chi^{(3)}_{xxyy} + \chi^{(3)}_{xyyx}) - \chi^{(3)}_{xyxy} = 0$ and $\chi^{(3)}_{xxyy} = \chi^{(3)}_{xyyx} = \chi^{(3)}_{xyxy}$

$$4P^0_{\sigma+}/\varepsilon_0 = (3/2)|E^0_{\sigma+}|^2E^0_{\sigma+}(\chi^{(3)}_{xxxx} + \chi^{(3)}_{xyxy}) + 3E^0_{\sigma+}|E^0_{\sigma-}|^2(\chi^{(3)}_{xxxx} + \chi^{(3)}_{xyxy}). \tag{4.69}$$

Introducing $A = \chi^{(3)}_{xxxx} + \chi^{(3)}_{xyxy}$ we arrive at

$$(4/3\varepsilon_0)P^0_{\sigma+} = \frac{1}{2}\left(A|E^0_{\sigma+}|^2 + 2A|E^0_{\sigma-}|^2\right)E^0_{\sigma+} \tag{4.70}$$
$$= \mathcal{C}_{1111}|E^0_{\sigma+}|^2E^0_{\sigma+} + \mathcal{C}_{1221}|E^0_{\sigma-}|^2E^0_{\sigma+}.$$

From where we identify: $\mathcal{C}_{1111} = (1/2)(\chi^{(3)}_{xxxx} + \chi^{(3)}_{xyxy})$ and $\mathcal{C}_{1221} = 2\mathcal{C}_{1111}$.

Alternatively we may also find the susceptibility tensor in the new frame based on the original (x, y, z) frame. The nonlinear induced polarization that we are interested in is

$$(\mathbf{P}^0)^{(3)} = \varepsilon_0\chi^{(3)}(-\omega_s; \omega_s, -\omega_s, \omega_s)\frac{3}{4}\mathbf{E}^0(\mathbf{E}^0)^*\mathbf{E}^0. \tag{4.71}$$

From the above the susceptibility tensor is transformed according to

$$\mathcal{C}_{\mu\alpha\beta\gamma} = r_{\mu k} r_{l\alpha}^{-1} (r_{m\beta}^{-1})^* r_{n\gamma}^{-1} \chi_{klmn}^{(3)}. \tag{4.72}$$

From this we find the tensor element

$$\mathcal{C}_{1111} = r_{1k} r_{l1}^{-1} (r_{m1}^{-1})^* r_{n1}^{-1} \chi_{klmn}^{(3)}. \tag{4.73}$$

Using the transfer matrix Eqn. (4.60) and its inverse Eqn. (4.61), we find

$$\begin{aligned} \mathcal{C}_{1111} &= \frac{1}{4}\left(\chi_{1111}^{(3)} + \chi_{1122}^{(3)} - \chi_{1212}^{(3)} + \chi_{1221}^{(3)} + \chi_{2112}^{(3)} - \chi_{2121}^{(3)} + \chi_{2211}^{(3)} + \chi_{2222}^{(3)}\right) \\ &= \frac{1}{2}\left(\chi_{1111}^{(3)} + \chi_{1221}^{(3)}\right). \end{aligned} \tag{4.74}$$

Further calculations show that $\mathcal{C}_{1111} = 0$ corresponds to a vanishing contribution from $(E_{\sigma-}^0)^2 (E_{\sigma+}^0)^*$ and in addition $\mathcal{C}_{1111} = \mathcal{C}_{1221} = \mathcal{C}_{1122}$.

—— ■ ——

4.5 Contracted notation for second-order susceptibility tensors*

As a consequence of permutation symmetry, the second-order induced polarization is often expressed using a contracted notation, in which the susceptibility tensor having 27 elements reduces to a tensor having 18 elements.

4.5.1 Second-harmonic generation

In the case of second-harmonic generation, the induced polarization is expressed as

$$(P_{2\omega}^0)^{(2)} = \varepsilon_0 \frac{1}{2} \chi^{(2)}(-2\omega; \omega, \omega) \mathbf{E}_\omega^0 \mathbf{E}_\omega^0, \tag{4.75}$$

due to intrinsic permutation symmetry,

$$\chi_{ijk}^{(2)}(-2\omega; \omega, \omega) = \chi_{ikj}^{(2)}(-2\omega; \omega, \omega), \tag{4.76}$$

hence contributions from jk and kj are identical and may be expressed using a single index. Typically, a numerical index is used to represent the contributions from jk as shown in table 4.2.

The second-order induced polarization is then expressed in matrix form as follows

$$\begin{bmatrix} (P_{2\omega}^0)_x \\ (P_{2\omega}^0)_y \\ (P_{2\omega}^0)_z \end{bmatrix} = \varepsilon_0 \begin{bmatrix} d_{11} & d_{12} & d_{13} & d_{14} & d_{15} & d_{16} \\ d_{21} & d_{22} & d_{23} & d_{24} & d_{25} & d_{26} \\ d_{31} & d_{32} & d_{33} & d_{34} & d_{35} & d_{36} \end{bmatrix} \begin{bmatrix} (E_\omega^0)_x \\ (E_\omega^0)_y \\ (E_\omega^0)_z \\ 2(E_\omega^0)_y (E_\omega^0)_z \\ 2(E_\omega^0)_x (E_\omega^0)_z \\ 2(E_\omega^0)_x (E_\omega^0)_y \end{bmatrix}. \tag{4.77}$$

If the material shows Kleinman symmetry, then the d-tensor reduces even further. In addition, if Kleinman symmetry holds, the contracted notation may be expanded to other second-order processes.

Table 4.2: Contracted indices.

m	1	2	3	4	5	6
jk	xx	yy	zz	zy yz	zx xz	xy yx

4.6 Summary

- Transformation of tensor

$$T'_{\mu\alpha_1\cdots\alpha_n} = R_{\mu u}R_{\alpha_1 a_1}\cdots R_{\alpha_n a_n}T_{u a_1\cdots a_n}. \quad (4.4)$$

- Internal symmetry: Intrinsic permutation symmetry, Overall permutation symmetry, Kleinman symmetry. These symmetry rules describe how arguments and corresponding indices may be interchanged.

- Spatial symmetry: Photonic materials are all characterized by a spatial symmetry of crystal class. The spatial symmetry causes various tensor elements to be related to others or even zero. The relations between tensor elements is evaluated using Neumann principle.

- Neumann Principle: The spatial symmetry of a given material describes that the material is invariant under certain rotations, translations etc. Consequently the tensor has to reflect these symmetry operations. This is described by the Neumann Principle

$$T'_{\mu\alpha_1\cdots\alpha_n} = T_{\mu\alpha_1\cdots\alpha_n}, \quad (4.7)$$

  where T is the material tensor in a frame and T' is the material in a new frame, where the new and the original frames are correlated by given spatial symmetries that apply to the material. To find the tensor in the new frame, the original is transformed using generating matrices as described in Appendix D.

- Cubic and isotropic material like silicon and glass, from Section 4.3: The non-zero elements are $\chi^{(3)}_{iijj}$, $\chi^{(3)}_{ijji}$, $\chi^{(3)}_{ijij}$ and $\chi^{(3)}_{iiii}$. From the Born-Oppenheimer approximation the nonlinear response may be written as

$$d_{ijkl}(t) = a(t)\delta_{ij}\delta_{kl} + \frac{1}{2}b(t)\left(\delta_{il}\delta_{jk} + \delta_{ik}\delta_{jl}\right) + c(t)\delta_{ijkl} \quad (4.31)$$

$$\sigma_{ijkl} = \frac{1}{6}\sigma\left(\delta_{ij}\delta_{kl} + \delta_{ik}\delta_{jl} + \delta_{il}\delta_{jk}\right) + \frac{1}{2}\tau\delta_{ijkl}, \quad (4.32)$$

  where $c(t) \neq 0$, $\tau \neq 0$ for cubic materials, for example silicon, and $c(t) = \tau = 0$ for isotropic materials, for example glass.

- Cyclic coordinate system

$$\sigma^+ = \frac{1}{\sqrt{2}}(\mathbf{e}_x + i\mathbf{e}_y) \quad \sigma^- = \frac{1}{\sqrt{2}}(\mathbf{e}_x - i\mathbf{e}_y) \quad \boldsymbol{C_3} = \mathbf{e}_z. \quad (4.59)$$

- Contracted notation of d tensors rather than $\chi^{(2)}$ tensors

$$\begin{bmatrix} d_{11} & d_{12} & d_{13} & d_{14} & d_{15} & d_{16} \\ d_{21} & d_{22} & d_{23} & d_{24} & d_{25} & d_{26} \\ d_{31} & d_{32} & d_{33} & d_{34} & d_{35} & d_{36} \end{bmatrix}. \quad (4.77)$$

Chapter 5

The nonlinear wave equation

Contents

In this chapter, the induced polarization is inserted into Maxwell's equations, and the nonlinear wave equation is derived. We use this to look at second-order nonlinear effects, for example second-harmonic generation and the generation of sum- and difference-frequencies of light, in Chapter 6. In the Chapters 7 and 8 we consider third-order nonlinear effects in silica glass based waveguides. Chapter 7 focuses on Raman scattering, whereas Chapter 8 focuses on the Kerr effect. Chapter 9 describes propagation of short pulses, whereas Chapter 10 deals with four-wave mixing in optical fibers.

From Chapter 1, the propagation of electromagnetic waves is governed by Maxwell's equations for a charge and current free ($\mathbf{J}_f(\mathbf{r},t) = 0$, and $\rho_f(\mathbf{r},t) = 0$)

medium,

$$\nabla \times \mathbf{H}(\mathbf{r}, t) = \frac{\partial \mathbf{D}(\mathbf{r}, t)}{\partial t}, \tag{5.1a}$$

$$\nabla \times \mathbf{E}(\mathbf{r}, t) = -\frac{\partial \mathbf{B}(\mathbf{r}, t)}{\partial t}, \tag{5.1b}$$

$$\nabla \cdot \mathbf{D}(\mathbf{r}, t) = 0, \tag{5.1c}$$

$$\nabla \cdot \mathbf{B}(\mathbf{r}, t) = 0. \tag{5.1d}$$

In addition, the constitutive relations for non-magnetic materials are

$$\mathbf{D}(\mathbf{r}, t) = \varepsilon_0 \mathbf{E}(\mathbf{r}, t) + \mathbf{P}(\mathbf{r}, t), \tag{5.2a}$$

$$\mathbf{B}(\mathbf{r}, t) = \mu_0 \mathbf{H}(\mathbf{r}, t), \tag{5.2b}$$

where $\mathbf{P}(\mathbf{r}, t)$ is the macroscopic induced polarization density (electric dipole moment per unit volume).

Taking the curl of the second Maxwell equation Eqn. (5.1b) and inserting $\nabla \times \mathbf{H}$ from Eqn. (5.1a) and the displacement vector field $\mathbf{D}$ from Eqn. (5.2a) we get

$$\nabla \times \nabla \times \mathbf{E}(\mathbf{r}, t) + \frac{1}{c^2} \frac{\partial^2 \mathbf{E}(\mathbf{r}, t)}{\partial t^2} = -\mu_0 \frac{\partial^2 \mathbf{P}(\mathbf{r}, t)}{\partial t^2}, \tag{5.3}$$

where the induced polarization may be written in terms of a perturbation series as

$$\mathbf{P}(\mathbf{r}, t) = \sum_{k=1}^{\infty} \mathbf{P}^{(k)}(\mathbf{r}, t). \tag{5.4}$$

Note, this series includes the linear induced polarization $\mathbf{P}^{(1)}$. The induced polarization is used to describe the interaction between light and matter as discussed in the previous chapters. If this is absent the equation describes the propagation of an electromagnetic wave in vacuum. In effect the induced polarization acts as a source term in the mathematical description of wave propagation, making the otherwise homogeneous equation an inhomogeneous differential equation.

For completeness it is noted that when the induced polarization is calculated from the Bloch equations (outside the scope of this book) as opposed to the above perturbation series, Maxwells equations and the above wave equation is denoted the Maxwell-Bloch equations.

By inserting the perturbation series for the induced polarization, which applies to arbitrary electric field distributions and field intensities of the light, we find

$$\nabla \times \nabla \times \mathbf{E}(\mathbf{r}, t) + \frac{1}{c^2} \frac{\partial^2}{\partial t^2} \int_{-\infty}^{\infty} (1 + \chi^{(1)}) \mathbf{E}(\mathbf{r}, \omega) e^{-i\omega t} d\omega = -\mu_0 \frac{\partial^2 \mathbf{P}^{(\mathrm{NL})}(\mathbf{r}, t)}{\partial t^2}, \tag{5.5}$$

where $\mathbf{P}^{(\mathrm{NL})}(\mathbf{r},t) = \sum_{k=2}^{\infty} \mathbf{P}^{(k)}(\mathbf{r},t)$ is the nonlinear induced polarization. Eqn. (5.5) is the nonlinear wave equation in the time domain.

Alternatively to the above wave equation in the time domain, we often study nonlinear propagation in the frequency domain, especially when the electromagnetic problem is related to monochromatic waves (continuous waves) in which case the frequency dependence is simply reduced to the interaction between discrete frequencies in a spectrum.

In the frequency domain, the wave equation is

$$\nabla \times \nabla \times \mathbf{E}(\mathbf{r},\omega) = \frac{\omega^2}{c^2}\mathbf{E}(\mathbf{r},\omega) + \mu_0\omega^2\mathbf{P}(\mathbf{r},\omega), \tag{5.6}$$

where we have used the property of the Fourier transform of plane waves

$$f'(t) \leftrightarrow -i\omega F(\omega) \quad \text{and} \quad f''(t) \leftrightarrow -\omega^2 F(\omega). \tag{5.7}$$

Note, these Fourier transformation pairs are only valid under the Fourier transformation convention chosen in this text, see Chapter 3 as well as Appendix C.

The integral on the left-hand side of Eqn. (5.5) includes the first-order, or linear, susceptibility tensor $\chi^{(1)}$. An often made assumption is that the considered material is isotropic i.e. the first-order susceptibility has identical non-zero components in the 'diagonal' only. The relative permittivity ε_r, defined as $\varepsilon_r = (1+\chi^{(1)})$, may then be expressed as $\varepsilon_{r,\mu\alpha} = n_0^2\delta_{\mu\alpha}$, where n_0 is the first-order, or linear, contribution to the refractive index of the medium. In Chapter 6 we discuss propagation in non-isotropic materials, i.e. uniaxial or biaxial materials.

5.1 Mono and quasi-monochromatic beams

Before we move on to discuss the transverse problem, we consider two subcases

- sum of monochromatic fields
- quasi-monochromatic light

5.1.1 Sum of monochromatic fields

We initially remind ourselves that if an electric field is a sum of **monochromatic** waves, then the time dependence is described by an exponential term as

$$\mathbf{E}(\mathbf{r},t) = \frac{1}{2}\sum_{\omega_\sigma \geq 0}\left[\mathbf{E}^0_{\omega_\sigma}(\mathbf{r})e^{-i\omega_\sigma t} + c.c.\right], \tag{5.8}$$

where ω_σ denotes any of the interacting frequencies.

In the frequency domain, this is represented by two delta functions

$$\mathbf{E}(\mathbf{r},\omega) = \frac{1}{2}\sum_{\omega_\sigma \geq 0}\left[\mathbf{E}^0_{\omega_\sigma}(\mathbf{r})\delta(\omega-\omega_\sigma) + (\mathbf{E}^0_{\omega_\sigma})^*(\mathbf{r})\delta(\omega+\omega_\sigma)\right]. \tag{5.9}$$

Since the electric field is a sum of monochromatic fields, the induced polarization also ends up being a sum of monochromatic waves and may be written as

$$\mathbf{P}^{(\mathrm{NL})}(\mathbf{r},t) = \frac{1}{2}\sum_{\omega_\sigma \geq 0}\left[(\mathbf{P}^0_{\omega_\sigma})^{(\mathrm{NL})}(\mathbf{r})e^{-i\omega_\sigma t} + c.c.\right] \tag{5.10}$$

in the time domain, where 'NL' is the order of the nonlinearity as described for example in Chapter 3. In the frequency domain, this becomes

$$\mathbf{P}^{(\mathrm{NL})}(\mathbf{r},\omega) = \frac{1}{2}\sum_{\omega_\sigma \geq 0}\left[(\mathbf{P}^0_{\omega_\sigma})^{(\mathrm{NL})}(\mathbf{r})\delta(\omega-\omega_\sigma) + (\mathbf{P}^0_{\omega_\sigma})^{(\mathrm{NL})*}(\mathbf{r})\delta(\omega+\omega_\sigma)\right]. \tag{5.11}$$

With this inserted into Eqn. (5.5) we get, for each frequency component of the oscillating monochromatic wave,

$$\nabla\times\nabla\times\mathbf{E}^0_{\omega_\sigma}(\mathbf{r}) - \frac{\omega_\sigma^2}{c^2}(1+\chi^{(1)})\mathbf{E}^0_{\omega_\sigma}(\mathbf{r}) = \mu_0\omega_\sigma^2(\mathbf{P}^0_{\omega_\sigma})^{(\mathrm{NL})}(\mathbf{r}). \tag{5.12}$$

5.1.2 Quasi-monochromatic light

If the field is **quasi-monochromatic** i.e. the electric field and the induced polarization consist in the time domain of one or more carrier waves multiplied by a slowly varying envelope, we have

$$\mathbf{E}(\mathbf{r},t) = \frac{1}{2}\sum_{\omega_\sigma \geq 0}\left[\mathbf{E}^0_{\omega_\sigma}(\mathbf{r},t)e^{-i\omega_\sigma t} + c.c.\right], \tag{5.13a}$$

$$\mathbf{P}(\mathbf{r},t) = \frac{1}{2}\sum_{\omega_\sigma \geq 0}\left[\mathbf{P}^0_{\omega_\sigma}(\mathbf{r},t)e^{-i\omega_\sigma t} + c.c.\right]. \tag{5.13b}$$

As opposed to Eqn. (5.8) through Eqn. (5.11) the amplitudes of the electric field as well as of the induced polarization are now functions of time. In the frequency domain this is represented by

$$\mathbf{E}(\mathbf{r},\omega) = \frac{1}{2}\sum_{\omega_\sigma \geq 0}\left[\mathbf{E}^0_{\omega_\sigma}(\mathbf{r},\omega-\omega_\sigma) + (\mathbf{E}^0_{\omega_\sigma})^*(\mathbf{r},\omega+\omega_\sigma)\right], \tag{5.14a}$$

$$\mathbf{P}(\mathbf{r},\omega) = \frac{1}{2}\sum_{\omega_\sigma \geq 0}\left[\mathbf{P}^0_{\omega_\sigma}(\mathbf{r},\omega-\omega_\sigma) + (\mathbf{P}^0_{\omega_\sigma})^*(\mathbf{r},\omega+\omega_\sigma)\right]. \tag{5.14b}$$

Considering only the positive frequencies and using the notation for quasi-monochromatic light, the integral in Eqn. (5.5) equals

$$\frac{1}{c^2}\frac{\partial^2}{\partial t^2}\int_{-\infty}^{\infty}(1+\chi^{(1)})\mathbf{E}(\mathbf{r},\omega)e^{-i\omega t}d\omega = -\frac{1}{2}\sum_{\omega_\sigma \geq 0}\int_{-\infty}^{\infty}k^2\mathbf{E}^0_{\omega_\sigma}(\mathbf{r},\omega-\omega_\sigma)e^{-i\omega t}d\omega, \tag{5.15}$$

where we have inserted the electrical field from Eqn. (5.14a) and furthermore used $k^2 = (\omega n_0/c)^2$ where $n_0 = \sqrt{1+\chi^{(1)}}$. Here we may use that the electric field is nearly monochromatic, hence we only need to know $k(\omega)$ in a narrow region around ω_σ, i.e. $k(\omega)$ may be approximated through

$$\begin{aligned} k^2(\omega) &\approx \left(k_\sigma + \left.\frac{dk}{d\omega}\right|_{\omega_\sigma}(\omega-\omega_\sigma) + \frac{1}{2}\left.\frac{d^2k}{d\omega^2}\right|_{\omega_\sigma}(\omega-\omega_\sigma)^2\right)^2 \\ &\approx k_\sigma^2 + 2k_\sigma \left.\frac{dk}{d\omega}\right|_{\omega_\sigma}(\omega-\omega_\sigma) + k_\sigma \left.\frac{d^2k}{d\omega^2}\right|_{\omega_\sigma}(\omega-\omega_\sigma)^2, \end{aligned} \tag{5.16}$$

where $k_\sigma = k(\omega_\sigma)$, and where the first approximation is due to the expansion of k to second order only. The second approximation is due to the fact that all terms not including k_σ are assumed to be negligible. From this we find the integral in Eqn. (5.5)

$$\begin{aligned} &\frac{1}{c^2}\frac{\partial^2}{\partial t^2}\int_{-\infty}^{\infty}(1+\chi^{(1)})\mathbf{E}^0(\mathbf{r},\omega)e^{-i\omega t}d\omega = \qquad (5.17)\\ &-\frac{1}{2}\sum_{\omega_\sigma\geq 0} e^{-i\omega_\sigma t}\left(k_\sigma^2 - i2k_\sigma\left.\frac{dk}{d\omega}\right|_{\omega_\sigma}\frac{\partial}{\partial t} - k_\sigma\left.\frac{d^2k}{d\omega^2}\right|_{\omega_\sigma}\frac{\partial^2}{\partial t^2}\right)\mathbf{E}^0_{\omega_\sigma}(\mathbf{r},t), \end{aligned}$$

where we have used the Fourier transformation rule $e^{-i\omega_0 t}f'(t) \leftrightarrow -i(\omega-\omega_0)F(\omega-\omega_0)$, see Appendix C.2. Inserting this into Eqn. (5.5) we obtain

$$\begin{aligned} &\sum_{\omega_\sigma\geq 0} e^{-i\omega_\sigma t}\left[\nabla\times\nabla\times\mathbf{E}^0_{\omega_\sigma}(\mathbf{r},t) - \left(k_\sigma^2 - i2k_\sigma\left.\frac{dk}{d\omega}\right|_{\omega_\sigma}\frac{\partial}{\partial t} - k_\sigma\left.\frac{d^2k}{d\omega^2}\right|_{\omega_\sigma}\frac{\partial^2}{\partial t^2}\right)\mathbf{E}^0_{\omega_\sigma}(\mathbf{r},t)\right] \\ &= -\mu_0\frac{\partial^2(\mathbf{P}^0_{\omega_\sigma})^{(\mathrm{NL})}(\mathbf{r},t)}{\partial t^2}. \end{aligned} \tag{5.18}$$

Assuming that the nonlinear induced polarization may be written as a quasi-monochromatic wave e.g. Eqn. (5.14a), we finally arrive at

$$\begin{aligned} &\nabla\times\nabla\times\mathbf{E}^0_{\omega_\sigma}(\mathbf{r},t) - \qquad (5.19)\\ &\left(k_\sigma^2 - i2k_\sigma\left.\frac{dk}{d\omega}\right|_{\omega_\sigma}\frac{\partial}{\partial t} - k_\sigma\left.\frac{d^2k}{d\omega^2}\right|_{\omega_\sigma}\frac{\partial^2}{\partial t^2}\right)\mathbf{E}^0_{\omega_\sigma}(\mathbf{r},t) \\ &= -\mu_0\left[-{\omega_\sigma}^2(\mathbf{P}^0_{\omega_\sigma})^{(\mathrm{NL})}(\mathbf{r},t) - 2i\omega_\sigma\frac{\partial(\mathbf{P}^0_{\omega_\sigma})^{(\mathrm{NL})}(\mathbf{r},t)}{\partial t} + \frac{\partial^2(\mathbf{P}^0_{\omega_\sigma})^{(\mathrm{NL})}(\mathbf{r},t)}{\partial t^2}\right], \end{aligned}$$

where we solve for each frequency component ω_σ. At this point we need to treat the transverse problem i.e., how to deal with the term

$$\nabla\times\nabla\times\mathbf{E}^0_{\omega_\sigma}(\mathbf{r},t). \tag{5.20}$$

5.2 Plane waves—the transverse problem

Until now we have used the plane transverse wave approximations

$$\mathbf{E}(\mathbf{r},t) = \mathbf{E}(z,t) \quad \text{and} \quad \mathbf{E}(z,t)\cdot\mathbf{e}_z = \mathbf{0}. \tag{5.21}$$

From this

$$\nabla \times \nabla \rightarrow -\frac{\partial^2}{\partial z^2}. \tag{5.22}$$

However, this description is not always sufficient since for example a Gaussian beam may be strongly focussed into a crystal or a waveguide to form a mode where the electric field is strongly confined. Examples of the latter include photonic crystal wave guides, planar wave guides and optical fibers. We return to the latter two cases in the following sections. However, for the remainder of this section we use the plane transverse wave approximation.

Our starting point is a monochromatic plane wave propagating along the z-direction, that is

$$\mathbf{E}(z,t) = \frac{1}{2}\left[\mathbf{E}^0_{\omega_\sigma} e^{-i(\omega_\sigma t - kz)} + c.c\right]. \tag{5.23}$$

It is noted, that we have used the frequency ω_σ rather than simply ω to remind the reader that we are considering a fixed frequency. In addition, we use the wave equation in the frequency domain, Eqn. (5.12), i.e.

$$\nabla \times \nabla \times \mathbf{E}^0_{\omega_\sigma}(z) - \frac{\omega_\sigma^2}{c^2}(1+\chi^{(1)})\mathbf{E}^0_{\omega_\sigma}(z) = \mu_0\omega_\sigma^2(\mathbf{P}^0_{\omega_\sigma})^{(\mathrm{NL})}(z). \tag{5.24}$$

The first term on the left-hand side is straightforward, using Eqn. (5.22)

$$\nabla \times \nabla \times \mathbf{E}^0_{\omega_\sigma}(z) = k^2\mathbf{E}^0_{\omega_\sigma}. \tag{5.25}$$

To solve the nonlinear wave equation we now need to evaluate the relevant nonlinear induced polarization, i.e. the induced polarization at the same frequency as the electric field and solve for the electric field. In the following, some important nonlinear phenomena and relations based on plane wave propagation are discussed.

5.2.1 The intensity dependent refractive index $\chi^{(3)}(-\omega;\omega,-\omega,\omega)$

First we consider the intensity dependent refractive index, the so-called optical Kerr effect, as opposed to the DC Kerr effect originating from second-order nonlinear effects $\chi^{(2)}(-\omega;0,\omega)$. Consequently, we need to identify the induced polarization at the frequency ω. This can only originate from terms of the type $\chi^{(3)}(-\omega;\omega,-\omega,\omega)E^0_\omega E^0_{-\omega} E^0_\omega$ and thus the amplitude of the third-order induced polarization at frequency ω is written as

$$(P^0_\omega)^{(3)}_i = \frac{3}{4}\varepsilon_0\chi^{(3)}_{ijkl}E^0_j(E^0_k)^*E^0_l. \tag{5.26}$$

Note that the prefactor K introduced in Chapter 3, Table 3.2 is written explicitly as $\frac{3}{4}$. Assuming that the electric field is linearly polarized, for example along the x-axis, then the third-order induced polarization equals

$$(P^0_\omega)^{(3)}_x = \frac{3}{4}\varepsilon_0\chi^{(3)}_{xxxx}|E^0_x|^2E^0_x. \tag{5.27}$$

Inserting this into the wave Eqn. (5.24) for a monochromatic wave, we get

$$k^2E^0_x - \frac{\omega_\sigma^2}{c^2}(1+\chi^{(1)}_{xx})E^0_x = \mu_0\omega_\sigma^2\frac{3}{4}\varepsilon_0\chi^{(3)}_{xxxx}|E^0_x|^2E^0_x. \tag{5.28}$$

From this we identify the dispersion relation

$$k^2 - \frac{\omega_\sigma^2}{c^2}(1+\chi^{(1)}_{xx}) = \mu_0\omega_\sigma^2\frac{3}{4}\varepsilon_0\chi^{(3)}_{xxxx}|E^0_x|^2. \tag{5.29}$$

Since $n_0^2 = \left(1+\chi^{(1)}_{xx}\right)$ and $k = n\omega_\sigma/c$ we arrive at the identity

$$n^2 = n_0^2 + \frac{3}{4}\chi^{(3)}_{xxxx}|E^0_x|^2. \tag{5.30}$$

Impact of the state of polarization of the electric field

From Chapter 3, Eqn. (3.54), the induced polarization may be expressed in terms of two parameters $\tilde{A}$ and $\tilde{B}$ as

$$\mathbf{P}^{(3)} = \varepsilon_0\tilde{A}\mathbf{E}(\mathbf{E}\cdot\mathbf{E}^*) + \varepsilon_0\tilde{B}\mathbf{E}^*(\mathbf{E}\cdot\mathbf{E}). \tag{5.31}$$

- Linear polarization

$$\mathbf{E} = E^0\begin{pmatrix}\cos\psi\\ \sin\psi\end{pmatrix}, \tag{5.32}$$

from this

$$\mathbf{P}^{(3)} = \varepsilon_0\mathbf{E}|E^0|^2\left(\tilde{A}+\tilde{B}\right). \tag{5.33}$$

That is, the state of polarization is unchanged and the intensity dependent refractive index equals

$$n^2 = (n_0^2 + |E^0|^2\left(\tilde{A}+\tilde{B}\right)). \tag{5.34}$$

That is

$$n \approx n_0 + \frac{1}{2n_0}|E^0|^2\left(\tilde{A}+\tilde{B}\right). \tag{5.35}$$

- Circular polarization

If we now would like to consider the impact of propagating circularly polarized light we write the electric field as a sum of left-hand circularly polarized light and right-hand circularly polarized light, i.e.

$$\mathbf{E}^0 = E^+\boldsymbol{\sigma}^+ + E^-\boldsymbol{\sigma}^-, \tag{5.36}$$

where the circular-polarization unit vectors are defined by

$$\boldsymbol{\sigma}^\pm = \frac{\mathbf{e}_x \pm i\mathbf{e}_y}{\sqrt{2}}. \tag{5.37}$$

It is noted that a beam of arbitrary polarization propagating in the positive z-direction may be written as a sum of a right- and left-hand circular polarized field. From Eqn. (5.36) we find

$$(\mathbf{E}^0)^* \cdot \mathbf{E}^0 = |E^+|^2 + |E^-|^2 \tag{5.38}$$

and

$$\mathbf{E}^0 \cdot \mathbf{E}^0 = 2E^+E^-. \tag{5.39}$$

The induced polarization is then

$$(\mathbf{P}^0)^{(3)} = \varepsilon_0 \left[\tilde{A}(|E^+|^2 + |E^-|^2)\mathbf{E}^0 + \tilde{B}2(E^+E^-)(\mathbf{E}^0)^*\right]. \tag{5.40}$$

Using that $(\boldsymbol{\sigma}^+)^* = \boldsymbol{\sigma}^-$ and $(\boldsymbol{\sigma}^-)^* = \boldsymbol{\sigma}^+$, we may write the induced polarization as a vector in terms of $\boldsymbol{\sigma}^+$ and $\boldsymbol{\sigma}^-$ as

$$\begin{pmatrix} P^+ \\ P^- \end{pmatrix} = \varepsilon_0 \begin{pmatrix} \left(\tilde{A}|E^+|^2 + (\tilde{A} + 2\tilde{B})|E^-|^2\right) E^+ \\ \left(\tilde{A}|E^-|^2 + (\tilde{A} + 2\tilde{B})|E^+|^2\right) E^- \end{pmatrix}. \tag{5.41}$$

By writing the wave equation in the coordinates of $\boldsymbol{\sigma}^+$ and $\boldsymbol{\sigma}^-$ the refractive index and its modification due to the induced polarization become [1]

$$n^\pm \approx n_0 + \frac{\tilde{A}(|E^\pm|^2) + (\tilde{A} + 2\tilde{B})|E^\mp|^2}{2n_0}, \tag{5.42}$$

where n_0 is the linear refractive index.

Thus, for circularly polarized light the intensity dependent refractive index equals $\tilde{A}(|E^\pm|^2)/2n_0$. For comparison the intensity dependent refractive index that linearly polarized light experiences equals $(\tilde{A} + \tilde{B})(|E^\pm|^2)/2n_0$, i.e. larger than the intensity dependent refractive experienced by a circularly polarized beam.

Considering the example of a single mode optical fiber supporting two orthogonal states of polarization with identical transverse distribution, that in an ideal fiber has identical effective refractive index, n_{eff} related to the propagation constant β through $\beta = n_{\text{eff}} 2\pi/\lambda$, for each polarization mode. In practice though, all fibers exhibit some modal birefringence and the beat length is typically on the order of 10 m. In polarization maintaining fibers, birefringence is purposely created, and the resulting fiber is called a linearly birefringent fiber. When the intensity of the light increases, an additional birefringence occurs that depends on the intensity of the light.

Consider an electric field consisting of a single frequency, i.e.

$$\mathbf{E} = \frac{1}{2}\{(\mathbf{e}_x E_x^0 + \mathbf{e}_y E_y^0)e^{-i\omega_0 t} + c.c\}, \tag{5.43}$$

where $\mathbf{e}_x$ and $\mathbf{e}_y$ are unit vectors along the x- and y-axis respectively. In this case the induced third-order polarization at frequency ω_0 along the x-axis, according to Eqn. (3.58), equals

$$\begin{aligned} P_x^0 = \varepsilon_0 \frac{3}{4}\big[&\chi^{(3)}_{xxxx} E_x^0 (E_x^0)^* E_x^0 + \chi^{(3)}_{xxyy} E_x^0 (E_y^0)^* E_y^0 \\ &+ \chi^{(3)}_{xyyx} E_y^0 (E_y^0)^* E_x^0 + \chi^{(3)}_{xyxy} E_y^0 (E_x^0)^* E_y^0\big]. \end{aligned} \tag{5.44}$$

An analogous expression for the polarization along the y-axis is easily found by replacing x and y in the equation for P_x^0. We assume that the nonlinearity is of pure electronic origin, and that the fiber has two principal axes along which linearly polarized light remains linearly polarized in the absence of nonlinear effects. This is referred to as linearly birefringent. P_x^0 and P_y^0 are then

$$\begin{aligned} P_x^0 &= \varepsilon_0 \frac{3}{4}\chi^{(3)}_{xxxx}\left[\left(|E_x^0|^2 + \frac{2}{3}|E_y^0|^2\right)E_x^0 + \frac{1}{3}\left(E_y^0\right)^2\left(E_x^0\right)^*\right], \\ P_y^0 &= \varepsilon_0 \frac{3}{4}\chi^{(3)}_{xxxx}\left[\left(|E_y^0|^2 + \frac{2}{3}|E_x^0|^2\right)E_y^0 + \frac{1}{3}\left(E_x^0\right)^2\left(E_y^0\right)^*\right]. \end{aligned} \tag{5.45}$$

From these equations the nonlinear refractive index along the x- and y-axis may be evaluated. Along the x-axis, the nonlinear refractive index equals [3]

$$\Delta n_x = n_2\left\{|E_x^0|^2 + \frac{2}{3}|E_y^0|^2\right\}, \tag{5.46}$$

and similarly for Δn_y by replacing index x and y by y and x in the expression for Δn_x. Assuming that light propagates through a very long fiber, then it will evolve through all possible states of polarization and experience an average nonlinear refractive index. This may be shown to equal 8/9 of the often quoted value of n_2. Figure 5.1 shows a result of a numerical simulation of the impact of an arbitrary rotation of the state of polarization of the electric field as a function of distance of propagation through an optical fiber.

5.2.2 Manley–Rowe relations

An important feature of frequency conversion in lossless nonlinear media was described in 1956 by J.M. Manley and H.E. Rowe, who proposed some general relations

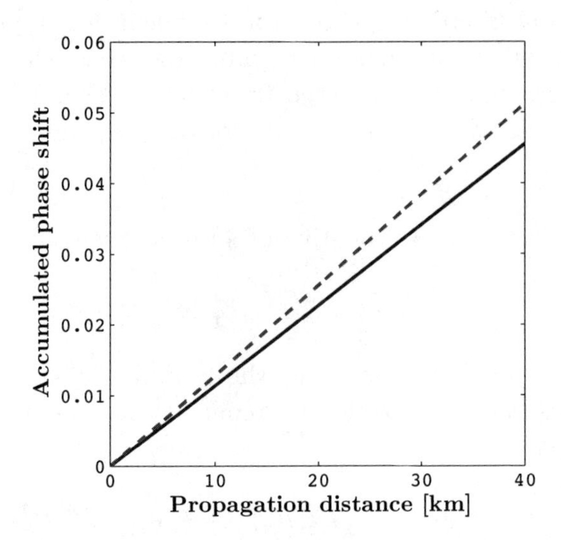

Figure 5.1: Accumulation of nonlinear phase shift as a function of propagation distance in an optical fiber. The dashed line shows predicted nonlinear phase shift assuming that the launched electric field is linearly polarized and remains in the same state of polarization during propagation. The solid line represents the accumulated phase shift assuming that the state of polarization of the electric field varies stochastically throughout propagation.

describing the energy transfer among different frequencies of an electric field [40]. In the general case of second-order nonlinear propagation, where $\omega_3 = \omega_1 + \omega_2$, and under the assumption of overall permutation symmetry, i.e. all frequencies can be permutated if the Cartesian indices are simultaneously permutated, then

$$\begin{aligned}\chi^{(2)}_{\mu\alpha\beta}(-\omega_3;\omega_1,\omega_2) &= \chi^{(2)}_{\alpha\mu\beta}(\omega_1;-\omega_3,\omega_2) \\ &= \chi^{(2)}_{\alpha\beta\mu}(\omega_1;\omega_2,-\omega_3) \\ &= \chi^{(2)}_{\beta\alpha\mu}(\omega_2;\omega_1,-\omega_3),\end{aligned} \tag{5.47}$$

that is to say that there is invariance under the $3! = 6$ possible permutations. It is noted that in the prefactor, the K factor, there are only two input frequencies and assuming ω_1, ω_2, and ω_3 are all different, the K factor equals 1 $(= (1/2)^2 \cdot 2 \cdot 2)$, $1/2^2$ since it is a second-order nonlinearity, 2 since we are interested in the amplitude of a non DC component of the induced second-order polarization and the last factor of two because there are two permutations among the two input frequencies, see also Table 3.2. This symmetry applies when all optical frequencies appearing in the formula for the susceptibility are removed far from any transition frequencies of the medium in other words the medium is transparent, or **lossless** at all relevant frequencies.

For simplicity we treat only the scalar case. The wave equation is divided into separate equations, one for each frequency. In the case of three-wave mixing we find

$$\frac{\partial E_1^0}{\partial z} = i\frac{\omega_1}{2n_1c}\chi_{\text{eff}}^{(2)} E_3^0 (E_2^0)^* e^{-i\Delta k z}, \tag{5.48a}$$

$$\frac{\partial E_2^0}{\partial z} = i\frac{\omega_2}{2n_2c}\chi_{\text{eff}}^{(2)} E_3^0 (E_1^0)^* e^{-i\Delta k z}, \tag{5.48b}$$

$$\frac{\partial E_3^0}{\partial z} = i\frac{\omega_3}{2n_3c}\chi_{\text{eff}}^{(2)} E_1^0 E_2^0 e^{i\Delta k z}, \tag{5.48c}$$

where $\Delta k = k_1+k_2-k_3$. Since the intensity of an electric field E_i is $I_i = \frac{1}{2}\varepsilon_0 c n_i |E_i^0|^2$, we find

$$\frac{\partial I_i}{\partial z} = \frac{1}{2}\varepsilon_0 c n_i \frac{\partial}{\partial z}(E_i^0(E_i^0)^*). \tag{5.49}$$

Inserting Eqn. (5.48) into Eqn. (5.49), we find

$$\frac{\partial I_1}{\partial z} = \frac{1}{2}\varepsilon_0\omega_1 \text{Re}\left[i\chi_{\text{eff}}^{(2)}(E_1^0)^*(E_2^0)^* E_3^0 e^{-i\Delta k z}\right], \tag{5.50a}$$

$$\frac{\partial I_2}{\partial z} = \frac{1}{2}\varepsilon_0\omega_2 \text{Re}\left[i\chi_{\text{eff}}^{(2)}(E_1^0)^*(E_2^0)^* E_3^0 e^{-i\Delta k z}\right], \tag{5.50b}$$

$$\frac{\partial I_3}{\partial z} = \frac{1}{2}\varepsilon_0\omega_3 \text{Re}\left[i\chi_{\text{eff}}^{(2)} E_1^0 E_2^0 (E_3^0)^* e^{i\Delta k z}\right]. \tag{5.50c}$$

Using that $\omega_3 = \omega_1 + \omega_2$, and assuming $\chi^{(2)}$ to be real, we note by comparison from these equations that

$$\frac{1}{\omega_1}\frac{\partial I_1}{\partial z} = \frac{1}{\omega_2}\frac{\partial I_2}{\partial z} = -\frac{1}{\omega_3}\frac{\partial I_3}{\partial z}. \tag{5.51}$$

That is, the rate of change in photon number per unit length at frequency ω_1 and ω_2 equals the rate of change in photon number per unit length at frequency ω_3 but with opposite sign.

Since the total intensity $I_{tot} = I_1 + I_2 + I_3$ we find

$$\frac{\partial I_{\text{tot}}}{\partial z} = \sum_{i=1,2,3}\frac{\partial I_i}{\partial z} = (\omega_1+\omega_2-\omega_3)\varepsilon_0 \text{Re}\left[i\chi_{\text{eff}}^{(2)}(E_1^0)^*(E_2^0)^* E_3^0 e^{-i\Delta k z}\right], \tag{5.52}$$

which equals zero since $\omega_3 = \omega_1 + \omega_2$.

This shows that there is no transfer of energy from the waves to the material. If the intensity at one frequency drops it has to appear at one of the other frequencies. These equations, more specifically Eqn. (5.51), are known as the Manley-Rowe relations.

5.3 Waveguides

In this section, we derive the nonlinear wave equation for a waveguide. Particular emphasis is given to the cases when the nonlinear induced polarization can be written as an effective first-order induced polarization.

The starting point is to treat the effect due to the nonlinear induced polarization as a perturbation to the linear wave equation. This equation may be solved in some simple cases, one of them being the circular cylindrical case, i.e. the optical fiber. From Appendix E.2 the solution to the wave equation in an optical fiber defines the so-called modes, i.e. the radial distribution of the electric field with corresponding propagation constants. These solutions are obtained by assuming that the electric field can be separated into a transverse part and a longitudinal part. Assuming furthermore that the electrical field is linearly polarized, for example along the x-axis, the wave equation reduces to the scalar wave equation under the assumption the refractive index contrast responsible for the wave guiding is weak. From this, we further assume that the electrical field is in the form of a quasi-monochromatic wave, i.e. in the form

$$E(r,t) = \frac{1}{2}\left(E^0_{\omega_0}(r,t)e^{-i\omega_0 t} + c.c.\right). \tag{5.53}$$

By using our previous results from quasi-monochromatic waves we know that the Fourier transform of the quasi-monochromatic wave is

$$E(r,\omega) = \frac{1}{2}\left(E^0_{\omega_0}(r,\omega-\omega_0) + (E^0_{\omega_0})^*(r,\omega+\omega_0)\right). \tag{5.54}$$

We now assume that the amplitude of the electric field can be separated into an envelope $A(\omega-\omega_0)$, a radial distribution $R(r)$, and a factor that gives propagation (together with $e^{-i\omega_0 t}$), i.e.

$$E_{\omega_0}(r,\omega-\omega_0) = \mathcal{A}(\omega-\omega_0)R(r)\exp(i\beta_0 z). \tag{5.55}$$

Also the linear induced polarization in the time and the frequency domain can be written as

$$\begin{aligned} P^{(\mathrm{L})}(r,t) &= \frac{1}{2}\left((P^0_{\omega_0})^{(\mathrm{L})}(r,t)e^{-i\omega_0 t} + c.c\right) \\ &\longleftrightarrow \quad P^{(\mathrm{L})}(r,\omega) = \frac{1}{2}\left((P^0_{\omega_0})^{(\mathrm{L})}(r,\omega-\omega_0) + (P^0_{\omega_0})^{(\mathrm{L})*}(r,\omega+\omega_0)\right). \end{aligned} \tag{5.56}$$

The nonlinear induced polarization in the time- and the frequency-domain can be written as

$$\begin{aligned} P^{(\mathrm{NL})}(r,t) &= \frac{1}{2}\left((P^0_{\omega_0})^{(\mathrm{NL})}(r,t)e^{-i\omega_0 t} + c.c\right) \\ &\longleftrightarrow \quad P^{(\mathrm{NL})}(r,\omega) = \frac{1}{2}\left((P^0_{\omega_0})^{(\mathrm{NL})}(r,\omega-\omega_0) + (P^0_{\omega_0})^{(\mathrm{NL})*}(r,\omega+\omega_0)\right). \end{aligned} \tag{5.57}$$

Inserting this into the wave equation in the frequency domain and considering only the positive frequencies gives

$$\begin{aligned}\nabla \times \nabla \times E^0_{\omega_0}(r, \omega - \omega_0) \\ = \frac{\omega^2}{c^2} E^0_{\omega_0}(r, \omega - \omega_0) + \mu_0 \omega^2 \left((P^0_{\omega_0})^{(\mathrm{L})}(r, \omega - \omega_0) + (P^0_{\omega_0})^{(\mathrm{NL})}(r, \omega - \omega_0) \right).\end{aligned} \tag{5.58}$$

Assuming that the used material is homogeneous, the left-hand side may be rewritten using $\nabla \times \nabla \times E = -\nabla^2 E$, and we arrive at the wave equation

$$\left(\nabla^2 + \frac{\omega^2}{c^2} \left(1 + \chi^{(1)} \right) \right) E^0_{\omega_0}(r, \omega - \omega_0) = -\mu_0 \omega^2 (P^0_{\omega_0})^{(\mathrm{NL})}(r, \omega - \omega_0), \tag{5.59}$$

which is the most general expression for a wave equation.

5.3.1 Effective linear induced polarization

In many cases, the induced polarization may be written as an effective linear polarization. Examples include Stokes shifted Raman and Brillouin scattering, Optical Kerr effect, i.e. self-phase modulation and induced birefringence, degenerate four-wave mixing, self-focusing, cross-phase modulation, and two-photon absorption. In these cases the induced polarization may be written as

$$(P^0_{\omega_0})^{(\mathrm{NL})}(r, \omega - \omega_0) = \varepsilon_0 K \chi^{(3)} E^0_{\omega_1} E^0_{\omega_2} E^0_{\omega_0} = \varepsilon_0 \tilde{\chi}^{(1)} E^0_{\omega_0}, \tag{5.60}$$

where $\tilde{\chi}^{(1)}$ has been introduced as an effective first-order susceptibility. Using the above electric field from Eqn. (5.55), we arrive at the wave equation

$$\left[-\nabla^2_{\perp} R(r) - \frac{\omega^2}{c^2} \left(1 + \left(\chi^{(1)} + \tilde{\chi}^{(1)} \right) \right) R(r) \right] \mathcal{A} e^{i\beta_0 z} = \nabla^2_z \left(R(r) \mathcal{A} e^{i\beta_0 z} \right). \tag{5.61}$$

To solve this, we use separation of variables i.e. we separate the radial and the longitudinal dependence

$$\frac{1}{R(r)} \left[-\nabla^2_{\perp} R(r) - \frac{\omega^2}{c^2} \left(1 + \left(\chi^{(1)} + \tilde{\chi}^{(1)} \right) \right) R(r) \right] = \frac{1}{\mathcal{A} e^{i\beta_0 z}} \nabla^2_z \left(\mathcal{A} e^{i\beta_0 z} \right). \tag{5.62}$$

Now the left-hand side depends only on r whereas the right-hand side depends only on z, which means each side must equal the same constant. From this, we get two equations

$$\frac{1}{R(r)} \left[\nabla^2_{\perp} R(r) + \frac{\omega^2}{c^2} \left(1 + \left(\chi^{(1)} + \tilde{\chi}^{(1)} \right) \right) R(r) \right] = \tilde{\beta}^2, \tag{5.63}$$

$$\frac{-1}{\mathcal{A} e^{i\beta_0 z}} \nabla^2_z \left(\mathcal{A} e^{i\beta_0 z} \right) = \tilde{\beta}^2. \tag{5.64}$$

The reason for choosing $\tilde{\beta}^2$ will become obvious in a moment. These two equations are important intermediate results. From Eqn. (5.63) the radial dependence, i.e. the transverse mode, is defined through the solution to the equation

$$\left[\nabla_\perp^2 R(r) + \frac{\omega^2}{c^2}\left(1 + \left(\chi^{(1)} + \tilde{\chi}^{(1)}\right)\right) R(r)\right] - \tilde{\beta}^2 R(r) = 0. \tag{5.65}$$

To solve this we assume that the nonlinearity does not affect the mode $R(r)$ i.e. we neglect the nonlinearity $\tilde{\chi}^{(1)}$ and solve the mode problem as in Appendix E to find $R(r)$ and $\beta(\omega)$. This is justified in the example below, and in Section 5.3.2, we show that the propagation constant due to a perturbation in the relative permittivity is changed by

$$\Delta\beta = \frac{k_0^2}{2\beta_0}\frac{\int_A \tilde{\chi}^{(1)}|R(r)|^2 da}{\int_A |R(r)|^2 da}, \tag{5.66}$$

where the integral is over the cross section of the fiber, A. Using this, Eqn. (5.64) provides the propagation equation. Performing the differentiation results in

$$\left(\frac{\partial^2 \mathcal{A}}{\partial z^2} + 2i\beta_0\frac{\partial \mathcal{A}}{\partial z}\right) - \left({\beta_0}^2\mathcal{A} - \tilde{\beta}^2\mathcal{A}\right) = 0. \tag{5.67}$$

Example 5.1. *Perturbation of the relative permittivity*
To justify the approximation of neglecting the nonlinearity when evaluating the transverse mode in a waveguide, even when the launched power is high, let us consider an optical fiber where a 1 W signal is launched into the fiber. For simplicity we assume that the transverse intensity profile follows a Gaussian distribution, which is a good approximation when the beam is launched into the fundamental mode guided by the fiber [3]. That is, we assume the intensity profile

$$I = I_0\left\{\exp(-(r/w_0)^2)\right\}^2, \tag{5.68}$$

where I_0 is the intensity at the center of the optical waveguide and w_0 is the radius of the the spot size of the transverse mode. Since the power carried by the mode is

$$P = 2\pi\int_0^\infty I_0\exp(-2(r/w_0)^2)r dr, \tag{5.69}$$

we find the intensity at the center of the fiber to be related to the power through

$$I_0 = P\frac{2}{\pi w_0^2}. \tag{5.70}$$

As a realistic number for the radius of the transverse spot, we use $w_0 = 5\cdot 10^{-6}$ m, and a power of 1 W. This results in a center intensity $I_0 \approx 25\cdot 10^9$ W/m^2. Using the intensity dependent refractive index n_2^I of silica glass, $n_2^I = 2.3\cdot 10^{-20}$ [m^2/W] then the refractive index change due to the transverse field is $\Delta n \approx 5.75\cdot 10^{-10}$. This should be compared against the refractive index step of a single mode optical fiber which is in the order of 10^{-2} to 10^{-3}, i.e. six to seven orders of magnitude higher than the perturbation caused by the 1 W optical beam.

— ■ —

Slowly varying envelope approximation

If a wave is propagating in the z-direction, it may be written as a complex-valued slowly varying envelope function $A_{\omega_\sigma}(z,t)$ of the electric field. The criteria for a slowly varying envelope function is

$$\left|\frac{\partial^2 \mathbf{A}_{\omega_\sigma}}{\partial z^2}\right| \ll \left|k_\sigma \frac{\partial \mathbf{A}_{\omega_\sigma}}{\partial z}\right|, \tag{5.71}$$

showing that the amplitude changes very little over a propagation length comparable to the wavelength. Using the slowly varying envelope approximation and setting $\tilde{\beta} = \beta(\omega) + \Delta\beta \approx \beta_0 + \beta_1(\omega - \omega_0) + \frac{1}{2}\beta_2(\omega - \omega_0)^2 + \Delta\beta$, where $\Delta\beta$ is as in Eqn. (5.66) and $\beta_0 = \beta(\omega_0)$ and $\beta_1 = \left.\frac{\partial \beta}{\partial \omega}\right|_{\omega_0}$ and $\beta_2 = \left.\frac{\partial^2 \beta}{\partial \omega^2}\right|_{\omega_0}$ we arrive at

$$2i\beta_0 \frac{\partial \mathcal{A}}{\partial z} + 2\beta_0 \left(\beta_1(\omega - \omega_0) + \frac{1}{2}\beta_2(\omega - \omega_0)^2 + \Delta\beta\right) \mathcal{A} = 0, \tag{5.72}$$

where we have also used the approximation $\beta_0^2 - \tilde{\beta}^2 \approx 2\beta_0(\beta_0 - \tilde{\beta})$. This may be translated into a propagation equation in the time domain,

$$\frac{\partial A}{\partial z} + \beta_1 \frac{\partial A}{\partial t} + i\frac{1}{2}\beta_2 \frac{\partial^2 A}{\partial t^2} - i\Delta\beta A = 0, \tag{5.73}$$

where we have used that $\omega F(\omega) \leftrightarrow i\frac{\partial}{\partial t}$, for details see Appendix C.2.

Example 5.2. *Moving frame*
To solve Eqn. (5.73) it is often useful to introduce a coordinate system that moves with the group velocity, i.e. introducing

$$\tau = t - z/v_g, \tag{5.74}$$

where $1/v_g = \beta_1$. Consequently, we need to express the partial derivatives of the amplitude A in the new coordinate system, i.e. we shift from coordinates (z,t) to (x,τ). The derivative of the amplitude, A, in the new frame, (x,τ), is then

$$\begin{aligned} \frac{\partial A}{\partial z} &= \frac{\partial A}{\partial x}\frac{\partial x}{\partial z} + \frac{\partial A}{\partial \tau}\frac{\partial \tau}{\partial z} \\ &= \frac{\partial A}{\partial x} - \frac{\partial A}{\partial \tau}\frac{1}{v_g}, \end{aligned} \tag{5.75}$$

and

$$\begin{aligned} \frac{\partial A}{\partial t} &= \frac{\partial A}{\partial x}\frac{\partial x}{\partial t} + \frac{\partial A}{\partial \tau}\frac{\partial \tau}{\partial t} \\ &= \frac{\partial A}{\partial \tau}. \end{aligned} \tag{5.76}$$

From this also $\partial^2 A/\partial t^2 = \partial^2 A/\partial \tau^2$. Inserting this into the propagation equation, Eqn. (5.73) it reduces to

$$\frac{\partial A}{\partial x} + i\frac{1}{2}\beta_2 \frac{\partial^2 A}{\partial \tau^2} - i\Delta\beta A = 0. \tag{5.77}$$

This may be even further reduced by introducing yet another coordinate system according to

$$\tau = (t - z/v_g)/T_0 \quad \text{and} \quad \xi = x|\beta_2|/T_0^2, \tag{5.78}$$

which eliminates β_2 in Eqn. (5.77). Finally, the amplitude function A may also be normalized, which results in a dimensionless propagation equation that is simpler to solve mathematically [3].

— ■ —

5.3.2 Perturbation*

In the above, we have assumed that the nonlinearity does not impact the transverse mode distribution nor the propagation constant. This is justified in the following.

We consider a waveguide structure that includes a perturbation to an unperturbed simple waveguide structure from which the eigenmode shape R and corresponding propagation constant β are known. The perturbation is expressed in terms of a change in the relative dielectric constant $\Delta\varepsilon_r$, i.e. we replace ε_r by $\varepsilon_r + \Delta\varepsilon_r$ where ε_r is the unperturbed dielectric constant ($\varepsilon_r = 1 + \chi^{(1)}$) in Eqn. (5.63) and $\Delta\varepsilon_r$ is the uniform perturbation along the z-direction, $\Delta\varepsilon_r = \tilde{\chi}^{(1)}$ in Eqn. (5.63). Assuming that only one mode is guided,the impact due to a perturbation in the permittivity may be expanded to include multiple modes [41]. However, in this treatment we keep it to the single mode case. Then in response to the perturbation of ε_r, the propagation constant and the transverse field is modified to $\beta + \Delta\beta$ and $R + \Delta R$ respectively.

Neglecting the higher-order perturbation terms that we get from the scalar wave equation, compare to Eqn. (5.63)

$$\begin{gathered} \nabla_\perp^2 (R + \Delta R) + [(\varepsilon_r + \Delta\varepsilon_r)k^2 - (\beta + \Delta\beta)^2](R + \Delta R)] = 0 \quad \Longleftrightarrow \\ \nabla_\perp^2 \Delta R + \varepsilon_r k^2 \Delta R + \Delta\varepsilon_r k^2 R - 2\beta\Delta\beta R - \beta^2 \Delta R = 0. \end{gathered} \tag{5.79}$$

Multiplying by the complex conjugate of the transverse mode R and integrating over the entire waveguide cross section, we get

$$\begin{aligned} 2\beta\Delta\beta \int_A |R|^2 da = &\int_A \Delta\varepsilon_r k^2 |R|^2 da + \\ &\int_A [(\nabla_\perp^2 \Delta R)R^* + \varepsilon_r k^2 \Delta R R^* - \beta^2 \Delta R R^*] da. \end{aligned} \tag{5.80}$$

We need to show that the second integral on the right-hand side equals zero.

To do so, we start from the wave equation

$$\nabla_\perp^2 R + [\varepsilon_r k^2 - \beta^2]R = 0, \tag{5.81}$$

we multiply the complex conjugate of the transverse wave equation by ΔR to get

$$\Delta R(\nabla_\perp^2 R^* + [\varepsilon_r^* k^2 - \beta^{*2}]R) = 0. \tag{5.82}$$

With this, Eqn. (5.80) is replaced by

$$\begin{aligned} 2\beta\Delta\beta \int_A |R|^2 da = \int_A \Delta\varepsilon_r k^2 |R|^2 da + \int_A [(\nabla_\perp^2 \Delta R)R^* \\ - \Delta R(\nabla_\perp^2 R^*) + \Delta R k^2(\varepsilon_r - \varepsilon_r^*)R^* - \Delta R(\beta^2 - \beta^{*2})R^*]da. \end{aligned} \tag{5.83}$$

Note that since ε_r and β are purely real then the integral reduces to

$$\begin{aligned} 2\beta\Delta\beta \int_A |R|^2 da &= \int_A \Delta\varepsilon_r k^2 |R|^2 da \\ &+ \int_A \left[(\nabla_\perp^2 \Delta R)R^* - \Delta R(\nabla_\perp^2 R^*)\right] da. \end{aligned} \tag{5.84}$$

To evaluate the integral on the right hand side we use Green's theorem

$$\int_A \left[(\nabla_\perp^2 \Delta R)R^* - \Delta R(\nabla_\perp^2 R^*)\right] da = 0, \tag{5.85a}$$

$$\Delta\beta = \frac{k^2}{2\beta} \frac{\int_A \Delta\varepsilon_r |R|^2 da}{\int_A |R|^2 da}, \tag{5.85b}$$

where the reader is reminded that $k = \omega/c$ and $\beta = k n_{\text{eff}}$ where n_{eff} is the effective refractive (or mode index) of the waveguide.

From this, we conclude that under the assumption of a purely real permittivity and hence real propagation constant, the change in propagation constant is solely determined from $\Delta\varepsilon$. In other words the small change in the transverse mode shape ΔR has no effect on the propagation constant.

5.4 Vectorial approach

In the method described above, the problem was to reduce the vectorial problem to a scalar problem. The approach used above is often pursued for simplicity, and it does provide significant insight. However, it may not always be sufficient, for example when the material has complicated symmetry, or if the associated waveguide supports very strong guidance, or if the propagation involves multiple modes. Several different approaches have been pursued to describe nonlinear pulse propagation

[42][43][44][45][46][47]. In the following we briefly highlight a method proposed by Afshar and Monro [42].

The method relies on the use of the so-called **reciprocal theorem**. This theorem relates a perturbed electromagnetic field to an unperturbed electromagnetic field. The perturbed field may for example be short pulses as opposed to a CW in the unperturbed case, the perturbed field may experience a completely different refractive index, and/or be subject to group velocity dispersion as opposed to a dispersion free material and/or be subject to nonlinearity. Denoting the unperturbed electromagnetic field $\mathbf{E}_0$ and $\mathbf{H}_0$, respectively, and the perturbed electromagnetic field $\mathbf{E}$ and $\mathbf{H}$, respectively then a vector function $\mathbf{F}_C$ is introduced as

$$\mathbf{F}_C = \mathbf{E}_0 \times \mathbf{H}^* + \mathbf{E}^* \times \mathbf{H}_0. \tag{5.86}$$

Using the reciprocal theorem the perturbed and the unperturbed fields are related through

$$\frac{\partial}{\partial z}\int_A \mathbf{F}_C \cdot \mathbf{z}da = \int_A \nabla\mathbf{F}_C da. \tag{5.87}$$

Reminding ourselves of Maxwell's equations from Chapter 1,

$$\nabla \times \mathbf{E}^0 = i\mu_0\omega\mathbf{H}^0 \tag{5.88a}$$

$$\nabla \times \mathbf{H}^0 = -i\varepsilon_0(1+\chi^{(1)})\omega\mathbf{E}^0 - i\omega(\mathbf{P^0})^{(\mathrm{NL})} \tag{5.88b}$$

we can consider two scenarios, the perturbed and the unperturbed case.

The unperturbed fields $\mathbf{E}_0^0(r,\omega_0)$ and $\mathbf{H}_0^0(r,\omega_0)$, which represent the electromagnetic fields of narrowband pulses at frequency ω_0 for which the dispersion, loss, and nonlinearity terms are zero, and the linear susceptibility is $\chi^{(1)}$.

The perturbed fields $\mathbf{E}^0(r,\omega)$ and $\mathbf{H}^0(r,\omega)$, represent electromagnetic fields of frequency ω associated with wideband pulses centered at ω_0, where the dispersion, loss, and nonlinearity terms are non-zero, and the effective linear susceptibility is $\tilde{\chi}^{(1)}$.

Applying equations (5.88a) to the unperturbed and perturbed fields, i.e. $\nabla \times \mathbf{E}_0^0 = i\mu_0\omega_0\mathbf{H}_0^0$ and $\nabla \times \mathbf{H}_0^0 = -i\varepsilon_0(1+\chi^{(1)})\omega_0\mathbf{E}_0^0$ for the unperturbed fields and $\nabla \times \mathbf{E}^0 = i\mu_0\omega\mathbf{H}^0$ and $\nabla \times \mathbf{H}^0 = -i\varepsilon_0(1+\tilde{\chi}^{(1)})\omega\mathbf{E}^0 - i\omega(\mathbf{P}^0)^{(\mathrm{NL})}$ for the perturbed fields, we find

$$\nabla\mathbf{F}_C = -i\mu_0(\omega-\omega_0)(\mathbf{H}^0)^*\cdot\mathbf{H}_0^0 - i\varepsilon_0(\omega\tilde{\varepsilon}_r - \omega_0\varepsilon_r)(\mathbf{E}^0)^*\cdot\mathbf{E}_0^0 - i\omega\mathbf{E}_0^0(\mathbf{P}^0)^{(\mathrm{NL})*}, \tag{5.89}$$

where $\varepsilon_r = 1+\chi^{(1)}$ for the unperturbed case and $\tilde{\varepsilon}_r = 1+\tilde{\chi}^{(1)}$ for the perturbed case.

Next we expand the perturbed fields $\mathbf{E}^0$ and $\mathbf{H}^0$ according to the set of forward and backward propagating modes and what does not fit the modes, appear as radiation RM, that is

$$\mathbf{E}^0 = \sum_\mu \frac{a_\mu(z)}{\sqrt{N_\mu}}\mathbf{e}_\mu e^{i\beta_\mu z} + \frac{a_{-\mu}(z)}{\sqrt{N_{-\mu}}}\mathbf{e}_{-\mu} e^{-i\beta_\mu z} + \mathrm{RM} \tag{5.90a}$$

$$\mathbf{H}^0 = \sum_\mu \frac{a_\mu(z)}{\sqrt{N_\mu}}\mathbf{h}_\mu e^{i\beta_\mu z} + \frac{a_{-\mu}(z)}{\sqrt{N_{-\mu}}}\mathbf{h}_{-\mu} e^{-i\beta_\mu z} + \mathrm{RM}. \tag{5.90b}$$

Note the radial dependence of the modes is in the vectors $\mathbf{e}_\mu$, $\mathbf{h}_\mu$ which are functions of the transverse coordinates (x, y) in Cartesian coordinates, whereas the amplitude $a_\mu(z)$ is a scalar. This is important for example when studying a circular state of polarization.

The unperturbed field is a superposition of guided modes that is

$$\mathbf{E}_0^0 = \sum_\nu \frac{\mathbf{e}_\nu}{\sqrt{N_\nu}} e^{i\beta_\nu z} \tag{5.91a}$$

$$\mathbf{H}_0^0 = \sum_\nu \frac{\mathbf{h}_\nu}{\sqrt{N_\nu}} e^{i\beta_\nu z}. \tag{5.91b}$$

Where the modes are fully orthogonal

$$\int_A (\mathbf{e}_\mu \times \mathbf{h}_\nu^*) \cdot \hat{\mathbf{z}} da = N_\mu \delta_{\mu\nu}, \tag{5.92}$$

and the coefficient N_μ defines the intensity in a modes, that is

$$N_\mu = \frac{1}{2}\int_A (\mathbf{e}_\mu \times \mathbf{h}_\mu^*) \cdot \hat{\mathbf{z}} da. \tag{5.93}$$

Example 5.3. *Kerr effect*
We will now consider the perturbed fields Eqn. (5.90) and the unperturbed fields Eqn. (5.91) for a single propagating mode ν, i.e.

$$\mathbf{E}_0^0 = \frac{\mathbf{e}_\nu}{\sqrt{N_\nu}} e^{i\beta_\nu z}, \ \mathbf{H}_0^0 = \frac{\mathbf{h}_\nu}{\sqrt{N_\nu}} e^{i\beta_\nu z}. \tag{5.94}$$

In order to evaluate the left-hand side of the reciprocal theorem Eqn. (5.87), we first calculate the field vector $\mathbf{F}_C$

$$\begin{aligned} \mathbf{F}_C &= \mathbf{E}_0 \times \mathbf{H}^* + \mathbf{E}^* \times \mathbf{H}_0 \qquad (5.95) \\ &= \frac{\mathbf{e}_\nu}{\sqrt{N_\nu}} e^{i\beta_\nu z} \times \sum_\mu \frac{a_\mu^*(z)}{\sqrt{N_\mu}} \mathbf{h}_\mu^* e^{-i\beta_\mu z} + \sum_\mu \frac{a_\mu^*(z)}{\sqrt{N_\mu}} \mathbf{e}_\mu^* e^{-i\beta_\mu z} \times \frac{\mathbf{h}_\nu}{\sqrt{N_\nu}} e^{i\beta_\nu z}. \end{aligned}$$

Using the orthogonality requirement Eqn. (5.92), then the left-hand side of the reciprocal theorem may be rewritten as

$$\frac{\partial}{\partial z}\int_A \mathbf{F}_C\cdot\mathbf{z}da = 4\frac{\partial a_\nu^*(z)}{\partial z}, \tag{5.96}$$

and the integral of the divergence of the vector function $\mathbf{F}_C$, i.e. the right-hand side of the reciprocal theorem

$$\begin{aligned}\int_A \nabla\mathbf{F}_C da = &- i\mu_0(\omega-\omega_0)\int_A\sum_\mu \frac{a_\mu^*(z)}{\sqrt{N_\mu}}\mathbf{h}_\mu^* e^{-i\beta_\mu z}\cdot\frac{\mathbf{h}_\nu}{\sqrt{N_\nu}}e^{i\beta_\nu z}da \\ &- i\varepsilon_0(\omega\tilde{\epsilon}_r-\omega_0\epsilon_r)\int_A\sum_\mu\frac{a_\mu^*(z)}{\sqrt{N_\mu}}\mathbf{e}_\mu^* e^{-i\beta_\mu z}\cdot\frac{\mathbf{e}_\nu}{\sqrt{N_\nu}}e^{i\beta_\nu z}da \\ &- i\omega\int_A\frac{\mathbf{e}_\nu}{\sqrt{N_\nu}}e^{i\beta_\nu z}(\mathbf{P}^0)^{(\mathrm{NL})^*}da.\end{aligned} \tag{5.97}$$

Introducing

$$A_{\nu\mu} = -i\mu_0(\omega-\omega_0)\frac{e^{-i(\beta_\mu-\beta_\nu)z}}{4\sqrt{N_\mu N_\nu}}\int_A \mathbf{h}_\mu^*\cdot\mathbf{h}_\nu da \tag{5.98a}$$

$$B_{\nu\mu} = -i\varepsilon_0(\omega\tilde{\epsilon}_r-\omega_0\epsilon_r)\frac{(z)e^{-i(\beta_\mu z-\beta_\nu)z}}{4\sqrt{N_\mu N_\nu}}\int_A \mathbf{e}_\mu^*\cdot\mathbf{e}_\nu da \tag{5.98b}$$

we rewrite the reciprocal theorem as

$$\frac{da_\nu(z)}{dz} = \sum_\mu A_{\nu\mu}^* a_\mu(z) + \sum_\mu B_{\nu\mu}^* a_\mu + i\frac{\omega e^{-i\beta_\nu z}}{4\sqrt{N_\nu}}\int_A \mathbf{e}_\nu^*(\mathbf{P}^0)^{(\mathrm{NL})}da. \tag{5.99}$$

This is the main result, which provides a governing equation for the mode amplitude. It is noted that the nonlinear induced polarization appears as

$$i\frac{\omega e^{i\beta_\nu z}}{4\sqrt{N_\nu}}\int_A \mathbf{e}_\nu(\mathbf{P}^0)^{(\mathrm{NL})^*}da. \tag{5.100}$$

As noted initially, the above method may be used to analyze the Kerr effect [42] and Raman scattering [43]. However, we will not discuss any application of this method any further in this book.

— ■ —

5.5 Nonlinear birefringence

We now return to the problem of single mode optical fibers that support two orthogonal states of polarization with identical spatial distribution, as discussed in

Section 5.2.1, and we will discuss polarization coupling of electric field at a single frequency. If the electric field is separated into a transverse field distribution and an amplitude $A_i(z,t)$, $i \in (x,y)$, which is a function that depends on time and position, a propagation equation for the field amplitude may be derived

$$\begin{aligned}\frac{\partial A_x}{\partial z} &+ \beta_{1x}\frac{\partial A_x}{\partial t} + i\frac{\beta_2}{2}\frac{\partial^2 A_x}{\partial t^2} \\ &= i\gamma\frac{3}{4}\left[\left(|A_x|^2 + \frac{2}{3}|A_y|^2\right)A_x + \frac{1}{3}\left(A_y\right)^2 A_x^* e^{(-2i\Delta\beta z)}\right],\end{aligned} \tag{5.101}$$

where $\Delta\beta$ is the difference in the propagation constant between the x- and y-component $\Delta\beta = \beta_{0x} - \beta_{0y}$. This method works well to describe birefringence in optical fibers.

Assuming that the fiber is elliptically birefringent, then the electric field may be written as

$$\mathbf{E} = \frac{1}{2}\left\{(\boldsymbol{\sigma}_1 E_x^0 + \boldsymbol{\sigma}_2 E_y^0)e^{-i\omega_0 t} + c.c.\right\}, \tag{5.102}$$

where

$$\boldsymbol{\sigma}_1 = \frac{\mathbf{e}_x + ir\mathbf{e}_y}{\sqrt{1+r^2}}, \quad \boldsymbol{\sigma}_2 = \frac{r\mathbf{e}_x - i\mathbf{e}_y}{\sqrt{1+r^2}}. \tag{5.103}$$

If $r = 0$ the field corresponds to linearly polarized light while $r = 1$ corresponds to circularly polarized light. Inserting this into the wave equation and assuming that $\Delta\beta$ is large, it is approximated by [3]

$$\frac{\partial A_x}{\partial z} + \beta_{1x}\frac{\partial A_x}{\partial t} + i\frac{\beta_2}{2}\frac{\partial^2 A_x}{\partial t^2} = i\gamma\frac{9}{4}\left[\left(|A_x|^2 + B|A_y|^2\right)A_x\right], \tag{5.104}$$

where

$$B = \frac{2 + 2\sin^2\theta}{2 + \cos^2\theta}. \tag{5.105}$$

Example 5.4. *How to treat birefringence in optical fibers*
Even though standard conventional optical fibers are circular cylinder symmetrical, they always possess some deviations both from the ideal theoretical geometry and also from the ideal cylindrical symmetrical refractive index profile. Consequently the fiber possesses some small birefringence that is varying along the z-axis.

To model this, we consider light that is propagating along the z-axis. When the light is propagating along the slow and fast axes, it is described as

$$\mathbf{E}(z) = E_x \exp(-i\beta_x z)\mathbf{e}_x + E_y \exp(-i\beta_y z)\mathbf{e}_y. \tag{5.106}$$

Assuming the loss is polarization independent, then propagation through an element which is rotated an angle right-handed rotation θ is described through

$$\mathbf{E}_{\text{out}}(z) = R(-\theta)T(\omega)R(\theta)\mathbf{E}_{\text{in}}. \tag{5.107}$$

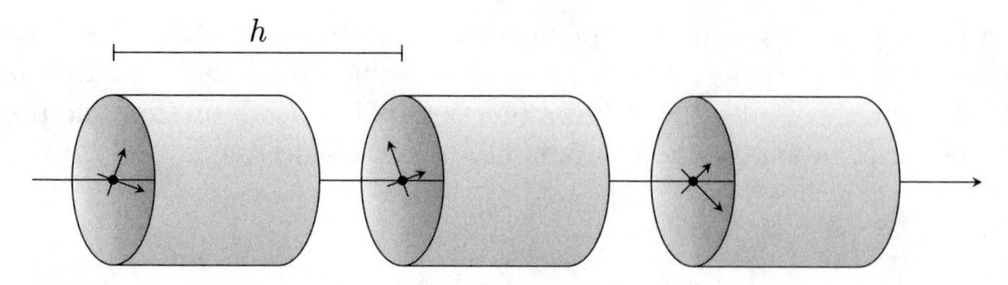

Figure 5.2: Illustration of how propagation through a long optical fiber is treated numerically, where h is the numerical step size. Figure adapted from [48].

$R(-\theta)$ re-expresses the field in terms of x- and y-components. A long fiber involving n segments each with a different angle offset with respect to x is evaluated through the general matrix product

$$\mathbf{E}_{\text{out}}(z) = [R(-\theta_n)T_n(\omega)R(\theta_n)]\dots[R(-\theta_2)T_2(\omega)R(\theta_2)]\,[R(-\theta_1)T_1(\omega)R(\theta_1)]\,\mathbf{E}_{\text{in}}. \tag{5.108}$$

— ■ —

5.6 Summary

- Starting from Maxwell's equations, the purpose of Chapter 5 has been to derive and discuss wave equations that can be solved to predict propagation of an electric field influenced by nonlinearities. In the time domain we have derived the wave equation

$$\nabla \times \nabla \times \mathbf{E}(\mathbf{r},t) = -\frac{1}{c^2}\frac{\partial^2 \mathbf{E}(\mathbf{r},t)}{\partial t^2} - \mu_0 \frac{\partial^2 \mathbf{P}(\mathbf{r},t)}{\partial t^2}, \tag{5.3}$$

and in the frequency domain we have shown the wave equation

$$\nabla \times \nabla \times \mathbf{E}(\mathbf{r},\omega) = \frac{\omega^2}{c^2}\mathbf{E}(\mathbf{r},\omega) + \mu_0\omega^2 \mathbf{P}(\mathbf{r},\omega). \tag{5.6}$$

- We most often describe nonlinear problems by using a sum of monochromatic waves or a slowly varying amplitude multiplied onto a monochromatic wave, the so-called quasi monochromatic approximation. Consequently we need to solve the nonlinear wave equation

$$\nabla \times \nabla \times \mathbf{E}^0_{\omega_\sigma}(\mathbf{r}) - \frac{\omega_\sigma^2}{c^2}(1+\chi^{(1)})\mathbf{E}^0_{\omega_\sigma}(\mathbf{r}) = \mu_0\omega_\sigma^2(\mathbf{P}^0_{\omega_\sigma})^{(\mathrm{NL})}(\mathbf{r}) \tag{5.12}$$

and we are concerned only with amplitudes indicated with superscript 0 in most equations.

- The Manley-Rowe relations

$$\frac{\partial}{\partial z}\left(\frac{I_1}{\omega_1}\right) = \frac{\partial}{\partial z}\left(\frac{I_2}{\omega_2}\right) = -\frac{\partial}{\partial z}\left(\frac{I_3}{\omega_3}\right) \tag{5.51}$$

state that the sum of intensity flux or photon number is constant, i.e. if one photon is added at one frequency it has to be removed from another frequency component.

- In optical wave guides, the challenge of propagating an electric field influenced by nonlinear effects is addressed by separating the electric field into a transverse part and a longitudinal part. In the transverse part, nonlinearities are typically neglected, and it requires solution of the eigenvalue problem

$$\frac{1}{R(r)}\left[\nabla_\perp^2 R(r) + \frac{\omega^2}{c^2}\left(1+\left(\chi^{(1)}+\tilde{\chi}^{(1)}\right)\right)R(r)\right] = \tilde{\beta}^2, \tag{5.63}$$

from which the propagation constant $\tilde{\beta}$ as well as the transverse field distribution $R(r)$ is evaluated. The longitudinal problem addresses the nonlinear propagation by solving the problem

$$\frac{-1}{\mathcal{A}e^{i\beta_0 z}}\nabla_z^2\left(\mathcal{A}e^{i\beta_0 z}\right) = \tilde{\beta}^2, \tag{5.64}$$

where the nonlinearity is now included in an expansion in $\tilde{\beta}$.

Chapter 6

Second-order nonlinear effects

Contents

Second-order nonlinear effects leading to optical frequency conversion was first demonstrated in 1961 by Franken *et al.* [5] shortly after the demonstration of the first working laser by Mainman in 1960 [49]. Many applications of second-order nonlinear effects have proven successful in various fields; for example as phase or amplitude modulators e.g. in optical communication systems, or to generate almost arbitrary wavelengths from existing laser systems.

Most applications of second-order nonlinear effects rely on the use of nonlinear crystals, where the applied field perturbs the dielectric material in order to generate a second-order nonlinear response. The electrical field can be either a purely optical field or a combination of an optical field and a constant electric field applied to the material.

The material response is described by a second-order nonlinear susceptibility tensor [50].

In addition to a general discussion of second-order nonlinear effects, and an introduction to coupled wave theory, this chapter provides examples of specific second-order nonlinear phenomena:

- Second-harmonic generation, used to generate light at a frequency of twice the launched frequency; $\chi^{(2)}_{ijk}(-2\omega_1;\omega_1,\omega_1)$.
- Non-degenerate parametric frequency conversion, for example used to generate sum- or difference-frequencies; $\chi^{(2)}_{ijk}(-\omega_3;\omega_1,\pm\omega_2)$.
- Electro optic modulators, used for modulation of optical signals; $\chi^{(2)}_{ijk}(-\omega_1;0,\pm\omega_1)$.

The above list is not a complete list of applications of second-order nonlinear effects, but it shows some important examples that are considered in this chapter.

6.1 General theory

The induced second-order nonlinear polarization can, according to Eqn. (3.31), be written as

$$(\mathbf{P}^0_{\omega_\sigma})^{(2)} = \varepsilon_0 \sum_{\omega_\sigma} K\left(-\omega_\sigma;\omega_1,\omega_2\right)\chi^{(2)}\left(-\omega_\sigma;\omega_1,\omega_2\right)\mathbf{E}^0_{\omega_1}\mathbf{E}^0_{\omega_2}, \tag{6.1}$$

where ω_σ is the sum of the input frequencies ω_1 and ω_2. The summation sign indicates that all contributions originating from distinct frequencies ω_1 and ω_2 in the applied optical field which satisfy the relation $\omega_\sigma = \omega_1 + \omega_2$ must be included. The prefactor K for the various processes is listed in Table 3.2.

Assuming monochromatic fields, we adopt the nonlinear wave equation from Chapter 5, using Eqn. (5.24) as a starting point we get

$$\nabla^2\mathbf{E}^0_{\omega_\sigma} + \frac{\omega_\sigma^2}{c^2}\left(1+\chi^{(1)}\left(-\omega_\sigma;\omega_\sigma\right)\right)\mathbf{E}^0_{\omega_\sigma} = -\mu_0\omega_\sigma^2(\mathbf{P}^0_{\omega_\sigma})^{(2)}, \tag{6.2}$$

where the tensor $\chi^{(1)}(-\omega_\sigma;\omega_\sigma)$ is the linear susceptibility at frequency ω_σ. Rewriting the linear susceptibility in the form of the refractive index n, and assuming the system to be lossless and far from resonant transitions we get the wave equation for second-order nonlinear effects

$$\left(\nabla^2 + \frac{\omega_\sigma^2 n^2}{c^2}\right)\mathbf{E}^0_{\omega_\sigma} = -\mu_0\omega_\sigma^2(\mathbf{P}^0_{\omega_\sigma})^{(2)}. \tag{6.3}$$

The wave equation has a form similar to the Helmholtz equation, but with a driving term on the right-hand side given by the induced second-order polarization.

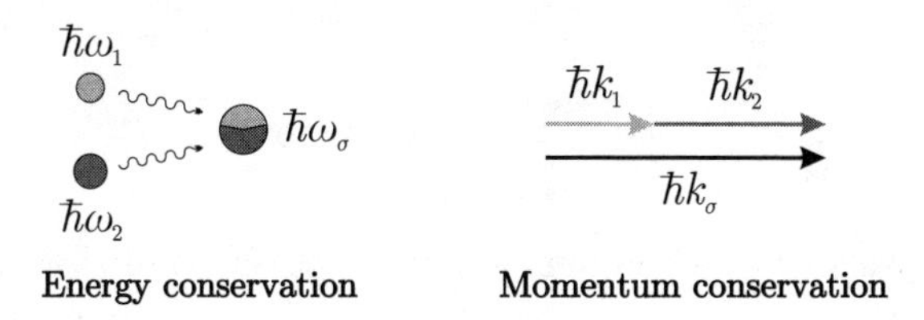

Figure 6.1: Two important aspects of second-order nonlinear interactions are the conservation of energy and momentum.

As usual for instantaneous lossless processes e.g. known from mechanics, energy and momentum have to be conserved through the interaction. In nonlinear frequency conversion **energy conservation** yields the following relation between the involved frequencies

$$\hbar\omega_\sigma = \hbar\omega_1 + \hbar\omega_2. \tag{6.4}$$

Momentum conservation in nonlinear optics is typically termed **phase matching**, where k_i is the wave number in the material at frequency ω_i

$$\hbar k_\sigma = \hbar k_1 + \hbar k_2. \tag{6.5}$$

It is noted that momentum conservation (phase matching) in general is a vectorial condition, but in the following we will only consider a scalar description, assuming that the propagation direction of all interacting fields are the same.

As it will be shown in Section 6.3, based on the wave equation Eqn. (6.3), the efficiency of the nonlinear interaction critically depends on whether or not the wave vectors of the induced polarization and the generated fields are successfully matched and maintained along the direction of propagation. Basically this is a challenge because of dispersion in the material in which the light is propagating, i.e. the fact that the refractive index is a function of wavelength.

A number of different techniques exist to achieve phase matching. Birefringent phase matching is an approach, where the material dispersion between the interacting waves are compensated by an appropriate choice of direction of propagation through the nonlinear material and direction of polarization of the interaction fields. Another approach is to apply a periodic or aperiodic structure in the material in order to compensate for the natural dispersion. A more elaborate discussion of the importance of phase matching and techniques to accomplish phase matching follows in Sections 6.3.1 and 6.3.2.

6.2 Coupled wave theory

In order to find solutions to the nonlinear wave equation Eqn. (6.3) let us consider a scalar electric field $E(t)$ oscillating at three different frequencies ω_1, ω_2 and ω_3

given by

$$E(t) = \frac{1}{2} \sum_{q=\pm1,\pm2,\pm3} E^0_{\omega_q} \exp(-i\omega_q t), \tag{6.6}$$

where the frequencies satisfy the energy conservation condition: $\omega_1 + \omega_2 = \omega_3$.

Inserting the electric field into the wave equation, the second-order induced polarization can be written as

$$(P^0)^{(2)}(t) = \varepsilon_0 K(-\omega_q - \omega_r; \omega_q, \omega_r) \chi^{(2)}_{\text{eff}} \sum_{q,r=\pm1,\pm2,\pm3} E^0_{\omega_q} E^0_{\omega_r} \exp(-i(\omega_q + \omega_r)t), \tag{6.7}$$

where $\chi^{(2)}_{\text{eff}}$ is the effective second-order susceptibility for the chosen directions of polarization and propagation for the interacting frequencies.

The induced polarization acts as a source term on the right-hand side of the wave equation at all three wavelengths. This enables the separation of the induced polarization into three equations, giving a source term oscillating at each of the three coupled frequencies [24].

Using short notation for the prefactor K, the induced polarization at each of the frequencies can be written as

$$(P^0_{\omega_1})^{(2)}(t) = \varepsilon_0 K \chi^{(2)}_{\text{eff}} E^0_{\omega_3} (E^0_{\omega_2})^* \exp(-i(\omega_3 - \omega_2)t) \tag{6.8a}$$

$$(P^0_{\omega_2})^{(2)}(t) = \varepsilon_0 K \chi^{(2)}_{\text{eff}} E^0_{\omega_3} (E^0_{\omega_1})^* \exp(-i(\omega_3 - \omega_1)t) \tag{6.8b}$$

$$(P^0_{\omega_3})^{(2)}(t) = \varepsilon_0 K \chi^{(2)}_{\text{eff}} E^0_{\omega_1} E^0_{\omega_2} \exp(-i(\omega_1 + \omega_2)t). \tag{6.8c}$$

Consequently the wave equation Eqn. (6.3) can now be split into three coupled equations, one for each of the interacting frequencies

$$(\nabla^2 + k_1^2 n_1^2) E^0_{\omega_1} \exp(-i\omega_1 t) = -\varepsilon_0 \mu_0 K \chi^{(2)}_{\text{eff}} \omega_1^2 E^0_{\omega_3} (E^0_{\omega_2})^* \exp(-i(\omega_3 - \omega_2)t) \tag{6.9a}$$

$$(\nabla^2 + k_2^2 n_2^2) E^0_{\omega_2} \exp(-i\omega_2 t) = -\varepsilon_0 \mu_0 K \chi^{(2)}_{\text{eff}} \omega_2^2 E^0_{\omega_3} (E^0_{\omega_1})^* \exp(-i(\omega_3 - \omega_1)t) \tag{6.9b}$$

$$(\nabla^2 + k_3^2 n_3^2) E^0_{\omega_3} \exp(-i\omega_3 t) = -\varepsilon_0 \mu_0 K \chi^{(2)}_{\text{eff}} \omega_3^2 E^0_{\omega_2} E^0_{\omega_1} \exp(-i(\omega_1 + \omega_2)t). \tag{6.9c}$$

In order to simplify the equations a new variable $|a_\omega|$ is introduced defined as the square root of the photon flux ϕ_ω, where $\eta = \sqrt{\mu/\varepsilon}$ is the impedance parameter of the material

$$|a_\omega|^2 = \phi_\omega = \frac{I_\omega}{\hbar\omega} = \frac{|E^0_\omega|^2}{2\eta\hbar\omega}. \tag{6.10}$$

The electrical field can then be written as $E^0_\omega = a_\omega \sqrt{2\eta\hbar\omega} \exp(ik_\omega z)$ and inserted into the three coupled wave equations. Furthermore, a gain parameter

$g = \varepsilon_0 \chi_{\text{eff}}^{(2)} \sqrt{\frac{1}{2}\eta^3 \hbar\omega_1\omega_2\omega_3}$ will be used in the following derivation, where η is taken for each of the three frequencies.

Assuming that the field amplitude varies slowly (SVEA) with distance and time see Section 5.3.1, meaning that $\frac{\partial^2 a_\omega}{\partial t^2} \ll \omega_0 \frac{\partial a_\omega}{\partial t}$ and defining the phase mismatch as $\Delta k = k_3 - (k_1 + k_2)$, coupled equations describing the change in photon flux at the three frequencies is obtained

$$\frac{da_{\omega 1}}{dz} = iKga_{\omega_3}a^*_{\omega_2}\exp\left(i\Delta kz\right) \tag{6.11a}$$

$$\frac{da_{\omega 2}}{dz} = iKga_{\omega_3}a^*_{\omega_1}\exp\left(i\Delta kz\right) \tag{6.11b}$$

$$\frac{da_{\omega 3}}{dz} = iKga_{\omega_1}a_{\omega_2}\exp\left(-i\Delta kz\right). \tag{6.11c}$$

The advantage of using a_ω rather than E_ω is, that the gain coefficient is the same in three coupled equations. These coupled differential equations describes how the energy flows between the interacting frequencies as illustrated in Figure 6.2. Two input fields oscillating at frequencies ω_1 and ω_2 (gray) are seen to exchange energyleading to a third field at frequency ω_3 (black). The signal at ω_3 is generated with a phase given as the product of ω_1 and ω_2 shifted counterclockwise by $\pi/2$.

Similarly, the generated field mixes with each of the input fields, acting as a source term at the other input frequency. In the figure the input frequencies ω_1 and ω_2 are seen to generate the sum-frequency at ω_3, and the generated sum-frequency mixes with each of the input fields to generate a signal at the other one, but out of phase with the input, hence, the input signals are attenuated at the expense of the generated field, conserving the total power.

Considering the case of perfect phase matching, $\Delta k = 0$, it is straightforward to calculate the exchange of energy among the three fields, as they propagate through the nonlinear material, similar to Figure 6.2. Figure 6.3 shows the power at the input frequencies (gray) and the generated sum-frequency power (black). The total power is seen to be conserved (dashed), consistent with the assumption of a lossless material.

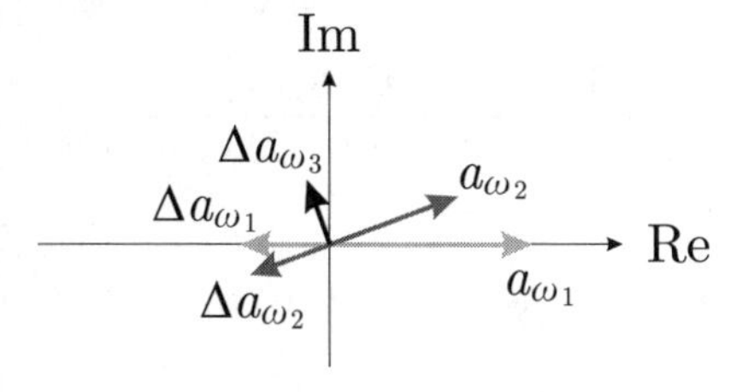

Figure 6.2: Three-wave mixing: the special case of sum-frequency generation.

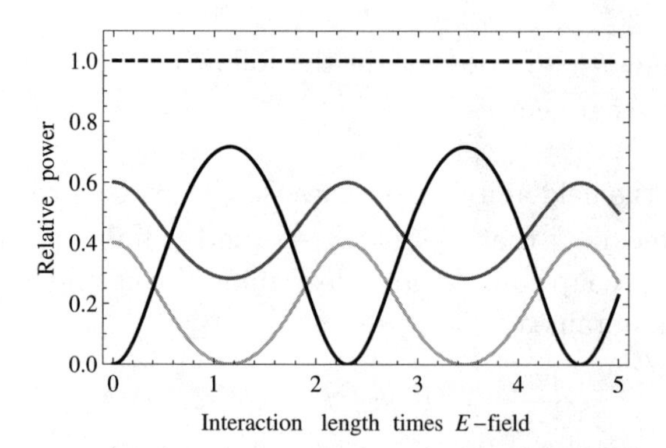

Figure 6.3: The power coupling between two input fields (gray) and their sum-frequency (black) are plotted as a function of the interaction length E-field product in the nonlinear material assuming perfect phase matching.

Figure 6.3 shows how the energy flows, first from the low frequencies ω_1 and ω_2 to the high frequency ω_3 through sum-frequency mixing, but then, as the power at ω_2 goes to zero, the phase shifts by π, and the power flows back to the lower frequencies through difference-frequency mixing. Note that the graphs are plotted as a function of crystal length E-field product.

Example 6.1. *Sum-frequency mixing of two lasers*
Considering two of the efficient laser transitions of Nd:YVO_4 at 1064 nm and 1342 nm, the sum-frequency wavelength ω_3 and gain parameter g^2 can be calculated. The sum-frequency wavelength is found according to the energy conservation Eqn. (6.4) rewritten in terms of wavelength

$$\frac{1}{\lambda_3} = \frac{1}{\lambda_1} + \frac{1}{\lambda_2} \Rightarrow \left(\frac{1}{1064 \text{ nm}} + \frac{1}{1342 \text{ nm}}\right)^{-1} = 593.5 \text{ nm}.$$

That is an up-converted wavelength in the yellow spectral range at $\lambda_3 = 593.5$ nm. Consider a nonlinear material with a second-order susceptibility of $\chi^{(2)}_{\text{eff}} = 5$ pm/V and a refractive index $n = 2.2$ then

$$\begin{aligned} g^2 &= \left(\varepsilon_0 \chi^{(2)}_{\text{eff}}\right)^2 \tfrac{1}{2}\eta^3 \hbar\omega_1\omega_2\omega_3 \\ &= (8.85 \text{ pF/m} \cdot 5 \text{ pm/V})^2 \frac{1}{2}\left(\frac{377\ \Omega}{2.2}\right)^3 1.05 \cdot 10^{-34} \\ &\quad \text{J} \cdot \text{s} \frac{\left(2\pi \cdot 3 \cdot 10^8 \text{ m/s}\right)^3}{1064 \text{ nm} \cdot 1342 \text{ nm} \cdot 593.5 \text{ nm}} \\ &= 4.1 \cdot 10^{-27} \text{ s}. \end{aligned}$$

This is the gain parameter to be used in the coupled wave equations, Eqn. (6.11).

— ■ —

6.3 Phase mismatch and acceptance bandwidths

Having introduced the phase mismatch parameter $\Delta k = k_3 - (k_1 + k_2)$ in the coupled wave equations Eqn. (6.11), this parameter describes the phase relation between the induced polarization and the real electric field at the same frequency, i.e. this is the phase relation between the nonlinear induced source term and the propagating field through the crystal [4]. The resulting photon flux at the output of the nonlinear material is calculated by integrating the contributions from the distributed source terms with the appropriate phase relation and using Eqn. (6.10)

$$\phi_3(L) = \left| \int_0^L iKga_{\omega_1}a_{\omega_2} \exp(-i\Delta kz)dz \right|^2 . \tag{6.12}$$

Assuming weak coupling between the interacting fields, i.e. a_{ω_1} and a_{ω_2} are assumed non-depleted through the crystal, these parameters can be moved outside the integral

$$\phi_3(L) = K^2g^2\phi_{\omega_1}\phi_{\omega_2} \left| \int_0^L \exp(-i\Delta kz)dz \right|^2 . \tag{6.13}$$

Evaluating the integral gives the conversion efficiency as a function of the phase mismatch relative to the perfect phase matched case, as seen in Figure 6.4

$$\left| \int_0^L \exp(-i\Delta kz)dz \right|^2 = \left| \frac{1 - \exp(-i\Delta kL)}{i\Delta k} \right|^2 = L^2 \frac{\sin^2(\Delta kL/2)}{(\Delta kL/2)^2}. \tag{6.14}$$

From Eqn. (6.14) the nonlinear frequency conversion efficiency is seen, for small Δk, to scale with the length of the nonlinear material squared, however, the wavelength acceptance bandwidth is inversely proportional to the length of the nonlinear material.

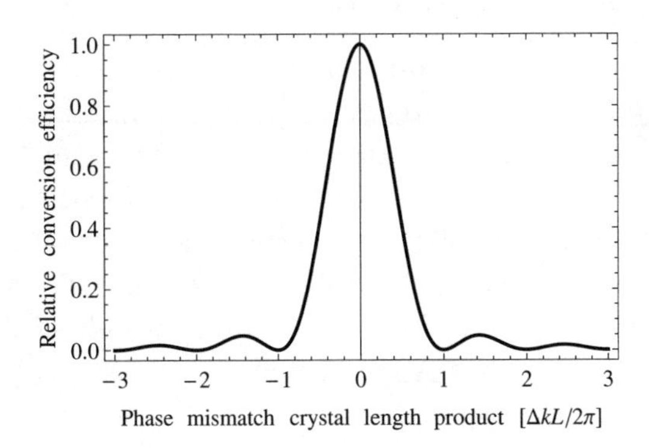

Figure 6.4: The relative conversion efficiency as a function of the phase mismatch crystal length product.

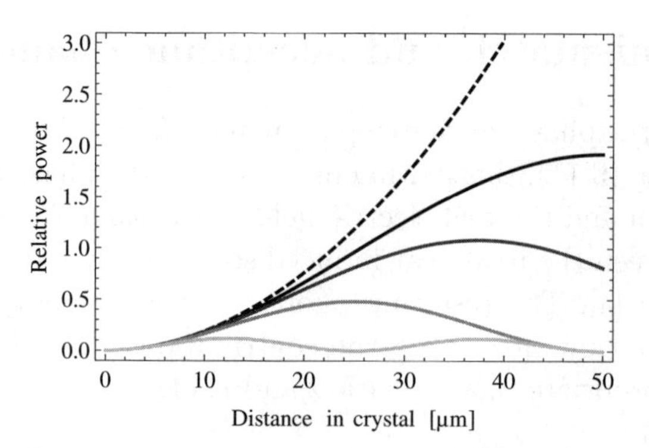

Figure 6.5: Plots of the power coupling between the three interacting frequencies. The dashed graph corresponds to perfect phase matching; lighter gray graphs correspond to coherence lengths of 100, 75, 50 and 25 μm respectively. The sum-frequency generated power is scaled by 10^9 compared to Figure 6.3.

From Eqn. (6.14) the so-called coherence length $l_{\text{coh}} = 2\pi/\Delta k$ is defined. This is the crystal length corresponding to a 2π phase shift between the induced polarization and the real field propagating through the nonlinear material, resulting in a cancellation of the converted power. Considering non-perfect phase matching, then the converted power in the first part of the nonlinear crystal is illustrated in Figure 6.5, where the dashed line is for perfect phase match and lighter gray graphs are for decreasing coherence lengths.

Figure 6.5 shows the generated power in the first part of the crystal, where the phase mismatch between the induced polarization and the real field causes back-conversion to occur after half a coherence length of propagation into the crystal, rather than a continuous build-up of power. In the perfect phase matched case all the power at one of the input frequencies needs to be converted before back-conversion occurs.

6.3.1 Birefringent phase matching

As discussed in the previous section, phase matching is an important concept in parametric three- and four-wave frequency conversion. Phase matching describes the relative phase of the induced polarization and the real field at the same frequency, $\Delta k = k_3 - k_1 - k_2$. In order to compensate for the natural dispersion of the material, the birefringence of the material can be used. More specifically the refractive index depends on the polarization of the electric field, i.e the ordinary and extra-ordinary refractive index of the material as well as the direction of propagation through the medium, see Appendix F.

Generally the refractive index depends on the wavelength of the optical field and on the temperature of the nonlinear material. In a birefringent medium the index

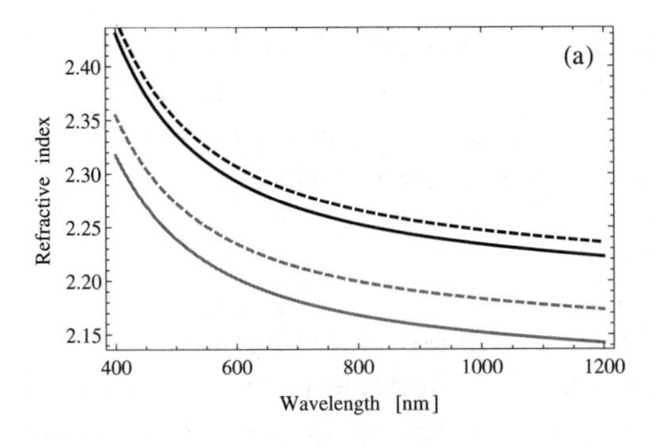

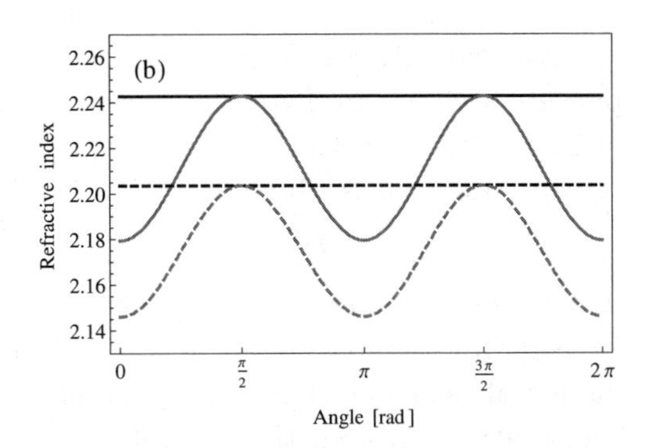

Figure 6.6: Considering MgO doped $LiNbO_3$ the graphs shows the ordinary (black) and extraordinary (gray) refractive index **(a)** as a function of wavelength for 25 °C (solid) and 125 °C (dashed), respectively. **(b)** Shows the refractive index at 125 °C as a function of direction of propagation relative to the optical axis at a wavelength of 1064 nm (dashed) and 532 nm (solid), respectively.

depends on polarization of the field, and the extraordinary index of refraction further depends on the angle between the optical axis and the direction of propagation as seen in Figure 6.6.

Example 6.2. *Non-critical phase matching*
Frequency mixing between fields propagating along one of the principle axis of the nonlinear material is particularly interesting, as beam walk-off is avoided in this case. Generally dispersion in the material makes phase matching impossible, however, one example where non-critical phase matching is possible, is second-harmonic generation of near infrared lasers propagating along the ordinary axis of $LiNbO_3$ crystals. With the fundamental field polarized along the ordinary axis, generation of a second-harmonic field along the extraordinary axis can be phase matched by the appropriate choice of temperature of the nonlinear material. Considering Figure 6.6 (a) it can be seen that the refractive index of an ordinary polarized field at 1031 nm and an extraordinary field at the second-harmonic frequency is identical at 25 °C, whereas the same is true for 1189 nm at 125 °C. In the same way, it can be shown that phase matching for second-harmonic generation of 1064 nm exists at 54 °C.

—— ■ ——

Birefringent phase matching relies on the coincidental occurrence of a direction through the nonlinear material that results in birefringence matching of the natural dispersion of the material. Hence, a direction leading to phase matching does not generally exist, and even if a phase matching direction exists, it generally does not match the direction and polarizations corresponding to the largest element of the second-order susceptibility tensor elements. Furthermore, beam propagation at an angle relative to the principle axis of the material results in a beam walk-off of the extraordinary field, reducing the mode overlap of the interacting fields.

Traditionally birefringent phase matching is divided into Type I and Type II phase matching. In Type I phase matching the two input beams are parallelly polarized orthogonal to the generated field, whereas Type II phase matching refers to the situation where the two driving fields are orthogonally polarized.

Example 6.3. *Birefringent phase matching*
Considering Figure 6.6 (b) it is seen that for Type I phase matching with the fundamental field polarized along the ordinary axis, it is possible to find an angle of propagation through the crystal resulting in phase matching for second-harmonic generation with the second-harmonic field polarized along the extraordinary axis.

Phase matched second-harmonic generation can, according to the conservation of momentum Eqn. (6.5) using index 1 for the fundamental and 2 for the second-harmonic field, be expressed as

$$0 = 2k_1 - k_2 = 2\frac{2\pi n_o(\lambda, T)}{\lambda} - \frac{2\pi n_e(\lambda/2, T, \theta)}{\lambda/2} = n_o(\lambda, T) - n_e(\lambda/2, T, \theta),$$

where $n_o(\lambda, T)$ and $n_e(\lambda, T)$ are the refractive indices along the principle axis of the crystal. The angular dependence of the extraordinary index can be expressed as

$$\frac{1}{n_e^2(\lambda, T, \theta)} = \frac{\sin^2(\theta)}{n_e^2(\lambda, T)} + \frac{\cos^2(\theta)}{n_o^2(\lambda, T)}.$$

Solving the above equation and inserting the phase match condition gives an angle of propagation $\theta = 0.68$ rad, giving a phase matched second-harmonic of a 1064 nm laser beam in MgO doped $LiNbO_3$ corresponding to the crossing point of the dashed black graph and the solid gray graph of Figure 6.6 (b).

—— ■ ——

6.3.2 Quasi-phase matching

From Figure 6.5 it is seen how phase mismatch results in back-conversion of the power. When the crystal length exceeds half a coherence length $L > l_{coh}/2$, the phase shift between the source terms and the real propagating fields in the wave equations exceeds π [51].

In some ferroelectric materials, like potassium titanyl phosphate (KTP), lithium niobate ($LiNbO_3$) and lithium tantalate ($LiTaO_3$) it is possible to invert the ferroelectric domains of the material. This can be done by applying an electric field across the nonlinear material above the coercive field strength (≈ 21 kV/mm in $LiNbO_3$). This results in a permanent inversion of the ferroelectric domains, this effectively corresponds to a change in sign of the effective nonlinear susceptibility [52]. Using electric field poling it is possible to make a patterned structure in the nonlinear material. This structure can be periodic with a length of Λ resulting in a

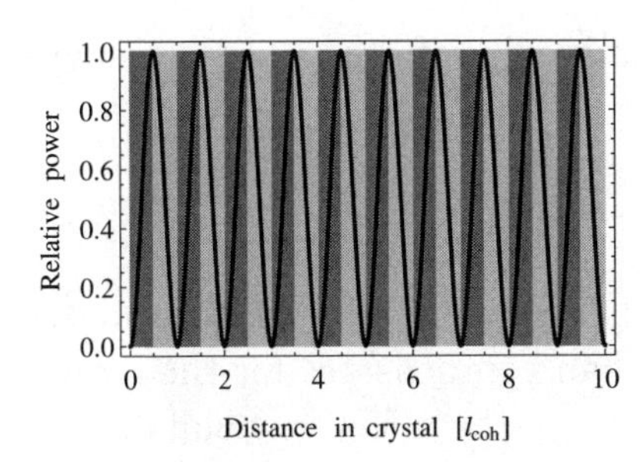

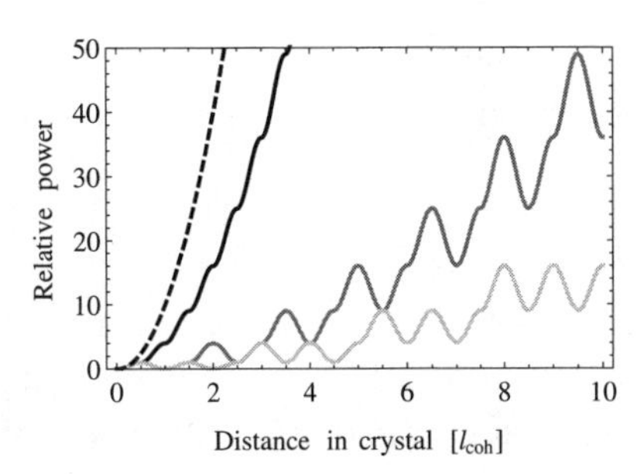

Figure 6.7: Quasi phase matching: **Left** graph without QPM structure. **Right** graphs show first-, third- and fifth-order QPM structures, respectively. Dashed graph corresponds to birefringent phase matching with a tensor element equal to that of the first-order poling.

wavevector for the structure $k_{QPM} = 2\pi/\Lambda$, that can compensate for the dispersion in the material, i.e. a new phase match condition can now be written as

$$\Delta k = k_3 - k_1 - k_2 \pm k_{qpm}. \tag{6.15}$$

As the periodicity of the structure is a designable parameter is it possible to invert the sign of the susceptibility at the point where the back-conversion would normally kick in, i.e. after propagation of half a coherence length, corresponding to a full domain length $\Lambda = l_{coh}$, see Figure 6.7.

The poling structure can be designed for different applications, using a periodic structure for a specific frequency mixing process or aperiodic structures to tailor the spectral response e.g. for broadband conversion. Periodic structures can be designed for uneven orders, first-, third- or fifth-order poling, corresponding to 1/2, 3/2 or 5/2 coherence lengths, respectively, as shown in Figure 6.7.

Using quasi phase matching, it is possible to design a system exploiting the highest possible nonlinear coefficient of the susceptibility tensor, as the phase match direction can be chosen according to the tensor element, typically along one of the principle axes of the material with all three fields polarized parallel to each other (Type III phase matching). The effective nonlinear coefficient is reduced by $2/(\pi m)$, where m is the order of the poling.

Example 6.4. *The effective nonlinear coefficient*
Non-critical phase matching is possible for second-harmonic generation of 1064 nm laser light using temperature tuning of $LiNbO_3$, having the two fundamental fields ordinarly polarized and the second-harmonic field extraordinarly polarized, see Figure 6.6, exploiting the $\chi^{(2)}_{xxz} = 4.35$ pm/V tensor element.

Moving to quasi-phase matching, all three fields are polarized along the extraordinary axis of the material giving the $\chi^{(2)}_{zzz} = 27$ pm/V tensor element. The nonlinear coupling is reduced due to the quasi-phase matching, for first-order poling structure the efficient nonlinear coefficient becomes

$$\chi_{\text{eff}}^{(2)} = \frac{2}{\pi m}\chi_{zzz}^{(2)} = 17.2 \text{ pm/V}.$$

Even though the coefficient is reduced due to the quasi-phase matching, the access to the largest tensor element more than compensates for the reduction. Significant variations exist in published values for the effective nonlinear coefficient. Comparing data from various commercial suppliers a typical value of the effective nonlinear coefficient of periodically poled lithium niobate (PP:LN) is approx. 14 pm/V.

—— ■ ——

6.4 Second-harmonic generation (SHG) $\chi_{ijk}^{(2)}(-2\omega;\omega,\omega)$

Second-harmonic generation is, as mentioned in some of the examples, a special case of three wave mixing, where the two input fields are degenerate, i.e. two photons of the same frequency ω, each with the energy $\hbar\omega$ combine to create one photon with energy $2\hbar\omega$

$$\hbar\omega + \hbar\omega \rightarrow 2\hbar\omega. \tag{6.16}$$

In order to evaluate second-harmonic generation using coupled wave theory, we use $\omega_1 = \omega_2 = \omega$ and $\omega_3 = 2\omega$. Assuming plane waves and zero input at the second-harmonic wavelength $|a_{2\omega}(0)|^2 = 0$, the Manley-Rowe relation Eqn. (5.50a) yields the following form

$$|a_{2\omega}(z)|^2 = \tfrac{1}{2}|a_\omega(0)|^2 - \tfrac{1}{2}|a_\omega(z)|^2 \Rightarrow \phi_{2\omega}(z) = \tfrac{1}{2}\phi_\omega(0) - \tfrac{1}{2}\phi_\omega(z). \tag{6.17}$$

Having only one fundamental input field a_ω, the coupled wave equations reduce to two equations

$$\frac{da_\omega}{dz} = iK(-\omega;2\omega,-\omega)\,g a_{2\omega}a_\omega^* \exp(i\Delta kz) \tag{6.18a}$$

$$\frac{da_{2\omega}}{dz} = iK(-2\omega;\omega,\omega)\,g a_\omega a_\omega \exp(-i\Delta kz). \tag{6.18b}$$

The gain parameter is defined as before: $g^2 = \left(\varepsilon_0\chi_{\text{eff}}^{(2)}\right)^2 \eta^3\hbar\omega^3$, now written in terms of the fundamental frequency, ω. It is, however, important to note that the prefactor K is different in the two equations, as the number of distinct permutations differ

$$K(-\omega;2\omega,-\omega) = 2^{l+m-n}\cdot p = 2^{1+0-2}\cdot 2 = 1 \tag{6.19a}$$

$$K(-2\omega;\omega,\omega) = 2^{l+m-n}\cdot p = 2^{1+0-2}\cdot 1 = \tfrac{1}{2}. \tag{6.19b}$$

The general form of the coupled wave equations for three-wave mixing Eqn. (6.11) is now cast into the special case of second-harmonic generation

$$\frac{da_\omega}{dz} = iga_{2\omega}a_\omega^* \exp(i\Delta kz) \tag{6.20a}$$

$$\frac{da_{2\omega}}{dz} = \tfrac{i}{2}ga_\omega a_\omega \exp(-i\Delta kz). \tag{6.20b}$$

6.4.1 Conversion efficiencies of second-harmonic generation

The coupled equations can now be solved in order to find the conversion efficiency of the second-harmonic generation process, first considering the special case of perfect phase matching, $\Delta k = 0$.

Taking the derivative of Eqn. (6.20a) and inserting this into Eqn. (6.20b) gives

$$\frac{d^2a_\omega}{dz^2} = iga_\omega^* \frac{da_{2\omega}}{dz} + iga_{2\omega}\frac{da_\omega^*}{dz}. \tag{6.21}$$

It is then straightforward to derive the following equation

$$\frac{d^2a_\omega}{dz^2} = -\frac{g^2}{2}|a_\omega(z)|^2 a_\omega(z) + g^2|a_{2\omega}(z)|^2 a_\omega(z). \tag{6.22}$$

Using the Manley-Rowe relations for the second-harmonic flux densities gives the following differential equation

$$\frac{d^2a_\omega}{dz^2} = -g^2 a_\omega(z)\left(|a_\omega(z)|^2 - \tfrac{1}{2}|a_{2\omega}(0)|^2\right). \tag{6.23}$$

A differential equation of the form $\frac{d^2f(x)}{dx^2} = -2\frac{b^2}{a^2}f^3(x) + b^2 f(x)$ is known to have a solutions of the form $f(x) = a\,\text{sech}(bx)$. Hence, the solution to Eqn. (6.23) can be written as

$$a_\omega(z) = a_\omega(0)\,\text{sech}\left(\sqrt{\tfrac{1}{2}}ga_\omega(0)\,z\right). \tag{6.24}$$

Multiplying by the complex conjugate, $a_\omega^*(z)$, the photon flux density $\phi_\omega(z) = |a_\omega(z)|^2$ is found

$$\phi_\omega(z) = \phi_\omega(0)\,\text{sech}^2\left(\sqrt{\tfrac{1}{2}}ga_\omega(0)\,z\right). \tag{6.25}$$

Applying the Manley-Rowe relations and using the identity, $\text{sech}^2(x) = 1 - \tanh^2(x)$, the following equation can be written for the generated second-harmonic flux density

$$\phi_{2\omega}(z) = \frac{\phi_\omega(0)}{2}\tanh^2\left(\sqrt{\tfrac{1}{2}}ga_\omega(0)\,z\right). \tag{6.26}$$

Multiplying with the photon energy the photon flux densities are converted to intensities. The second-harmonic conversion efficiency defined as the generated second-harmonic intensity divided by the fundamental input intensity, γ_{shg}, is now found according to

$$\gamma_{\mathrm{shg}} = \frac{I_{2\omega}(z)}{I_{\omega}(0)} = \tanh^2\left(\sqrt{\tfrac{1}{2}}g a_{\omega}(0)\, z\right). \tag{6.27}$$

6.4.2 SHG conversion efficiency in the weak coupling regime

When considering continuous wave frequency conversion, the weak coupling approximation is often applied, where the fundamental beam is to be considered non-depleted. In this limit a first-order Taylor expansion of the trigonometric function is used, $\tanh^2(x) \approx x^2$

$$\gamma_{\mathrm{shg}} = \tanh^2\left(\sqrt{\tfrac{1}{2}}g a_{\omega}(0)\, z\right) \approx \tfrac{1}{2}g^2 \phi_{\omega}(0)\, z^2. \tag{6.28}$$

Inserting the parametric gain g, into the equation for the conversion efficiency in the weak coupling regime, and considering a nonlinear crystal with a total interaction length L, gives

$$\gamma_{\mathrm{shg}} \approx \tfrac{1}{2}\left(\varepsilon_0 \chi_{\mathrm{eff}}^{(2)}\right)^2 \eta^3 \hbar \omega^3 \phi_{\omega}(0)\, L^2. \tag{6.29}$$

Rewriting in terms of intensities leads to the final equation for the conversion efficiency

$$\gamma_{\mathrm{shg}} \approx \frac{2\left(\chi_{\mathrm{eff}}^{(2)}\right)^2 \pi^2}{c\varepsilon_0 n_{\omega}^2 n_{2\omega} \lambda^2} I_{\omega}(0)\, L^2 = \frac{8 d_{\mathrm{eff}}^2 \pi^2}{c\varepsilon_0 n_{\omega}^2 n_{2\omega} \lambda^2} I_{\omega}(0)\, L^2. \tag{6.30}$$

It is noted that the right-hand side of Eqn. (6.30) refers to a commonly used notation where $d_{\mathrm{eff}} = \frac{1}{2}\chi_{\mathrm{eff}}^{(2)}$, see Section 4.5.1.

Example 6.5. *Weak coupling regime of second-harmonic generation*
The weak coupling assumption is validated considering second-harmonic generation in a 10 mm long $LiNbO_3$ crystal. Assuming a top-hat field (rectangular dist.), with a power of 10 W in a beam diameter of 100 μm, the conversion efficiency can be evaluated as

$$\begin{aligned}
\gamma_{\mathrm{shg}} &\approx \frac{2\left(\chi_{\mathrm{eff}}^{(2)}\right)^2 \pi^2}{c\varepsilon_0 n_{\omega}^2 n_{2\omega} \lambda^2} I_{\omega}(0)\, L^2 \\
&= \frac{2\left(\chi_{\mathrm{eff}}^{(2)}\right)^2 \pi^2}{3\cdot 10^8\ \mathrm{m/s}\cdot 8.85\cdot 10^{-12}\ \mathrm{F/m}\cdot 2.2^3\cdot (1064\cdot 10^{-9}\ \mathrm{m})^2} \\
&\quad \cdot \frac{10\ \mathrm{W}}{\pi\,(100\cdot 10^{-6}\ \mathrm{m})^2}\cdot (0.01\ \mathrm{m})^2 \\
&= \left(\chi_{\mathrm{eff}}^{(2)}\right)^2 \cdot 1.96\cdot 10^{19}\ \mathrm{V^2/m^2}.
\end{aligned}$$

Inserting the nonlinear coefficient for non-critical phase matching in bulk material ($\chi_{\text{eff}}^{(2)} = 4.35$ pm/V) and quasi-phase matching ($\chi_{\text{eff}}^{(2)} = 17.2$ pm/V) gives a conversion efficiency of $\gamma_{\text{shg}}^{\text{NCPM}} = 0.04$ % and $\gamma_{\text{shg}}^{\text{QPM}} = 0.58$ %, respectively. This proves the validity of the weak coupling (non-depleted) approximation for most continuous wave frequency conversion systems.

—— ■ ——

6.5 Non-degenerate parametric frequency conversion

Returning to the more general form of three wave mixing, where the three interaction wavelengths are non-degenerate $\omega_1 < \omega_2 < \omega_3$, we revert to the three coupled wave equations Eqn. (6.11). First the prefactor K needs to be calculated, in this case $K = 1$ in all three equations. Assuming perfect phase matching $\Delta k = 0$ gives the following equations

$$\frac{da_{\omega 1}}{dz} = i g a_{\omega_3} a_{\omega 2}^{*} \tag{6.31a}$$

$$\frac{da_{\omega 2}}{dz} = i g a_{\omega_3} a_{\omega 1}^{*} \tag{6.31b}$$

$$\frac{da_{\omega 3}}{dz} = i g a_{\omega_1} a_{\omega 2}, \tag{6.31c}$$

where the gain parameter is defined as $g = \varepsilon_0 \chi_{\text{eff}}^{(2)} \sqrt{\frac{1}{2}\eta^3 \hbar \omega_1 \omega_2 \omega_3}$.

In the following subsections sum-frequency mixing and difference-frequency generation will be considered.

6.5.1 Sum-frequency generation $\chi_{ijk}^{(2)}(-\omega_3; \omega_1, \omega_2)$

In the general form of sum-frequency generation where two fundamental fields are oscillating at different-frequencies, all three coupled equations need to be considered. However, one of the fields is often much stronger than the others. If that is the case the stronger field can be assumed non-depleted, i.e. the derivative with respect to propagation is set to zero. Defining the strong input field $a_{\omega_2}(0)$ to be real, the coupled equations can be written in the following form

$$\frac{da_{\omega 1}}{dz} = i g_{\text{sfg}} a_{\omega 3} \tag{6.32a}$$

$$\frac{da_{\omega 2}}{dz} = 0 \tag{6.32b}$$

$$\frac{da_{\omega 3}}{dz} = i g_{\text{sfg}} a_{\omega 1}, \tag{6.32c}$$

where $g_{\text{sfg}} = g a_{\omega_2}(0)$.

Taking the derivative of the Eqn. (6.32a) and inserting it into Eqn. (6.32c) the following second-order differential equation is obtained

$$\frac{d^2 a_{\omega_1}}{dz^2} = i g_{\mathrm{sfg}} \frac{d a_{\omega 3}}{dz} = -g_{\mathrm{sfg}}^2 a_{\omega_1}. \tag{6.33}$$

The solution to Eqn. (6.33) is known to be of the form $a_{\omega_1} = c_1 \cos(g_{\mathrm{sfg}} z) + c_2 \sin(g_{\mathrm{sfg}} z)$. Considering the boundary conditions, the derivative of a_{ω_1} at $z = 0$ needs to be zero, as $a_{\omega_3}(0) = 0$ i.e. $a_{\omega_1}(z) = a_{\omega_1}(0)\cos(g_{\mathrm{sfg}} z)$.

The photon flux through the nonlinear material can thus be described by the following set of equations

$$\phi_{\omega_1}(z) = \phi_{\omega_1}(0)\cos^2(g_{\mathrm{sfg}} z) \tag{6.34a}$$

$$\phi_{\omega_3}(z) = \phi_{\omega_1}(0)\sin^2(g_{\mathrm{sfg}} z). \tag{6.34b}$$

Multiplying with the photon energy the intensities of the two fields are found

$$\frac{I_{\omega_3}(z)}{I_{\omega_1}(0)} = \frac{\hbar\omega_3 \phi_{\omega_1}(0)\sin^2(g_{\mathrm{sfg}} z)}{\hbar\omega_1 \phi_{\omega_1}(0)} = \frac{\omega_3}{\omega_1}\sin^2(g_{\mathrm{sfg}} z). \tag{6.35}$$

Finally, considering the weak coupling regime the *sine*-function can be Taylor expanded and the final result for non-depleted phase matched sum-frequency mixing in the weak coupling regime is found

$$\gamma_{\mathrm{sfg}} = \frac{I_{\omega_3}(z)}{I_{\omega_1}(0)} = \frac{\omega_3}{\omega_1} g^2 \phi_{\omega_2}(0) L^2 = \frac{1}{2}\eta^3 \omega_3^2 \left(\varepsilon_0 \chi_{\mathrm{eff}}^{(2)}\right)^2 I_{\omega_2}(0) L^2. \tag{6.36}$$

It is seen that the conversion efficiency from ω_1 to ω_3 is proportional to the intensity at ω_2, $I_{\omega_3} = \gamma_{\mathrm{sfg}} I_{\omega_1}$. This is in contrast to second-harmonic generation, where the power at the up-converted wavelength scales quadratically with the fundamental power at ω_1.

Example 6.6. *Sum-frequency generation in periodically poled KTP*
A nonlinear material frequently used for quasi-phase matched sum-frequency mixing is KTP. Expanding on Example 6.1, yellow light generation has been demonstrated in a 9 mm long PP:KTP crystal by sum-frequency mixing of a strong 1342 nm laser (45 W) and a weaker 1064 nm laser (1.8 W).

Using the Sellmeier coefficients for KTP and with all three interacting fields polarized along the z-axis of the material, the first-order period of the QPM structure can be calculated to be 12.65 μm,. Using $\chi_{zzz}^{(2)} = 17$ pm/V, an efficient nonlinear coefficient of a first-order poling structure can be calculated: $\chi_{\mathrm{eff}}^{(2)} = \chi_{zzz}^{(2)} \cdot 2/\pi = 11$ pm/V. Using a refractive index of $n = 1.8$ the coupling coefficient can be calculated $g^2 = 36 \cdot 10^{-27}$ s, similar to Example 6.1.

Inserting this into Eqn. (6.36) using a top-hat beam with a radius of 35 μm and a length of the nonlinear material of $L = 9$ mm, gives

$$\begin{aligned}\gamma_{\text{sfg}} &= \frac{\omega_3}{\omega_1} g^2 \phi_{\omega_2}(0) L^2 = \frac{\lambda_1}{\lambda_3} \frac{g^2 P_2(0) L^2}{h\nu_2 \cdot A} \\ &= \frac{1064 \cdot 10^{-9} \text{ m}}{593.5 \cdot 10^{-9} \text{ m}} \cdot \frac{36 \cdot 10^{-27} \text{ s} \cdot 45 \text{ W} \cdot (0.009 \text{ m})^2}{6.62 \cdot 10^{-34} \text{J} \cdot \text{s} \cdot \frac{3 \cdot 10^8 \text{ m/s}}{1342 \cdot 10^{-9} \text{ m}} \cdot \pi (35 \cdot 10^{-6} \text{ m})^2} = 0.41.\end{aligned}$$

This means that 41 % of the 1064 nm laser power will be up-converted to the yellow, as long as the 1342 nm laser can be assumed non-depleted. A 1064 nm power of 1.8 W therefore results in approx. 0.74 W of 593.5 nm light, assuming perfect phase matching and overlap of top-hat waves [53].

— ■ —

Although sum-frequency mixing is commonly used with fundamental Gaussian beams, it can also be applied for up-conversion of both spatially coherent [54] and spatially incoherent [55] fields.

6.6 Difference-frequency generation $\chi^{(2)}_{ijk}(-\omega_3; \omega_1, -\omega_2)$

Difference-frequency generation or parametric down-conversion is a process very similar to sum-frequency mixing, except that the power flows from the high frequency (high energy) pump field to the low frequency fields, i.e. a high energy pump photon, $\hbar\omega_3$, is divided into two low energy photons, $\hbar\omega_1$ and $\hbar\omega_2$. The two low frequency fields are typically termed the "signal" and "idler" fields, respectively.

Parametric down-conversion exists in three different forms, difference-frequency mixing, optical parametric amplification, and optical parametric oscillation, depending on the setup and application.

In difference-frequency generation a pump and a signal field is mixed to generate an idler field. This process can be used for generation of coherent mid-infrared light. While the idler is generated, it is clear from the Manley-Rowe relations that the signal field is also being amplified. The generation of an idler photon requires the generation of a signal photon due to energy conservation, i.e. generation of the difference-frequency further results in amplification of the signal field. This is often called parametric amplification and is used to amplify signals where no traditional laser amplifiers are available, similar to e.g. Raman amplifiers as described in Chapter 7.

Using the same approach as in the previous section the starting point is the three coupled equations Eqn. (6.31). Assuming the pump field a_{ω_3} to be much stronger than the signal field, the pump can be assumed non-depleted

$$\frac{da_{\omega_1}}{dz} = ig_{\text{dfg}} a_{\omega_3} \tag{6.37a}$$

$$\frac{da_{\omega_2}}{dz} = ig_{\text{dfg}} a_{\omega_3} \tag{6.37b}$$

$$\frac{da_{\omega_3}}{dz} = 0, \tag{6.37c}$$

where $g_{\text{dfg}} = g a_{\omega_3}(0)$.

The coupled equations can once again be solved, leading to the following solutions for the signal and idler fluxes, respectively

$$\phi_{\omega_1}(z) = \phi_{\omega_1}(0) \cosh^2(g_{\text{dfg}} z) \tag{6.38a}$$

$$\phi_{\omega_2}(z) = \phi_{\omega_1}(0) \sinh^2(g_{\text{dfg}} z) . \tag{6.38b}$$

As seen from the above equations the photon flux at both the signal and idler wavelengths increases, as the fields propagate through the nonlinear material. Hence, parametric devices can be used to coherently amplify the input signal or generate a new signal at the difference-frequency of the two input fields.

Example 6.7. *Difference-frequency generation in periodically poled* $LiNbO_3$
As an example of difference-frequency generation a pump source at 1064 nm is considered mixed with a 1550 nm signal field. From the energy conservation it is evident that the generated idler signal is oscillating at a wavelength of $\lambda_2 = 1/(1/\lambda_3 - 1/\lambda_1) = 3393$ nm.

For this mixing process periodically poled $LiNbO_3$ can be used, with a QPM structure of 30.5 µm, giving an effective nonlinear coefficient of $\chi_{\text{eff}}^{(2)} = 17.2$ pm/V and a refractive index of $n = 2.2$, resulting in a coupling coefficient of

$$\begin{aligned} g^2 &= \left(\varepsilon_0 \chi_{\text{eff}}^{(2)}\right)^2 \tfrac{1}{2} \eta^3 \hbar \omega_1 \omega_2 \omega_3 \\ &= (8.85\ \text{pF/m} \cdot 17.2\ \text{pm/V})^2 \frac{1}{2} \left(\frac{377\ \Omega}{2.2}\right)^3 \frac{1.05 \cdot 10^{-34}\ \text{J} \cdot \text{s}\ \left(2\pi \cdot 3 \cdot 10^8\ \text{m/s}\right)^3}{1064\ \text{nm} \cdot 1550\ \text{nm} \cdot 3393\ \text{nm}} \\ &= 7.3 \cdot 10^{-27}\ \text{s}. \end{aligned}$$

Assuming a 100 W 1064 nm laser power with a top-hat beam radius of 50 µm the gain is calculated

$$\begin{aligned} g_{\text{dfg}}^2 &= g^2 \cdot \phi_{\omega_3}(0) = g^2 \cdot \frac{P_3(0)}{h\nu_2 \cdot A} \\ &= \frac{7.3 \cdot 10^{-27}\ \text{s} \cdot 100\ \text{W}}{6.62 \cdot 10^{-34} \text{J} \cdot \text{s} \cdot \frac{3 \cdot 10^8\ \text{m/s}}{1064 \cdot 10^{-9}\ \text{m}} \cdot \pi \cdot (50 \cdot 10^{-6}\ \text{m})^2} = 500\ \text{m}^{-2}. \end{aligned}$$

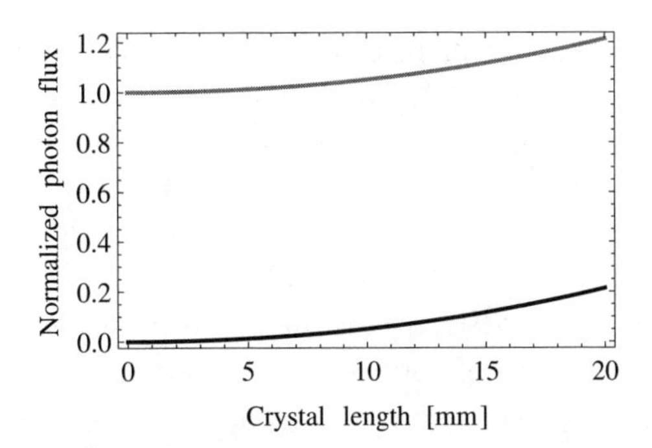

Figure 6.8: Considering difference-frequency generation and parametric amplification, the photon flux at the signal (gray) and idler (black) frequencies, respectively.

It is now possible to plot Eqn. (6.38a) and (6.38b). Assuming a signal power of 1 μW at 1550 nm, the change in photon flux throughout the crystal is plotted in Figure 6.8.

As seen from the figure, a gain of 20 % is obtained for the 1550 nm beam, while the same increase in photon flux is generated at the idler frequency, corresponding to a power of approx. 0.1 μW at 3.4 μm.

—— ■ ——

6.6.1 Optical parametric oscillators

So far we have considered systems with two inputs. However, it is known from quantum mechanics that vacuum fluctuations give rise to a field corresponding to having half a photon in any allowed state. This can be used to explain spontaneous emission in lasers, and similarly this is the origin of spontaneous parametric down-conversion in three-wave mixing, i.e. a pump photon can spontaneously be converted to a signal and an idler photon. Now having a non-zero field at all three wavelengths the process of parametric amplification described in the previous section can occur, and an output signal can be measured, even though only the pump frequency component is present at the input [56].

If furthermore a resonator provides feedback of the signal and/or idler field, a coherent signal is generated, if the parametric gain at the signal and idler wavelengths exceeds the round trip losses of the resonator. Such a system is commonly referred to as an optical parametric oscillator (OPO). An OPO is analogous to a laser, however, the gain process relies on the nonlinear interactions rather than population inversion and stimulated emission. The wavelength control of OPOs is obtained through control of the phase matching condition of the nonlinear interaction rather than by an atomic transition.

6.7 Frequency conversion of focused Gaussian beams*

In all of the above it has been assumed that the interacting fields can be described using plane-wave approximations. However, it has also been implicitly assumed that all fields are well confined in the transverse direction. When dealing with nonlinear interactions the limited aperture of the nonlinear material is important since the cross section of the field as well as the phase shift depends on the beam focus. In this section the interaction of focused beams with a transverse Gaussian field distribution will be considered. The field distribution of a Gaussian beam with its beam waist w_0 located at $z = 0$ propagating in the direction of the z-axis is defined as [24]

$$E^0(x,y,z) = E_0^0\left(\frac{w_0}{w(z)}\right)\exp\left(-\frac{x^2+y^2}{w^2(z)}\right)\exp\left(ikz + ik\frac{x^2+y^2}{2R(z)} - i\zeta(z)\right), \tag{6.39}$$

where the field amplitude E_0^0 is related to the on-axis intensity I_0 through Eqn. (6.10), which is related to the beam power as: $I_0 = \frac{2P}{\pi w_0^2}$. Furthermore, $\zeta(z)$ is the so-called Gouy phase shift. Using the Rayleigh range defined as $2z_0 = w_0^2 k$, the beam size and radius of curvature of the phase front is given as a function of the z-position as

$$w(z) = w_0\sqrt{1 + (z/z_0)^2} \tag{6.40a}$$

$$R(z) = z\left(1 + (z_0/z)^2\right). \tag{6.40b}$$

As seen in the previous sections, the frequency conversion efficiency depends on the fundamental electric fields, i.e. for a given fundamental power the conversion efficiency depends on the beam focusing. Intuitively it is clear that a small beam size in the nonlinear material increases the intensity, resulting in an increased conversion efficiency. However, there is a limit to the focusing, as a small spot size results in a short depth of focus of the beam, giving a small radius of curvature of the phase fronts of the field, i.e. a diverging pencil of wavevectors within the nonlinear material. This implies that an optimal beam size exists for a given length of the nonlinear material.

The converted power can be evaluated as an integral of the induced nonlinear polarization in the material. In 1968 Boyd and Kleinman [57] presented a theory for the dependence of the SHG power on the focussing parameter $\xi = L/b$, in a nonlinear uniaxial crystal. Assuming Type I phase matching in a lossless material, without depletion of the fundamental field, an efficiency factor $h(\sigma,\beta,\kappa,\xi,\mu)$ which is often called the Boyd-Kleinman factor can be calculated as

$$h(\sigma,\beta,\kappa,\xi,\mu) = \frac{1}{4\xi}\iint_{-\xi(1+\mu)}^{\xi(1+\mu)} \frac{e^{i\sigma(\tau-\tau')}e^{-\beta^2(\tau-\tau')^2}}{(1+i\tau)(1-i\tau)}d\tau d\tau', \tag{6.41}$$

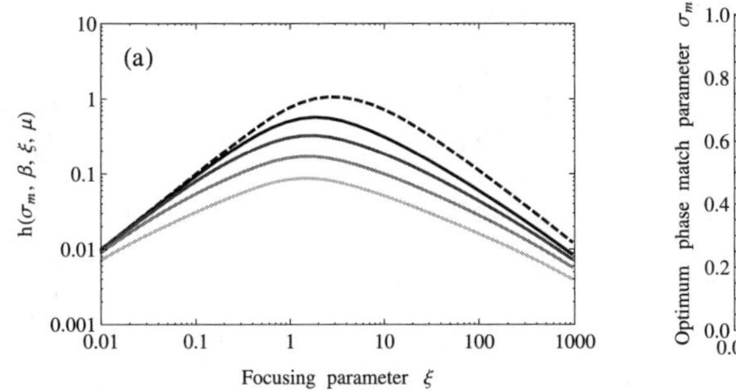

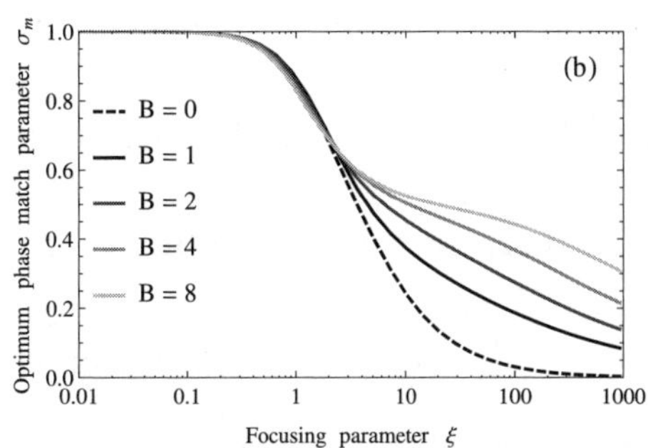

Figure 6.9: Plots of the Boyd-Kleinman efficiency and optimum phase match parameter. The curves are parameterized according to $B = 0, 1, 2, 4, 8$ from the top.

where $b = 2z_0 = \omega_0^2 k_1$ is the confocal parameter, $k_1 = 2\pi n_1/\lambda$ is the fundamental wave number and ω_0 is the beam waist in the nonlinear material. The walk-off parameter is defined as $B = \beta\sqrt{\xi} = \rho\sqrt{Lk_1}/2$, where ρ is the walk-off angle, i.e. the angle between the Poynting vectors corresponding to the ordinary and the extraordinary waves and L is the crystal length. μ is a parameter describing the position of the beam waist relative to the center of the nonlinear material. $\Delta k = 2k_1 - k_2$ is the phase mismatch as described in Section 6.3 and $\sigma = b\Delta k/2$ is a phase matching parameter which takes the local phase of the focused Gaussian fields into account.

Evaluation of the Boyd-Kleinman factor means solving the above integral while optimizing the phase mismatch parameter σ. Figure 6.9 shows the evaluated integral as a function of the focusing parameter (a) and the optimal phase mismatch parameter (b).

The highest efficiency is found for $B = 0$, i.e. no walk-off. As the walk-off increases the highest efficiency drops significantly and the crystal length for optimal efficiency decreases. For $B = 8$ the optimum focusing parameter is found to be $\xi = 1.39$ compared to $\xi = 2.84$ for $B = 0$, and the maximum efficiency has dropped by more than a decade. Systems are often designed for non-critical phase matching i.e. to avoid walk-off [58].

A better understanding of the origin of the optimum phase mismatch can be gained by inserting the Gaussian field from Eqn. (6.39) into the coupled wave equations Eqn. (6.20). This results in an induced polarization proportional to the square of the fundamental Gaussian field. Calculating the square of the Gaussian field in Eqn. (6.39) results in a new Gaussian field distribution which oscillates at twice the frequency, but with a beam waist reduced by a factor of $\sqrt{2}$ compared to the fundamental field and a phase term $\varphi_j(x, y, z)$, where the index j refers to the fundamental or second-harmonic field, respectively

$$[\exp(i\varphi_1(x,y,z))]^2 = \exp\left(i2k_1 z + i2k_1 \frac{x^2+y^2}{2R_1(z)} - i2\zeta_1(z)\right). \quad (6.42)$$

The corresponding term for the generated second-harmonic field can be written as

$$\exp(i\varphi_2(x,y,z)) = \exp\left(ik_2 z + ik_2 \frac{x^2+y^2}{2R_2(z)} - i\zeta_2(z)\right). \quad (6.43)$$

Considering the fundamental field and the generated second-harmonic field, it is clear that the beam waist is located at the same position. The second-harmonic field is oscillating at twice the frequency, which means that the beam size of the fundamental and the generated second-harmonic maintain a constant ratio: $w_1(z) = \sqrt{2}w_2(z)$. Furthermore, the wavefront curvatures for the squared fundamental field and the induced polarization are identical. This means that the electric field added to the second-harmonic field in any given slab of the crystal has the same beam propagation parameters as the second-harmonic field generated in other slabs of the nonlinear crystal, except for a possible phase mismatch originating from the Gouy phase shift difference between the induced polarization and the actual field at that second-harmonic frequency, $\zeta_1(z) = \zeta_2(z) \neq 2\zeta_2(z)$. This is the origin of the necessary optimization of phase mismatch $\sigma_m \neq 0$ for nonlinear interaction of focused Gaussian beams as seen in the Boyd-Kleinman integral of Figure 6.9.

Assuming that both the induced polarization and the second-harmonic field are Gaussian because their propagation parameters are identical except for a phase shift, it is enough to know the electric field on-axis in order to calculate the total generated second-harmonic field, which is the sum of the fields generated in each slab of the crystal propagated with the appropriate phase and beam size scaling to a given reference plan. This is illustrated in Figure 6.10.

Figure 6.10 (a) shows the on-axis generated field in each slab of the crystal, propagated to the center of the crystal; the graphs are normalized to the field for optimum focusing of $\xi = 2.84$ (black). The graph for stronger focusing $\xi = 11.36$ (lightgray) corresponds to half the spot size compared to optimal focusing and for weaker focusing $\xi = 0.71$ (darkgray) corresponds to a spot size of twice the optimal focusing. Figure 6.10 (b) shows the Gouy phase shift (solid) and the optimal phase shift σ_m (dashed). The graphs in Figure 6.10 (c) show the integrated power through the crystal for optimum phase match, whereas Figure 6.10 (d) shows the second-harmonic power generated in the crystal as a function of the phase mismatch.

As a consequence of different Gouy phase shifts, second-harmonic generation can be applied for mode conversion, i.e. a TEM_{10} pump beam can be converted to a TEM_{00} or a TEM_{02} at the second-harmonic frequency depending on the phase match condition [59].

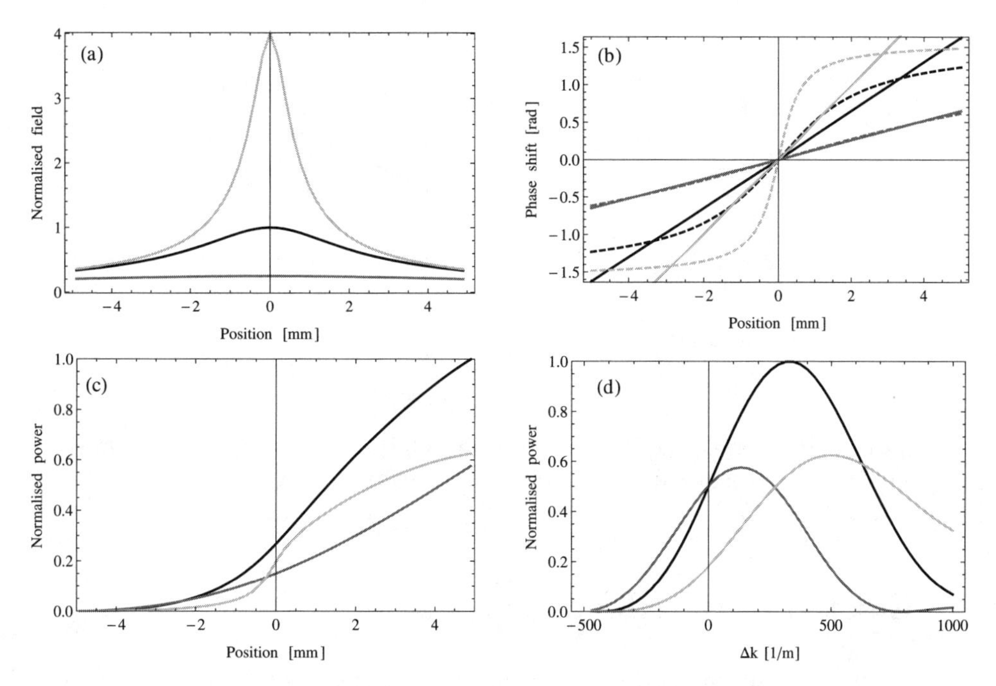

Figure 6.10: Considering different focusing ξ=11.36 (light gray), ξ=2.84 (black), ξ=0.71 (dark gray) (a) shows the on-axis generated field in each slab, (b) is the phase relation between the induced polarization and the field at the second-harmonic frequency, (c) is the power accumulated through the crystal and (d) is the power generated through the crystal as a function of the phase mismatch.

It is noted, that the analysis of second-harmonic generation is based on a non-depleted approximation, as depletion of the fundamental field would result in mode degradation, meaning that the fundamental field would not maintain its Gaussian distribution throughout the crystal length. The approach described here is, however, very general and can be applied even for higher-order transverse modes, taking the appropriate Gouy phase shifts into account. This means that pump depletion can be included in the calculation, but it would require the modes to be expanded in higher-order transverse modes after each slab of interaction in the nonlinear material.

In order to compare the plane-wave approximation and the Boyd-Kleinman approach, Figure 6.11 shows the generated power in a second-harmonic crystal as a function of crystal lengths for fixed beam size of $w_0 = 20$ μm in KTP ($n = 1.8$). The power is normalized to unity for optimum focusing $\xi = 2.84$. As seen, the two graphs give similar results for weak focusing, i.e. short crystal lengths compared to the depth of focus. However, the plane-wave approximation fails for longer crystals approaching the tightly focused regime.

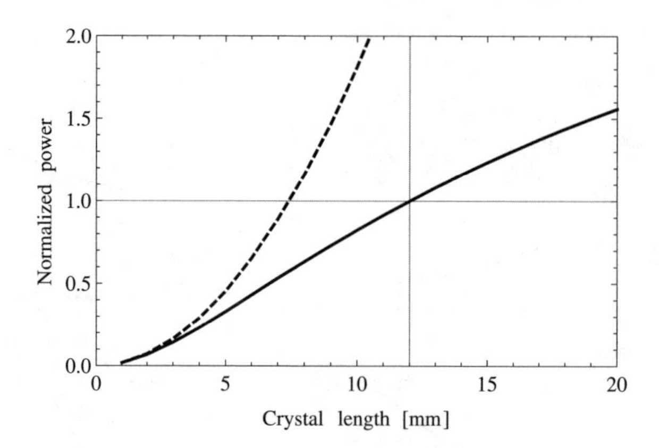

Figure 6.11: Comparing Boyd-Kleinmain (solid) and plane-wave approximation (dashed) normalized second-harmonic generation as a function of crystal length for fixed beam size.

6.8 Electro optic effects $\chi^{(2)}_{ijk}(-\omega_1; 0, \omega_1)$

Another important second-order nonlinear effect is the electro optic effect [4]. The electro optic effect is a phenomena where an external electrical field, a control signal, is applied across a crystal to modify the optical properties of the crystal as experienced by light propagating through the crystal. The frequency of the control signal is much lower than the optical frequency of the light.

The electro optic effect is sometimes called the linear electro optic effect or the Pockels effect. This is used for $\chi^{(2)}(-\omega; 0, \omega)$, and the term quadratic electro optic effects or DC Kerr effect is used in relation with $\chi^{(3)}(-\omega; 0, 0, \omega)$.

To describe electro optic effects it is customary to use the index ellipsoid. A review of linear propagation of light through an anisotropic crystal based on the index ellipsoid is provided in Appendix F. The index ellipsoid is described not through the susceptibility tensor but rather through the impermeability tensor which is defined through the reciprocal permittivity ($\varepsilon_r = 1 + \chi^{(1)}$).

In the absence of a control signal applied to the crystal the impermeability tensor is

$$\eta_{ij} = \varepsilon_0 / \varepsilon_{ij}. \tag{6.44}$$

When the control signal is applied to the crystal, the impermeability tensor can be expanded in terms of the electric field

$$\eta_{ij}\left(\mathbf{E}^0\right) = \eta_{ij} + \sum_k r_{ijk} E_k^0 + \sum_{kl} s_{ijkl} E_k^0 E_k^0, \tag{6.45}$$

where r_{ijk} is the linear electro optic coefficient. For completeness the quadratic electro optic tensor is included in Eqn. (6.45).

It is noted that rather than using $\chi^{(2)}(-\omega; 0, \omega)$ the electro optic effects are usually described using the electro optic tensor denoted by r_{ijk}. Clearly there is a relation between the second-order susceptibility tensor and the electro optic tensor, but due to material dispersion the relation is not simple.

Typically, when dealing with the electro optic effect, symmetry relations of the permittivity allow for a contracted notation reducing the tensor from 27 to 18 elements, as described in Table 4.2

$$r = \begin{pmatrix} r_{11} & r_{12} & r_{13} \\ r_{21} & r_{22} & r_{23} \\ r_{31} & r_{32} & r_{33} \\ r_{41} & r_{42} & r_{43} \\ r_{51} & r_{52} & r_{53} \\ r_{61} & r_{62} & r_{63} \end{pmatrix}. \tag{6.46}$$

Inserting Eqn.(6.45) into Eqn.(F.3), the index ellipsoid can be written in the principal coordinates of the material as

$$\begin{aligned} 1 = \left(\frac{1}{n_x^2} + r_{1k}E_k^0\right)x^2 + \left(\frac{1}{n_y^2} + r_{2k}E_k^0\right)y^2 + \left(\frac{1}{n_z^2} + r_{3k}E_k^0\right)z^2 \\ + 2yzr_{4k}E_k^0 + 2zxr_{5k}E_k^0 + 2xyr_{6k}E_k^0, \end{aligned} \tag{6.47}$$

where a summation over repeated indices is assumed i.e. a summation over k.

Example 6.8. *Electro optic effect in point group 3m - LiNbO$_3$*
LiNbO$_3$ belongs to the 3m point group of materials and its electro optic tensor elements are

$$r = \begin{pmatrix} 0 & -r_{12} & r_{13} \\ 0 & r_{22} & r_{13} \\ 0 & 0 & r_{33} \\ 0 & r_{51} & 0 \\ r_{51} & 0 & 0 \\ -r_{22} & 0 & 0 \end{pmatrix}.$$

Applying an electric field along the z-axis of the material, the linear term in the electric field can be written as

$$r \cdot \mathbf{E}^0 = \begin{pmatrix} 0 & -r_{12} & r_{13} \\ 0 & r_{22} & r_{13} \\ 0 & 0 & r_{33} \\ 0 & r_{51} & 0 \\ r_{51} & 0 & 0 \\ -r_{22} & 0 & 0 \end{pmatrix} \begin{pmatrix} 0 \\ 0 \\ E_z^0 \end{pmatrix} = \begin{pmatrix} E_z^0 r_{13} \\ E_z^0 r_{13} \\ E_z^0 r_{33} \\ 0 \\ 0 \\ 0 \end{pmatrix}.$$

Using Eqn. (6.44) and Eqn. (6.45) gives

$$\eta\left(\mathbf{E}^0\right) = \begin{pmatrix} 1/n_o^2 & 0 & 0 \\ 0 & 1/n_o^2 & 0 \\ 0 & 0 & 1/n_e^2 \end{pmatrix} + \begin{pmatrix} E_z^0 r_{13} & 0 & 0 \\ 0 & E_z^0 r_{13} & 0 \\ 0 & 0 & E_z^0 r_{33} \end{pmatrix}$$

$$= \begin{pmatrix} 1/n_o^2 + E_z^0 r_{13} & 0 & 0 \\ 0 & 1/n_o^2 + E_z^0 r_{13} & 0 \\ 0 & 0 & 1/n_e^2 + E_z^0 r_{33} \end{pmatrix}.$$

The new refractive indices are then found as a function the applied field

$$\frac{1}{n_o^2\left(E_z^0\right)} = \frac{1}{n_o^2} + r_{13} E_z^0 \tag{6.48a}$$

$$\frac{1}{n_e^2\left(E_z^0\right)} = \frac{1}{n_e^2} + r_{33} E_z^0, \tag{6.48b}$$

and by assuming small perturbations, approximative equations can be given for the perturbation of the ordinary and the extraordinary refractive indices

$$n_o\left(E_z^0\right) \approx n_o - \frac{1}{2} n_0^3 r_{13} E_z^0 \tag{6.49a}$$

$$n_e\left(E_z^0\right) \approx n_e - \frac{1}{2} n_e^3 r_{33} E_z^0. \tag{6.49b}$$

This shows that an electric field applied along the z-axis of $LiNbO_3$ changes the refractive index along all three axes according to the above equation, but the material remains uniaxial as indicated in Figure 6.12.

— ■ —

Example 6.9. *Electro optic effect in point group 42m - KDP*
KDP is in the 42m point group of materials and can be described by the following electro optic tensor elements

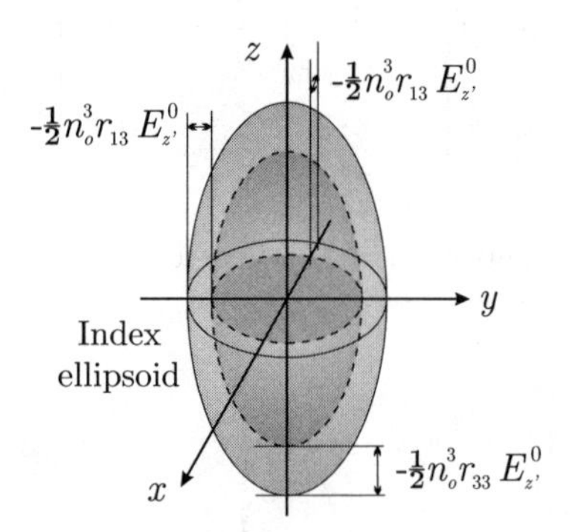

Figure 6.12: The index ellipsoid is shown with (dashed lines) and without (solid lines) an applied field along the principal z-axis for $LiNbO_3$.

$$r = \begin{pmatrix} 0 & 0 & 0 \\ 0 & 0 & 0 \\ 0 & 0 & 0 \\ r_{41} & 0 & 0 \\ 0 & r_{41} & 0 \\ 0 & 0 & r_{63} \end{pmatrix}.$$

KDP is a uniaxial material. Applying an electric field along the principal z-axis of the material gives

$$r \cdot \mathbf{E}^0 = \begin{pmatrix} 0 & 0 & 0 \\ 0 & 0 & 0 \\ 0 & 0 & 0 \\ r_{41} & 0 & 0 \\ 0 & r_{41} & 0 \\ 0 & 0 & r_{63} \end{pmatrix} \begin{pmatrix} 0 \\ 0 \\ E_z^0 \end{pmatrix} = \begin{pmatrix} 0 \\ 0 \\ 0 \\ 0 \\ 0 \\ E_z^0 r_{63} \end{pmatrix} \Rightarrow$$

$$\begin{aligned} \eta\left(\mathbf{E}^0\right) &= \begin{pmatrix} 1/n_o^2 & 0 & 0 \\ 0 & 1/n_o^2 & 0 \\ 0 & 0 & 1/n_e^2 \end{pmatrix} + \begin{pmatrix} 0 & E_z^0 r_{63} & 0 \\ E_z^0 r_{63} & 0 & 0 \\ 0 & 0 & 0 \end{pmatrix} \\ &= \begin{pmatrix} 1/n_o^2 & E_z^0 r_{63} & 0 \\ E_z^0 r_{63} & 1/n_o^2 & 0 \\ 0 & 0 & 1/n_e^2 \end{pmatrix}. \end{aligned}$$

It is now possible to find the new refractive indices as a function the applied electric field by rotating the principal axis by 45° around the z-axis, see Eqn. (4.6) for further details, $\mathbf{e}'_x = (\mathbf{e}_x + \mathbf{e}_y)/\sqrt{2}$, $\mathbf{e}'_y = (\mathbf{e}_x - \mathbf{e}_y)/\sqrt{2}$ and $\mathbf{e}'_z = \mathbf{e}_z$

$$\eta'\left(\mathbf{E}^0\right) = \begin{pmatrix} 1/n_{x'}^2 + E_{z'}^0 r_{63} & 0 & 0 \\ 0 & 1/n_{y'}^2 - E_{z'}^0 r_{63} & 0 \\ 0 & 0 & 1/n_{z'}^2 \end{pmatrix}.$$

The resulting refractive indices in the new coordinate system can now be expressed in terms of the applied electric field

$$\frac{1}{n_{x'}^2\left(E_{z'}^0\right)} = \frac{1}{n_o^2} + r_{63} E_{z'}^0 \tag{6.50a}$$

$$\frac{1}{n_{y'}^2\left(E_{z'}^0\right)} = \frac{1}{n_o^2} - r_{63} E_{z'}^0 \tag{6.50b}$$

$$\frac{1}{n_{z'}^2\left(E_{z'}^0\right)} = \frac{1}{n_e^2}. \tag{6.50c}$$

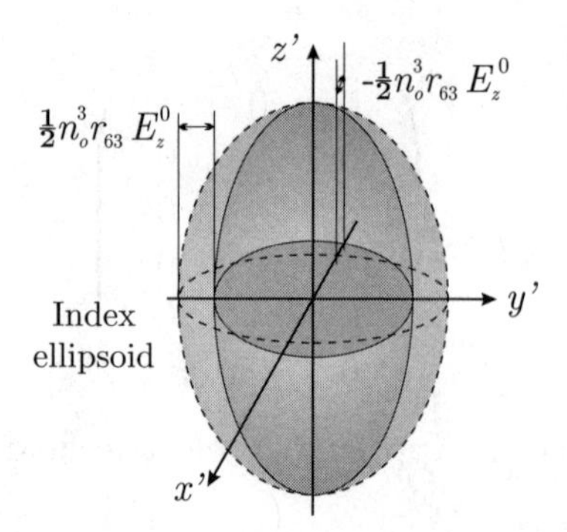

Figure 6.13: The index ellipsoid is shown with (dashed lines) and without (solid lines) an applied field along the principal z-axis for KDP.

A small perturbation approximation is made resulting in the following equations for the perturbation of the refractive indices, now leading to a biaxial material response as seen in Figure 6.13

$$n_{x'}\left(E_{z'}^0\right) \approx n_o - \frac{1}{2}n_0^3 r_{63} E_{z'}^0 \tag{6.51a}$$

$$n_{y'}\left(E_{z'}^0\right) \approx n_e + \frac{1}{2}n_e^3 r_{63} E_{z'}^0 \tag{6.51b}$$

$$n_{z'}\left(E_{z'}^0\right) = n_e. \tag{6.51c}$$

— ■ —

The electro optic effect can be used for optical phase modulators. Assuming an optical field E_ω, polarized along the x-axis of a material, the light will experience a phase shift of

$$\phi = \frac{\omega}{c} n_x\left(\mathbf{E}_m\right) L, \tag{6.52}$$

where L is the length of the electro optic material and $n_x\left(\mathbf{E}_m\right)$ is the refractive index along the x-axis as a function of the applied modulation field $\mathbf{E}_m$.

In a phase modulator the modulation field is not a DC but rather an applied electrical signal with a frequency ω_m, much lower than the optical frequency, i.e. $\mathbf{E}_m = \mathbf{E}_m^0 \sin(\omega_m t)$. To see the result of the applied modulation signal the propagating optical field is written as

$$E_\omega = E_\omega^0 \cos(\omega t - \frac{\omega}{c} n_x\left(n_x\left(\mathbf{E}_m\right)\right) L). \tag{6.53}$$

Introducing a new variable $\delta = \frac{\omega}{c}\Delta n_x\left(\mathbf{E}_m^0\right) L$, where $\Delta n_x\left(\mathbf{E}_m^0\right)$ is the perturbation of the refractive index introduced by the applied electrical signal, then Eqn. (6.53) can be written as

$$E_\omega = E_\omega^0 \cos(\omega t + \delta\, \sin(\omega_m t)). \tag{6.54}$$

The phase modulation is achieved by a modulation of the refractive index, with the modulation signal $\mathbf{E}_m$, resulting in a phase modulation of the optical output

$$\begin{aligned} E_\omega &= E_\omega^0 \left[\cos(\omega t + \delta \sin(\omega_m t))\right] \\ &= E_\omega^0 \left[\cos(\delta \sin(\omega_m t)) \cos(\omega t) - E_\omega^0 \sin(\delta \sin(\omega_m t)) \sin(\omega t)\right]. \end{aligned} \tag{6.55}$$

Example 6.10. *Electro optic phase modulation using $LiNbO_3$*
Returning to Example 6.8 it was found that with a modulation field applied along the principal z-axis of the material $(E_m)_z$, the material remains uniaxial. Using Eqn. (6.49) the perturbation of the ordinary and extraordinary refractive indices is found as a function of the applied field. Assuming an incoming optical field polarized along the x-axis and propagating along the z-axis, the light experiences a phase shift of

$$\phi = \frac{\omega}{c}\left(n_x - \frac{1}{2} n_x^3 r_{13} (E_m^0)_z\right) L,$$

and the modulation, to be inserted into Eqn.(6.55), becomes $\delta = \frac{\omega}{c}\frac{1}{2} n_x^3 r_{13} (E_m^0)_z L$.

— ■ —

6.9 Summary

- The wave equation for three-wave mixing is written as

$$\left(\nabla^2 + \frac{\omega_\sigma^2 n^2}{c^2}\right)\mathbf{E}^0_{\omega_\sigma} = -\mu_0\omega_\sigma^2(\mathbf{P}^0_{\omega_\sigma})^{(2)}. \tag{6.3}$$

Energy conservation Eqn. (6.4) has to be stringently fulfilled, whereas the conservation of momentum Eqn. (6.5) is important for efficient conversion

$$\hbar\omega_\sigma = \hbar\omega_1 + \hbar\omega_2, \tag{6.4}$$

$$\hbar k_\sigma = \hbar k_1 + \hbar k_2. \tag{6.5}$$

- Applying the SVEA and introducing an amplitude parameter proportional to the electric field, three coupled wave equations are derived

$$\frac{da_{\omega_1}}{dz} = iKga_{\omega_3}a^*_{\omega_2}\exp\left(i\Delta kz\right) \tag{6.11a}$$

$$\frac{da_{\omega_2}}{dz} = iKga_{\omega_3}a^*_{\omega_1}\exp\left(i\Delta kz\right) \tag{6.11b}$$

$$\frac{da_{\omega_3}}{dz} = iKga_{\omega_1}a_{\omega_2}\exp\left(-i\Delta kz\right). \tag{6.11c}$$

- In the weak coupling regime the second-harmonic conversion efficiency for plane waves can be written as

$$\gamma_{\text{shg}} \approx \frac{2\left(\chi^{(2)}_{\text{eff}}\right)^2\pi^2}{c\varepsilon_0 n_\omega^2 n_{2\omega}\lambda^2}I_\omega\left(0\right)L^2 = \frac{8d^2_{\text{eff}}\pi^2}{c\varepsilon_0 n_\omega^2 n_{2\omega}\lambda^2}I_\omega\left(0\right)L^2, \tag{6.30}$$

where $I_{2\omega} = \gamma_{\text{shg}} \cdot I_\omega$.

- Similarly for sum-frequency generation the conversion efficiency $I_{\omega_3} = \gamma_{\text{sfg}} \cdot I_{\omega_1}$, where

$$\gamma_{\text{sfg}} = \frac{I_{\omega_3}\left(z\right)}{I_{\omega_1}\left(0\right)} = \frac{\omega_3}{\omega_1}g^2\phi_{\omega_2}\left(0\right)L^2 = \frac{1}{2}\eta^3\omega_3^2\left(\varepsilon_0\chi^{(2)}_{\text{eff}}\right)^2 I_{\omega_2}\left(0\right)L^2, \tag{6.36}$$

and where I_{ω_2} is assumed to be much stronger than the other field i.e. non-depleted.

- Considering the electro optic effect, an applied electric field is seen to perturb the refractive index of the nonlinear material

$$\begin{aligned}1 = \left(\frac{1}{n_x^2} + r_{1k}E^0_k\right)x^2 + \left(\frac{1}{n_y^2} + r_{2k}E^0_k\right)y^2 + \left(\frac{1}{n_z^2} + r_{3k}E^0_k\right)z^2 \\ + 2yzr_{4k}E^0_k + 2zxr_{5k}E^0_k + 2xyr_{6k}E^0_k,\end{aligned} \tag{6.47}$$

where a summation over repeated indices is assumed i.e. a summation over k.

Chapter 7

Raman scattering

Contents

Raman scattering is interaction of light with the material that the light is propagating through, more specifically, interactions with microscopic vibrations and/or rotations within the material at the molecular level. There are several ways to describe and picture the process of Raman scattering. In a simple picture one may imagine that the light sets up dipoles and induces a macroscopic polarization that reradiates light.

An important application of Raman scattering is within spectroscopy, where Raman scattering has been used for many years to identify constituents of a given sample, more specifically the content of specific molecules that can be identified from a measurement of the Raman scattered light.

Within nonlinear fiber optics, the most significant applications of Raman scattering include optical amplifiers, Raman fiber lasers, and temperature sensors. However, Raman scattering also plays a significant role in the propagation of short pulses. Since pulses are getting shorter and shorter in the time domain, they get wider and wider in the frequency domain, and consequently frequency components within the pulse start to interact with each other through Raman scattering.

In the following, Raman scattering is first described from the point of view of physical mechanisms, then propagation equations in amplitude and power are discussed, and finally focus is directed towards applications within optical fiber amplifiers for optical communication, that is, for providing signal gain to signals located around the wavelength: 1555 nm.

7.1 Physical description

The process of Raman scattering may be described as a dipole that is excited through a so-called excitation or pump wave.

Rayleigh scattering is an interaction of light and bound electrons which oscillate at the optical frequency of the excitation wave. This induced oscillating dipole moment produces optical radiation at the same frequency as the incident optical field.

Raman scattering is an interaction between light and the molecular structure which in itself is oscillating. These oscillations are quantized as phonons. Therefore, the induced oscillating dipole moment contains the sum and difference frequency between the optical frequency and the rotational and or vibrational frequencies. These new frequencies are the Raman scattered light, the generation of a lower frequency component than the excitation light, and a phonon, is referred to as the Stokes process whereas generation of a higher frequency component than the excitation light, and absorption of a phonon, is referred to as the anti-Stokes process.

Physically, this may be described through a differential polarizability of the material. The excited dipole oscillates at the difference frequency between the excitation wave and the phonon wave. The radiated power from the dipole is recaptured by the fiber.

In a quantum mechanical approach the Raman process is described through a harmonic oscillator, and the creation of photons and phonons in the Stokes process or annihilation of photons and phonons in the anti-Stokes process. This approach is useful to describe spontaneous emission and the temperature dependence of the Raman process.

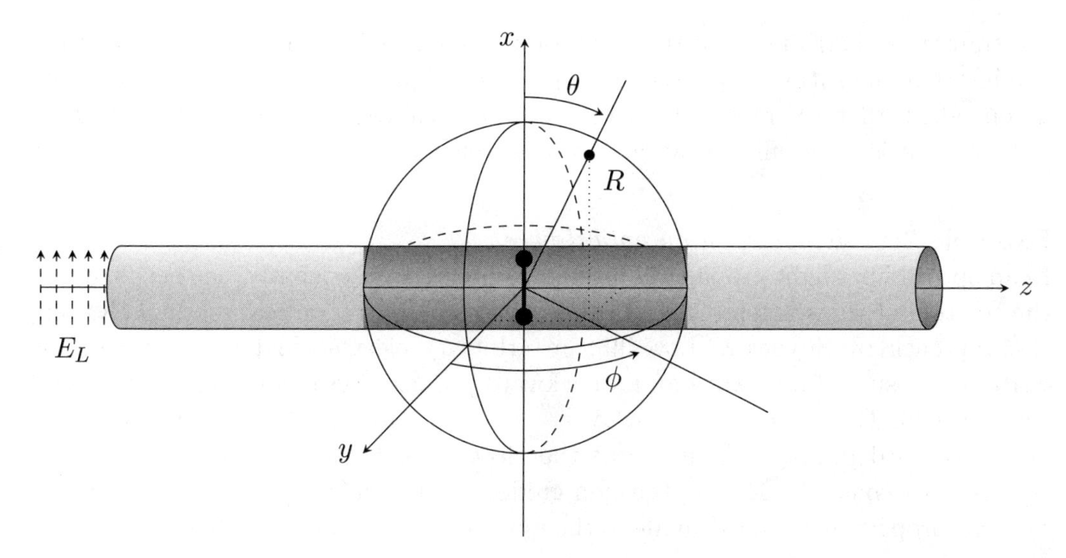

Figure 7.1: A dipole located inside a fiber at the origo and pointing in the x-direction; the electric field E_L interacts with the dipole. The black dot denotes a point on the surface of the sphere, and it has the spherical coordinates (R, θ, ϕ).

7.1.1 Electromagnetic description, applying dipole theory

Figure 7.1 illustrates the excitation of a dipole inside a fiber. The total power radiated by a dipole is

$$P_{\text{dip}} = \frac{\varepsilon_0 n_s \omega_s^4}{3\pi c^3} |\wp|^2 |E_L^0|^2, \tag{7.1}$$

where $|\wp|^2$ is the dipole strength and E_L^0 is the amplitude of the launched electric field. The use of the phrase: total power, reflects the fact that the radiated power P_{dip} has an angular distribution such that there is no radiation along the axis of the dipole and most of the radiation is at an angle 90° from the dipole axis. Since $\varepsilon_0 |E_L^0|^2 c n_s$ has dimension of intensity i.e [W/m^2] the constant $\omega_s^4/(c^4|\wp|^2)$ has dimension of an area and the notion of a cross section makes sense. The frequency dependence agrees with what is known from Rayleigh scattering.

The ensemble of dipoles is proportional to the intensity of the excitation field, and a material parameter $M(x, y, z)$, which equals the density of dipoles, i.e. the number of dipoles per unit volume. In an optical fiber it is fair to assume that M is independent of z and also naturally described in a circular cylindrical coordinate system, $M(r)$. The radiated power from an ensemble of dipoles is then

$$P = \frac{\omega_s^4 n_s}{6\pi c^4 n_L} P_L \frac{\int_r M(r)|R_L(r)|^2 |\wp|^2 r dr}{\int_r |R_L(r)|^2 r dr}, \tag{7.2}$$

where ω_s is the frequency of the scattered wave, n_s and n_L are the refractive indices at the wavelength of the scattered and the excitation wave, respectively. $R_L(r)$ is the transverse distribution of the excitation wave. It is important to note, that assuming that the ratio $(n_s \int_r M(r)|R_L(r)|^2|\wp|^2 rdr)/(n_L \int_r |R_L(r)|^2 rdr)$ is frequency independent, then the radiated power is proportional to the frequency to the power of four i.e. it is most efficient at the lower wavelengths.

Example 7.1. *Recapture in an optical fiber*
In an optical fiber light propagates as guided modes. Consequently, one needs to find the fraction of the scattered light that is recaptured in an optical mode. Following the approach by Snyder & Love [9], an arbitrary electric field in the fiber $\mathbf{E}$ is written as a sum of a forward and a backward propagating mode and a non-guided electric field, $\mathbf{E} = a^+\mathbf{e}^+e^{i\beta^+ z} + a^+\mathbf{e}^-e^{i\beta^- z} + \mathbf{E}_{\text{RC}}$, where $\mathbf{e}^+$ is the mode vector of the forward propagating mode of the electric field and β^+ the corresponding propagation constant, $\mathbf{E}_{\text{RC}}$ is the non-guided electric field. The amplitude of the forward propagating guided mode is then evaluated through the integral

$$a^+ = \frac{1}{P_n}\int_V (\mathbf{e}^+)^* \cdot \mathbf{J} e^{-i\beta^+ z} dv, \tag{7.3}$$

where $\mathbf{J}$ is the current density vector of the dipole excited inside the fiber. After some calculations, of which we only show and use the result, the recaptured power P_{RC} is found to be

$$P_{\text{RC}} = P_L \frac{\omega_s^2}{(4\pi c^2)^2 \varepsilon_0 n_s n_L} \int_z \frac{\int_r M(r)|R_L(r)|^2|R_s(r)|^2|\wp|^2 rdr}{\int_r |R_L(r)|^2 rdr \int_r |R_s(r)|^2 rdr} dz, \tag{7.4}$$

where P_L is the power in the launched electric field, and $R_s(r)$ the transverse distribution of the guided mode at the frequency, ω_s, of the scattered wave. It is noted that the frequency dependence is reduced to ω_s^2, under the assumption that the integral in Eqn. (7.4) is frequency independent. If $M(r)$ and $|\wp|^2$ are constant throughout the fiber cross section then the total recaptured power reduces to

$$P_{\text{RC}} = P_L \frac{\omega_s^2}{(4\pi c^2)^2 \varepsilon_0 n_s n_L} M(r)|\wp|^2 S_B \Delta z, \tag{7.5}$$

where $\Delta z = \int_z dz$ is the segment of the fiber that contributes to the recaptured power, and where the recapture fraction S_B is

$$S_B = \frac{\int_r |R_L(r)|^2|R_s(r)|^2 rdr}{\int_r |R_L(r)|^2 rdr \int_r |R_s(r)|^2 rdr}. \tag{7.6}$$

When the transverse distribution of the electric field in the guided mode at the excitation frequency equals the transverse distribution of the electric field at the frequency of the scattered wave, i.e. $R_L(r) = R_s(r)$ then S_B equals

$$S_B = \frac{3}{4k^2n^2}\frac{2\pi}{A_{\text{eff}}}. \tag{7.7}$$

—— ■ ——

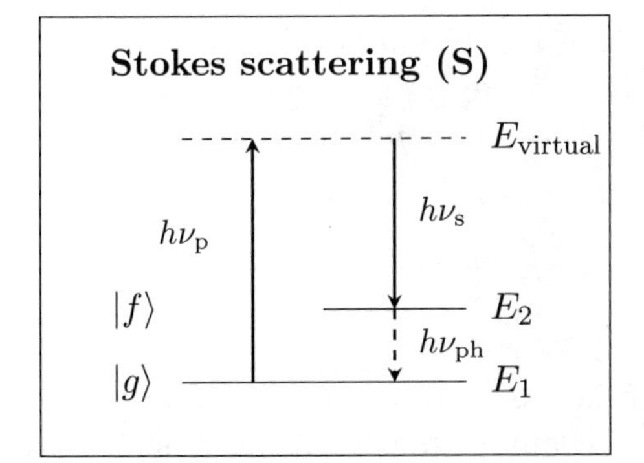

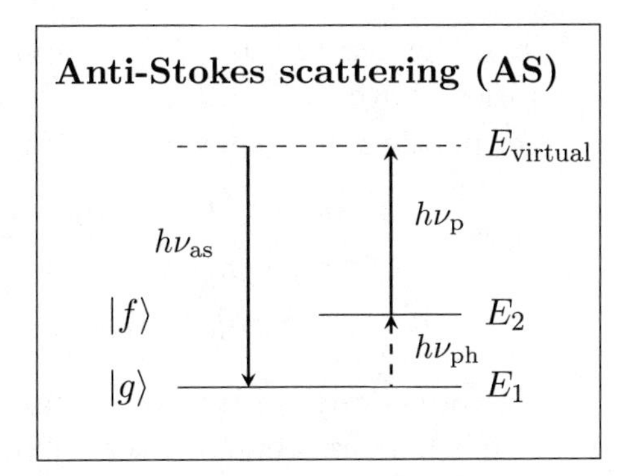

Figure 7.2: Energy level diagrams illustrating the Stokes process, emitting a photon at frequency ν_s (**left**) and the anti-Stokes process (**right**), emitting a photon at frequency ν_{as}. In both cases, the processes are excited by a so-called pump photon at frequency ν_{p}, $|g\rangle$ denotes a ground state, with energy E_1 while $|f\rangle$ denotes a final state with energy E_2.

7.1.2 Quantum mechanical description

In a quantum mechanical description one describes Raman scattering as a system that consists of a ground state and an excited state of a phonon. In the Stokes process two optical waves interact with a phonon and consequently change the state of the system from the ground state to the excited state.

Figure 7.2, left, shows the Stokes process whereas Figure 7.2, right, shows anti-Stokes. It is noted that the Stokes process may always happen but the anti-Stokes may not, since the latter requires the presence of phonons in their excited state.

Applying a quantum mechanical description of the harmonic oscillator with respect to creation and annihilation of a phonon, the rate for emission of photons at angular frequency ω_s (Stokes process) is proportional to [60]

$$\sigma(n_s + 1)n_p(n_\nu + 1), \tag{7.8}$$

where σ is the Raman cross section for the phonon energy $\hbar\Omega_{ph} = \hbar(\omega_p - \omega_s)$, n_s is the number of photons in the Stokes wave, n_p is the number of photons in the pump wave and n_ν is the number of phonons, $n_\nu = \{\exp(\hbar\Omega_{ph}/k_BT) - 1\}^{-1}$, which depends on the temperature T, Boltzmann constant k_B, and the phonon energy $\hbar\Omega_{ph}$, where Ω_{ph} is the frequency of the phonon and $\hbar$ Planck constant relative to 2π. The product $\sigma n_s n_p(n_\nu + 1)$ represents the stimulated emission and $\sigma n_p(n_\nu + 1)$ spontaneous emission. The rate of absorption of photons at ω_s (the anti-Stokes process) is proportional to

$$\sigma n_s(n_p + 1)n_\nu, \tag{7.9}$$

where $\sigma n_\nu n_s$ is the rate of spontaneous absorption, while $\sigma n_\nu n_p n_s$ is the rate of stimulated absorption.

To find the actual rate of change in photon number at the frequency ω_s, both of the ratios above in Eqn. (7.8) and Eqn. (7.9) need to be considered. The rate of change of photons at frequency ω_s, denoted W_s, due to emission and absorption is

$$\begin{aligned} W_s &= \sigma(n_s+1)n_p(n_\nu+1) - \sigma n_s(n_p+1)n_\nu \\ &= \sigma[n_s n_p(n_\nu+1) + n_p(n_\nu+1) - n_s n_p n_\nu - n_s n_\nu] \\ &\approx \sigma n_s n_p + \sigma n_p(n_\nu+1). \end{aligned} \tag{7.10}$$

An introduction to **phonons** is appropriate here. However, we return to this topic in much more detail in the following chapter. A phonon may be pictured as a collective vibration of atoms in a material. Due to the connections between atoms, the displacement of one or more atoms from their equilibrium positions gives rise to a vibration wave that propagates through the material. One such wave is shown in Figure 7.3. Typically a phonon is pictured as an excitation of a periodic, elastic arrangement of atoms or molecules, as in Figure 7.3. The collection of atoms oscillates uniformly at the same frequency and the collective motion defines the frequency of the collective motion i.e. the phonon. The collective motion is typically referred to as a normal mode. While normal modes are wave-like phenomena in classical mechanics, they have particle-like properties in quantum mechanics, similar to the wave particle duality for light versus photons. Thus where a photon may be pictured as the smallest quantum of light, a phonon may be pictured as the smallest quantum of vibrations.

The states defined by their frequency Ω_{ph} refer to the collective motion of atoms in the material. Each normal mode represents a single harmonic oscillator. Adding a phonon to the state referred to as the normal mode, increases the amplitude of the collective motions.

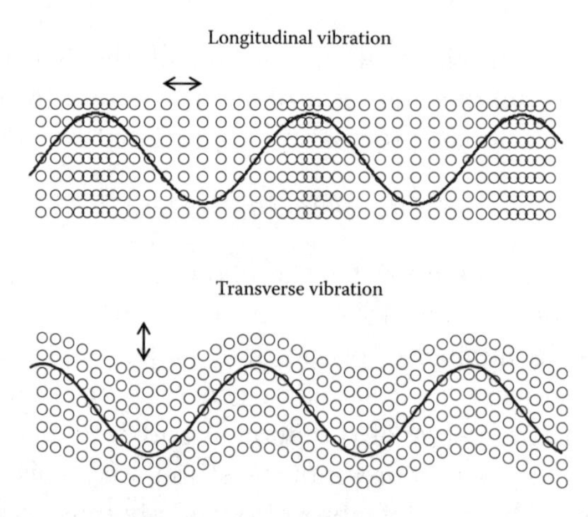

Figure 7.3: Generation of a phonon wave, the solid curve, created by moving atoms, the circles. Upper figure illustrates a longitudinal mode of vibration generating a phonon wave in the same direction as the movement of individual atoms, illustrated by the little double-arrow. The lower figure illustrates a transverse vibrational mode where the atoms move as illustrated by the little double arrow, and the phonon wave moves in a direction transverse to the movement of the atoms.

For the calculation we will do in the following we need the phonon population numbers as well as the mean number of phonons in a normal node. This is described in various text books on solid state physics for example [61] by C. Kittel. In accordance with the energy levels of a harmonic oscillator, which are $E_n = \hbar\Omega_{ph}(n+1/2)$ and assuming that the number of oscillators in state $|n+1\rangle$, where $\langle n\rangle$ is the thermal equilibrium occupation number of phonons, is related to those in state $|n\rangle$ by a Boltzman factor i.e. $\frac{N_{|n+1\rangle}}{N_{|n\rangle}} = e^{-\hbar\Omega_{ph}/(k_BT)}$, where k_B is Boltzman constant. That is, two adjacent energy levels are separated by $e^{-\hbar\Omega_{ph}/(k_BT)}$, and the population in state $|n\rangle$, N_n is: $N_n = N_0 e^{-n\hbar\Omega_{ph}/(k_BT)}$, where N_0 is the ground state population.

Let us evaluate the average number of phonons in a state, alternatively referred to as a normal mode. The probability of having n phonons in a state is proportional to

$$p(n) \propto e^{-n\hbar\Omega_{ph}/k_Bt}, \tag{7.11}$$

where $\hbar\Omega_{ph}$ is the separation between adjacent energy levels in the oscillator. Since the accumulated probability has to equal one i.e. $\sum_{n=0}^{\infty} p(n) = 1$ we can derive the accurate probability density distribution for the number of phonons in the normal mode

$$p(n) = \frac{1}{Z} e^{-n\hbar\Omega_{ph}/k_Bt}, \tag{7.12}$$

where $Z = \sum_{n=0}^{\infty} e^{-n\hbar\omega_{ph}/k_Bt} = \left(1 - e^{-\hbar\Omega_{ph}/(k_BT)}\right)$. With this probability distribution we can now find the mean number of phonons as

$$\langle n\rangle = \sum_{n=0}^{\infty} np(n) = \frac{1}{e^{\hbar\Omega_{ph}/(k_Bt)} - 1}, \tag{7.13}$$

which is the number that appear for example in equations Eqn. (7.8) and Eqn (7.9).

Example 7.2. *Peak frequencies in Raman spectrum*
Though in general very difficult, the peak frequencies of the Raman spectrum may be predicted if the material through which the light propagates is known, more specifically, if the bonds and bond-energies between various atoms in the molecules are known. Commercial software exists that allows one to evaluate bonds and bond-energies, and to predict Raman spectra numerically. However, in simple molecules the peak frequencies in the Raman spectrum may be predicted more easily using tabulated forces for example in simple diatomic molecules.

In the simplest cases of diatomic molecules where peak frequencies originate from vibrations, the Raman spectrum may be predicted by assuming that the two atoms are connected by a spring and that the natural frequency of the spring is identical to the frequency of the phonon. The spring forces for various diatomic molecules have been tabulated, see Table 7.1.

Table 7.1: The resonance frequency is $c/2\pi\sqrt{k/\mu}$ where μ is the reduced mass in atomic mass units which has the value of $1.6605 \cdot 10^{-27}$ kg.

Molecule	Force constant [N/m]	Reduced atomic mass unit	Frequency [cm^{-1}]
H_2	520	1/2	4159.2
D_2	530	1	2990.3
O_2	1140	8	1556.3
N_2	2260	7	2330.7

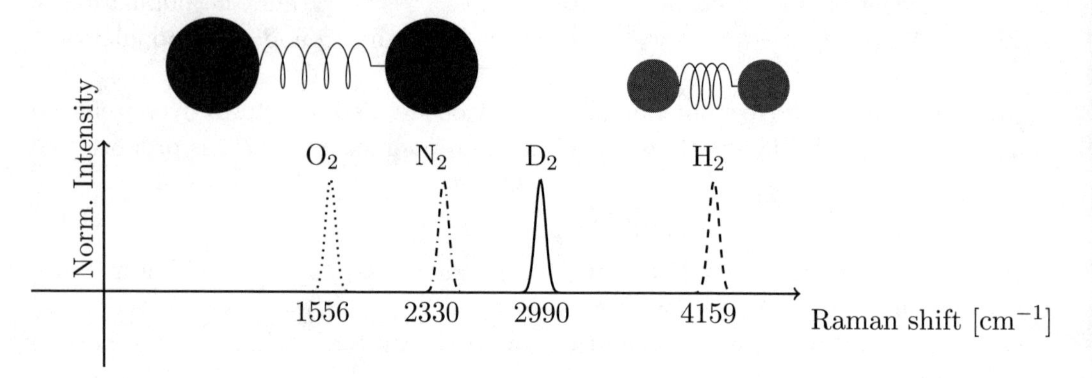

Figure 7.4: The smaller the reduced mass the larger the frequency. A smaller reduced mass corresponds to smaller atomic radius. amplitude chosen arbitrarily

Figure 7.4 illustrate the Raman spectra of the four diatomic gases in Table 7.1.

For completeness, it is noted that even some of these simple molecules may also have rotational energies in addition to the above calculated stretching modes.

The complexity is increased significantly if a molecule has three atoms rather than two. An important example is water.

— ■ —

7.2 Amplitude equations

With the use of Chapter 4 the relevant induced polarization due to Raman scattering is

$$(P_i^0)^{(3)} = \varepsilon_0 K(-\omega_s;\omega_p,-\omega_p,\omega_s)\sum_{jkl}\chi_{ijkl}^{(3)}(-\omega_s;\omega_p,-\omega_p,\omega_s)(E_p^0)_j(E_p^0)_k^*(E_s^0)_l \tag{7.14}$$

where $K(-\omega_s;\omega_p,-\omega_p,\omega_s) = 2\cdot 1/8\cdot 6 = 3/2$ is the prefactor discussed in Section 3.4 in Chapter 3.

In this chapter we focus on isotropic materials, more specifically silica glass. Consequently, the susceptibility tensor only has components where the indices appear in pairs or are all identical. Let us consider the x-component of the induced polarization, then the only non-zero components are $\chi^{(3)}_{xxxx}(-\omega_s;\omega_p,-\omega_p,\omega_s)$, $\chi^{(3)}_{xyyx}(-\omega_s;\omega_p,-\omega_p,\omega_s)$, $\chi^{(3)}_{xxyy}(-\omega_s;\omega_p,-\omega_p,\omega_s)$ and $\chi^{(3)}_{xyxy}(-\omega_s;\omega_p,-\omega_p,\omega_s)$. Out of these, especially the latter two are of interest. The contribution to the induced polarization from these two tensor elements is

$$\chi^{(3)}_{xxyy}(-\omega_s;\omega_p,-\omega_p,\omega_s)(E_p^0)_x(E_p^0)_y^*(E_s^0)_y, \tag{7.15}$$

and

$$\chi^{(3)}_{xyxy}(-\omega_s;\omega_p,-\omega_p,\omega_s)(E_p^0)_y(E_p^0)_x^*(E_s^0)_y. \tag{7.16}$$

Using the rules of intrinsic permutation symmetry, Chapter 4, the two tensor elements $\chi^{(3)}_{xxyy}(-\omega_s;\omega_p,-\omega_p,\omega_s)$ and $\chi^{(3)}_{xyxy}(-\omega_s;\omega_p,-\omega_p,\omega_s)$ are not identical. However, we note from physical arguments that the two susceptibility tensor elements must be identical since the Raman susceptibility only depends on the difference between ω_p and ω_s. Therefore $\chi^{(3)}_{xxyy}(-\omega_s;\omega_p,-\omega_p,\omega_s) = \chi^{(3)}_{xyxy}(-\omega_s;\omega_p,-\omega_p,\omega_s)$. Consequently the induced polarization due to Raman scattering is

$$\begin{aligned}(P_x^0)^{(3)} &= \left[\chi^{(3)}_{xxxx}(E_p^0)_x(E_p^0)_x^*(E_s^0)_x + \chi^{(3)}_{xyyx}(E_p^0)_y(E_p^0)_y^*(E_s^0)_x\right]e^{i(k_s)_xz} \\ &+ \left[(E_p^0)_y(E_p^0)_x^*(E_s^0)_y e^{i((k_p)_y-(k_p)_x)z}\right. \\ &\left.+(E_p^0)_x(E_p^0)_y^*(E_s^0)_y e^{i((k_p)_x-(k_p)_y)z}\right]\chi^{(3)}_{xyxy}e^{i(k_s)_yz}.\end{aligned} \tag{7.17}$$

We consider Raman scattering (Stokes case) where a pump at frequency ω_p is generating a Stokes wave at frequency ω_s, i.e the electrical field is

$$\mathbf{E} = \left[\begin{pmatrix}(E_p^0)_x\\(E_p^0)_y\\(E_p^0)_z\end{pmatrix}e^{-i\omega_pt} + \begin{pmatrix}(E_s^0)_x\\(E_s^0)_y\\(E_s^0)_z\end{pmatrix}e^{-i\omega_st} + c.c.\right]. \tag{7.18}$$

If we assume that $E_z^0 = 0$ for both the pump and signal, as valid in most guided structures, then the nonlinear polarization only has contributions from the x- and/or y-component, respectively. Considering terms in the susceptibility that generate the sum ω_s we see that these can only originate from $\omega_p - \omega_p + \omega_s$ or from $\omega_s - \omega_s + \omega_s$. The first one is responsible for Raman scattering whereas the latter is responsible for the intensity dependent refractive index.

To obtain a propagation equation we restrict the description to the most common fiber types, i.e. weakly guiding single mode fibers. In these fibers the electric field may be separated into a transverse part $R_i(r)$, $i = s,p$, and $r = (x,y)$ and a function of z $E_i^0(z)$, $i = s,p$. The total electric field is then

$$\begin{aligned}\mathbf{E} = \frac{1}{2}&\left[\mathbf{e}_pE_p^0(z)R_p(r)\exp\left(i(\beta_pz-\omega_pt)\right)\right. \\ &\left.+\ \mathbf{e}_sE_s^0(z)R_s(r)\exp\left(i(\beta_sz-\omega_st)\right) + c.c.\right],\end{aligned} \tag{7.19}$$

where $\mathbf{e}_i$, $i = s, p$ is a unit polarization vector, and β_i, $i = s, p$ the propagation constant as determined from the waveguide eigenvalue problem.

To obtain a propagation equation for the scattered field we identify the induced polarization at ω_s. The ith-coordinate of the induced third-order induced polarization may be written in terms of the third-order susceptibility tensor $\chi^{(3)}_{ijkl}$ through

$$(P_i^0)^{(3)} = \varepsilon_0 \sum_{jkl} K \chi^{(3)}_{ijkl} E_j^0 E_k^0 E_l^0, \tag{7.20}$$

where K is the prefactor to the induced polarization that accounts for all possible combinations of the involved frequencies. Assuming that the electric field is as in Eqn. (7.18), the amplitude of the induced polarization oscillating at frequency ω_s is

$$(P_i^0)^{(3)} = \frac{3}{2}\varepsilon_0 |R_p|^2 R_s \sum_{jkl} \chi^{(3)}_{ijkl} (E_p^0)_j (E_p^0)_k^* (E_s^0)_l. \tag{7.21}$$

In an optical fiber, i.e. an isotropic material, where the electric field is transverse, i.e $(i, j, k, l) \in (x, y)$, only four contributions appear in the sum on the right-hand side of Eqn. (7.21). Thus, the induced polarization equals

$$\begin{aligned}(P_i^0)^{(3)} = \frac{3}{2}\varepsilon_0 |R_p|^2 R_s \Big(&\chi^{(3)}_{iiii} (E_p^0)_i (E_p^0)_i^* (E_s^0)_i + \chi^{(3)}_{iijj} (E_p^0)_i (E_p^0)_j^* (E_s^0)_j \\ &+ \chi^{(3)}_{ijji} (E_p^0)_j (E_p^0)_j^* (E_s^0)_i + \chi^{(3)}_{ijij} (E_p^0)_j (E_p^0)_i^* (E_s^0)_j \Big),\end{aligned} \tag{7.22}$$

with a complex conjugate at $-\omega_s$. Alternatively, using Eqn. (3.72) the induced polarization may be expressed through the coefficients A and B, see Chapter 3, and is expressed as

$$(P_i^0)^{(3)}(\omega_s) = \varepsilon_0 \left[A (E_p^0)_i \left((\mathbf{E}_p^0)^* \cdot \mathbf{E}_s^0 \right) + \frac{1}{2} B \left\{ (E_s^0)_i \left| \mathbf{E}_p^0 \right|^2 + (E_p^0)_i^* (\mathbf{E}_p^0 \cdot \mathbf{E}_s^0) \right\} \right]. \tag{7.23}$$

In Chapter 4 two cases were described, one where $\mathbf{E}_p^0$ was parallel to $\mathbf{E}_s^0$ and one where $\mathbf{E}_p^0$ was orthogonal to $\mathbf{E}_s^0$. However, from the point of view of application as an amplifier, two cases are more interesting: one where the pump and signal have well defined states of polarization, and one where the pump or the signal is unpolarized. In the first cases the state of polarization evolves differently during propagation, which complicates the situation, unless a polarization maintaining fiber is used. These two cases are considered in the following subsections.

7.2.1 Pump and signal having well defined states of polarization

Relative to the description provided by Eqn. (7.22), Raman scattering may be further subdivided into a case using a polarization-maintaining fiber, and a case not using polarization-maintaining fiber.

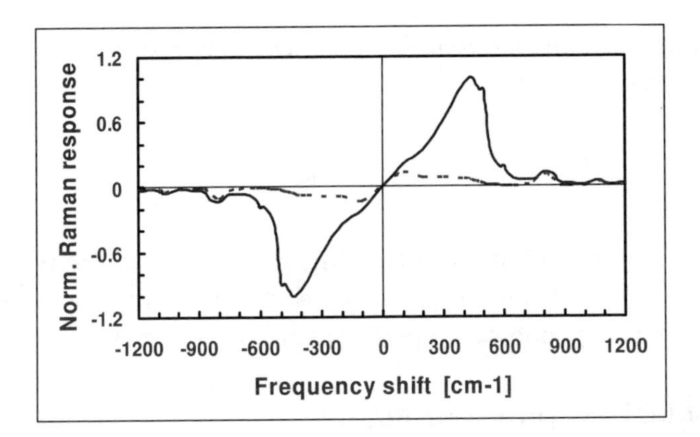

Figure 7.5: The Raman response normalized to the peak value of the parallel response versus the frequency shift between excitation wave and Stokes wave in wavenumbers (cm^{-1}). The data are valid for fused quartz. The solid curve is the parallel case whereas the dotted curve is the orthogonal case. Data from [17].

When using a **polarization-maintaining** fiber, the polarization states of the pump and signal are preserved during propagation, and it is possible to use Figure 7.5. Based on this figure, it is relevant to discuss a two examples.

Example 7.3. *Pump and signal, both oriented along the x-axis*
The induced polarization simplifies to $(P_x^0)^{(3)}(\omega_s) = (3/2)\varepsilon_0\chi_{xxxx}|E_p^0|^2E_s^0$. In this case the rate equation reduces to

$$\frac{\partial(E_s^0)_x}{\partial z}R_s = \frac{\omega_s^2}{2c^2\beta_s}iR_s|R_p|^2 6\chi_{xxxx}^{(3)}|E_p^0|^2E_s^0. \tag{7.24}$$

From this, the equation in power is written as

$$\frac{\partial P_s}{\partial z}R_s = 2\text{Re}[\frac{\omega_s^2}{2c^2\beta_s}i\chi_{xxxx}^{(3)}\int_A |R_s|^2|R_p|^2 6|E_p^0|^2|E_s^0|^2 da]. \tag{7.25}$$

The frequency domain Raman response corresponding to this situation is illustrated in Figure 7.5, as the solid curve. Note that the Raman response is normalized to the peak value of this parallel case. This setup gives the maximal gain.

— ■ —

Example 7.4. *Pump and signal orthogonal*
In Figure 7.5 the response corresponding to the scenario where the pump and signal are launched in orthogonal, yet linear, polarized states is displayed as the dashed curve. When comparing against the case when pump and signal are linearly polarized in the same state of polarization, it is clear that the orthogonal case is much weaker than the parallel case. A reduction in the peak values of the response of close to a factor of ten is seen. Equally important, the peak in the orthogonal case

occurs at a frequency shift different from the frequency shift that results in a peak for the parallel case.

— ■ —

Using a **non-polarization-maintaining fiber** the state of polarization of the signal and pump evolve during propagation, and as a result, the amplifier gain may become polarization dependent. A specific example is discussed in Section 7.4.3.

7.2.2 Un-polarized pump or signal

Assuming that $(E_s^0)_j$ is much smaller than $(E_s^0)_i$, contributions from $\chi_{ijij}^{(3)}$ and $\chi_{iijj}^{(3)}$ may be neglected and Eqn. (7.24) reduces to

$$\frac{\partial (E_s^0)_i}{\partial z} R_s = \frac{3\omega_s^2}{c^2 \beta_s} i R_s |R_p|^2 \left(\chi_{iiii}^{(3)} |(E_p^0)_i|^2 + \chi_{ijji}^{(3)} |(E_p^0)_j|^2 \right) (E_s^0)_i. \tag{7.26}$$

From this it is clear that $\chi_{iiii}^{(3)}$ is responsible for the Raman interaction when the pump and signal are along the same axis, whereas $\chi_{ijji}^{(3)}$ is responsible for Raman interaction when the pump and scattered wave are along orthogonal axis.

From Eqn. (7.26) the rate equation for the power $P_s = (1/2)\varepsilon_0 n_s c \int_A |E_s^0 R_s|^2 da$, where A is the cross-section of the fiber, is

$$\frac{dP_s}{dz} = 42\varepsilon_0 n_s c \int_A |R_s|^2 \mathrm{Re}\left[(E_s^0)^* \frac{dE_s^0}{dz} \right] da \equiv g_R P_s, \tag{7.27}$$

where the "≡" sign indicates that this is the equation that is used to define the gain coefficient.

Since the pump is unpolarized, the pump power is equally divided between the two components, represented with index i and j in Eqn. (7.26), at any point in time and at any point along the fiber, i.e. $(1/2)\varepsilon_0 n_p c \int_A |R_p (E_p^0)_i|^2 da = (1/2)\varepsilon_0 n_p c \int_A |R_p (E_p^0)_j| da = P_p/2$. Assuming furthermore that the susceptibilities are constant across the entire fiber cross section, the gain coefficient g_R is a simple average of the two contributions

$$g_R = -\frac{\omega_s^2}{2\varepsilon_0 n_p c^3 \beta_s} \frac{1}{A_{\mathrm{eff}}^{\mathrm{ps}}} \frac{6\mathrm{Im}[\chi_{iiii}^{(3)} + \chi_{ijji}^{(3)}]}{2}, \tag{7.28}$$

where $A_{\mathrm{eff}}^{\mathrm{ps}}$ is the effective area of the overlap between the pump and signal

$$A_{\mathrm{eff}}^{\mathrm{ps}} = \frac{\int_A |R_p|^2 da \int_A |R_s|^2 da}{\int_A |R_p|^2 |R_s|^2 da}. \tag{7.29}$$

In an optical single mode fiber with a core refractive index n_1 and a cladding refractive index n_2, the propagation constant β_s may be approximated by $\beta_s \approx \omega_s n_s/c$ where n_s is the effective refractive index for the signal as defined in Appendix F. The effective refractive index is between the core and cladding refractive index, i.e. $n_2 < n_s < n_1$, and the propagation constant β_s is proportional to ω_s. With this the gain coefficient equals

$$g_R = -\frac{3\omega_s}{\varepsilon_0 c^2 n_p n_s} \frac{\text{Im}[\chi_{iiii}^{(3)} + \chi_{ijji}^{(3)}]}{A_{\text{eff}}^{\text{ps}}}. \tag{7.30}$$

This illustrates that g_R scales linearly with the frequency of the Stokes shifted frequency. Since the optical frequencies are on the order of 10^{14} Hz, ω_s is often replaced with ω_p for convenience. In optical fibers g_R is typically on the order of 1 $(\text{Wkm})^{-1}$ and $A_{\text{eff}}^{\text{ps}} \approx 75 \cdot 10^{-12}$ m^2.

7.3 Fundamental characteristics of silica

In accordance with Chapter 2, both the electric field and the macroscopic polarization depend upon both position and time. With respect to position, we assume that the macroscopic polarization is local, and thus the induced third-order polarization is related to the response as described in Section 2.3, see also [17]. The tensor is a tensor of rank 4 that only depends on the differences among different times in the applied electric field and the actual time [17].

In the Raman scattering process, the response function consists of a nuclear response originating from vibrations and rotations of molecules in the glass network [28]. Since this has been described in more detail in Chapter 2, only a summary is given in the following.

7.3.1 Susceptibility

In correspondence with Chapter 2, the frequency domain Raman response is a complex quantity where the real part relates to a change in refractive index whereas the imaginary part relates to gain or depletion. Figure 7.6 shows the Raman susceptibility of silica.

The structure of the Raman spectrum is defined by the nature of the material that the light is being Raman scattered within. Silica, as in Figure 7.6, is an amorphous material, consequently the spectrum is relatively broad as compared to a crystalline material, as for example crystalline silicon. However, silica still has some structure in that it consists of tetrahedrons made of a silicon atom in the core surrounded by four oxygen atoms, located in the furthest corners of a cube. These tetrahedrons are then connected to each other in rings of for example six, four,

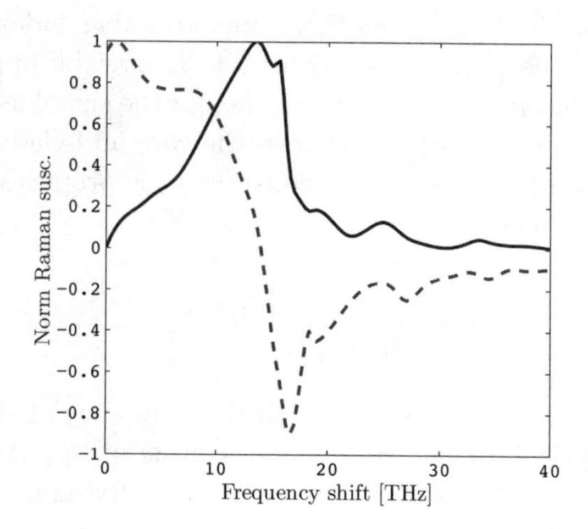

Figure 7.6: The normalized Raman susceptibility (dashed, real and solid, imaginary part). The scattered field is measured in the same plane of polarization as the pump light.

or three tetrahedrons. Figure 7.7 shows a three- and a four-fold ring. Each of these ring configurations results in different vibrational modes as illustrated in Figure 7.7.

As indicated above, the Raman spectrum is unique for any material. Each material has its individual set of rotational and/or vibrational modes. Consequently, the detailed structure of the Raman spectrum also differs from optical fiber to optical fiber. Figure 7.8 shows the Raman spectrum from different types of glass.

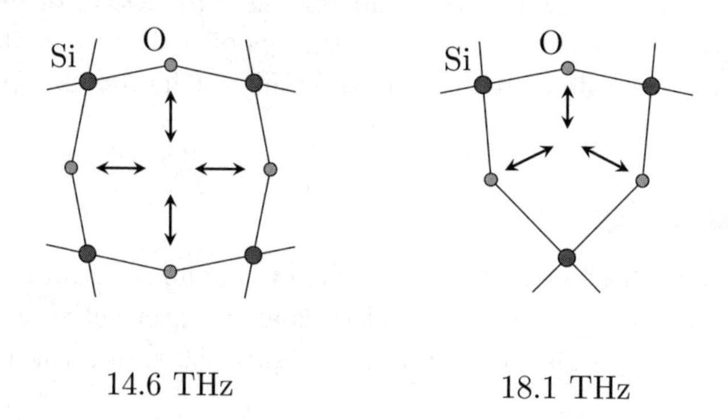

14.6 THz 18.1 THz

Figure 7.7: Stretching modes from Raman in silica from [62] and [63] for shifts of 14.6 THz corresponding to 485 cm^{-1} left and 18.1 THz, corresponding to 605 cm^{-1} right. For comparison against Figure 7.6 it is noted that the frequency of a specific mode corresponds to specific local peaks in the Raman susceptibility.

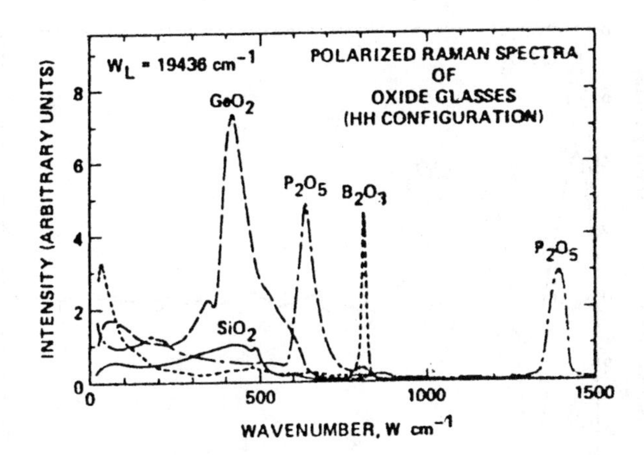

Figure 7.8: Raman response of various oxide glasses From [Galeener, F. L., et al. "The relative Raman cross sections of vitreous SiO_2, GeO_2, B_2O_3, and P_2O_5, Appl. Phys. Lett., Vol 32, p 34, (1978)]

7.3.2 Response function

The response time of the Raman process is related to the response of nuclear rotations and vibrations, and hence is longer than the electronic response. In [29] the Raman response time of fused silica is evaluated to be less than 100 fs. For most applications, this appears instantaneous, and especially in relation to the application as Raman amplifiers in optical communications systems. This is especially true when the propagation direction of the pump is chosen to be opposite to that of the signal, as for example when trying to mitigate unwanted polarization dependence in the amplifier gain. However, when the pump and signal propagate in the same direction, fluctuations in the pump translate directly to fluctuations in the signal gain.

When describing pulse propagation, the response time of the Raman scattering process is becoming an important parameter. This is because the pulse duration of a single bit in a single channel, as well as in a multiple wavelength channel communications system, is pushed close to a picosecond or below. In both cases the response time needs to be included even in a system that is not applying Raman amplification, but this is beyond the scope of this book.

Another important application of Raman scattering is within supercontinuum generation. This has received much attention, see for example [64]. Supercontinuum generation may be achieved by launching short pulses into an optical fiber with strong nonlinear effects obtained by reducing the effective area of the beam and a group velocity dispersion tailored to favor generation of light within a broad spectral range.

The response function varies from material to material similar to the Raman gain coefficient. Hence it is difficult to obtain accurate knowledge of the response

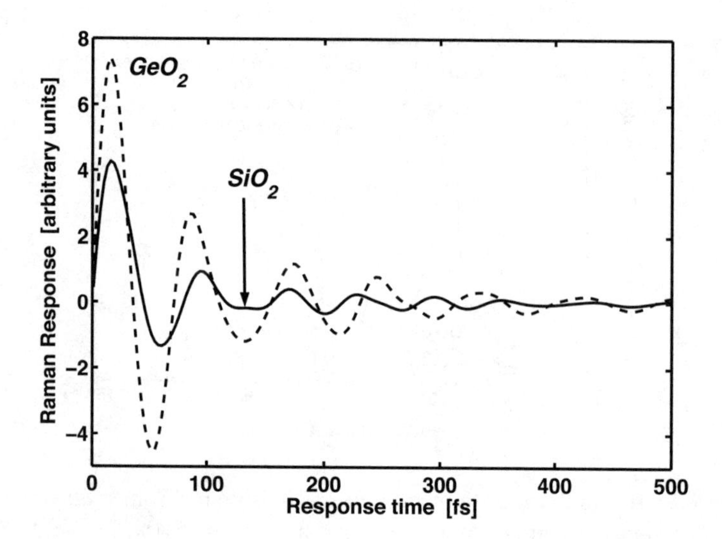

Figure 7.9: The Raman response function. The response is normalized according to $\int_{-\infty}^{\infty} h(t)dt = 1$.

function from fiber to fiber. However, the response function of the Raman scattering process, may be directly evaluated through the Fourier transform of the complex Raman susceptibility. Figure 7.9 displays both the response from SiO_2 and also the response from GeO_2.

In a simple picture the response time may be fitted to a single damped harmonic oscillator with two time constants, τ_ν related to the frequency of the 'phonon' and τ_s related to the attenuation of the network of vibrating atoms, i.e.

$$h(t) = ae^{(-t/\tau_s)} \sin(t/\tau_\nu). \tag{7.31}$$

For the two curves in Figure 7.9, we find the best fit to Eqn. (7.31) with the time constants listed in Table 7.2.

From the table there is no noticeable difference between the time constant related to the phonon frequency for the differ materials whereas there is a significant difference of the time constant related to the damping of the phonon. Both of the materials have a response time close to 100 fs with a difference of less than 10 %. The response time for germania is larger than for silica which may be attributed to the larger atomic weight of germanium relative to silicon.

Table 7.2: Time constants of the damped harmonic oscillator related to the Raman response function. The last column noted by * is adopted from [30].

Material	GeO_2	SiO_2	SiO_2*
τ_ν [fs]	12	13	12
τ_s [fs]	83	39	32

A comprehensive modelling of the response function requires more insight into the possible rotational and vibrational modes of the glass network. Such a modeling, provided by [34] and [31], has shown that it is necessary to combine an ensemble of damped oscillators to get an accurate fit to the response function. In this so-called intermediate broadening model, not homogeneously nor inhomogeneously broadened oscillators, the Raman response is modelled as an ensemble of 13 damped harmonic oscillators with a Gaussian distribution of the mean frequencies. As a result the gain spectrum is a convolution of a Gaussian and a Lorentz form, and the response function has a perfect match to a measured Raman spectrum from silica, as previously mentioned in Subsection.

7.4 The Raman fiber amplifier

Within optics communication, Raman scattering in optical fibers may be used to obtain amplification. This is achieved by simultaneously launching light at two frequencies separated by the Stokes shift. Hereby light is transferred from the high frequency beam to the low frequency beam. Figure 7.10 displays a generic fiber Raman amplifier. In principal the Raman amplifier configuration is defined by the amplifier fiber, the used pump laser(s), and their relative positions [65].

The benefits of Raman amplification include:

Gain at any wavelength. The Raman scattering process depends only upon the frequency shift between pump and signal. In silica this is close to 13 THz.

Pump beams at different wavelengths may be combined. By launching multiple pump wavelength a composite spectrum is obtained. In this way it is possible to broaden the gain bandwidth [66].

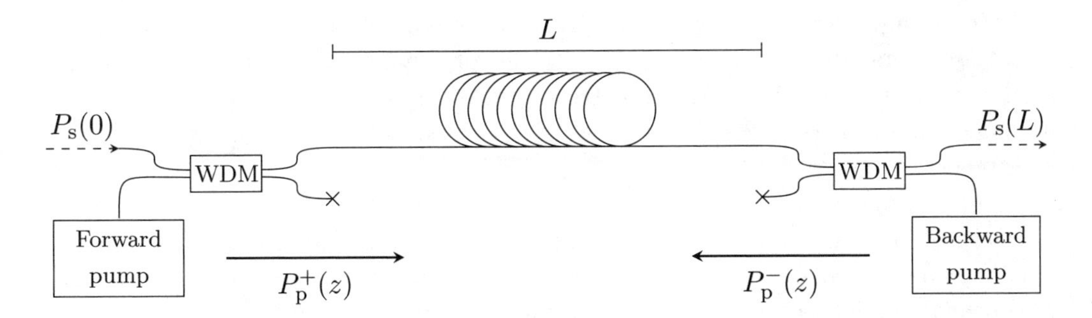

Figure 7.10: General setup of a Raman fiber amplifier. Forward and backward pumps are coupled together with the signal in Wavelength Division Multiplexers (WDM), i.e. wavelength dependent couplers. The signal propagates in one direction only; bi-directional signal transmission is not considered.

Suitable as distributed gain medium. The gain per unit length is relatively week, and hence it is possible to counterbalance the loss along the transmission line. This allows for distributed amplification which has favorable noise properties [67].

No special doping is necessary. Silica has a Raman efficiency that is sufficient to achieve gain, such that a transmission fiber may be turned into an amplifier in itself [68].

Simple. The amplifier is very simple compared to other schemes which require a large number of lasers and guard-bands to separate the signals into different bands so that signal bands can be amplified band by band.

The amplifier fiber is characterized by its Raman gain coefficient which is determined by the material constituents of the fiber and the waveguide design which defines the Raman effective area. In addition, the attenuation of the fiber at the pump and signal wavelengths are important parameters for the amplifier performance. Finally, the group velocity dispersion properties are essential both for the performance of the system and the amplifier performance when using multiple pump wavelengths, due to cross-coupling among individual pumps, and pumps and signals.

The individual pump lasers in Figure 7.10 are characterized by their wavelength, the emitted power, the polarization properties of the emitted light, and the linewidth of the laser. Finally, the noise properties of the pump lasers may also be important parameters for the amplifier performance.

The amplifier configuration is characterized by the location of pump-lasers and the possible use of other components in the amplifier. In the simplest configuration only one pump is used and the amplifier is either forward or backward pumped, see Figure 7.10. A further complication is added when using pumps propagating in both directions relative to the signal, i.e. a bi-directionally pumped amplifier. Especially in discrete amplifiers, the amplifier may consist of multiple stages, gain equalization filters, isolators, etc. In such cases the analysis of the amplifier performance is even further complicated. Figure 7.11 illustrates an example of a discrete Raman amplifier consisting of two stages. Each stage is separated from the others by an isolator [69].

Despite the simplicity of the architecture of a Raman amplifier, many factors must be considered in the design of the amplifier and the systems that use them. A thorough understanding of some key factors that determine the amplifier performance is required. These include pump to pump power transfer [70], signal to signal power transfer [71], pump depletion (saturation), Rayleigh scattering [72], multipath interference [73], and amplified spontaneous emission [74]. However, such detailed discussion is outside the scope of this book.

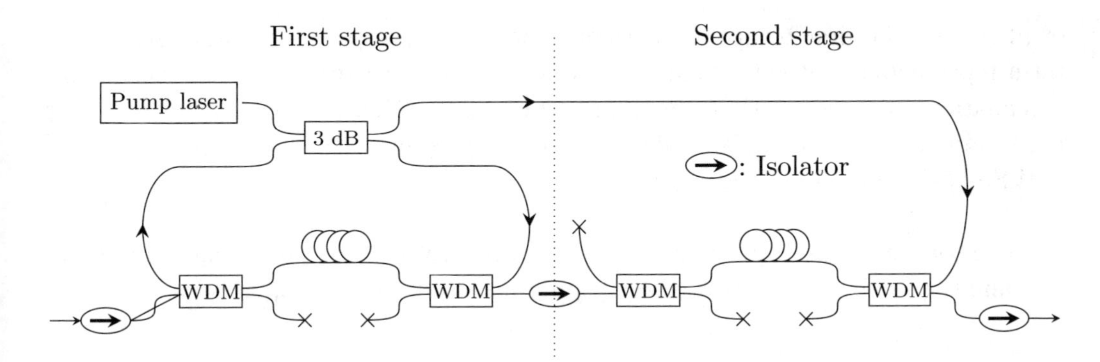

Figure 7.11: A two-stage discrete Raman amplifier configuration; each stage is separated by an isolator. The remaining pump power from the first stage is reused in the second. From [69]

In Section 7.4.1 the simplest model of gain and spontaneous emission is presented. In this model, the pump power is assumed to propagate independently from the signal power, and only one signal and one pump wavelength is considered [75]. This model may easily be expanded to include Rayleigh scattering at the signal wavelength. Although very simple, the model is powerful and provides excellent insight into Raman amplification. Additionally, effects such as depletion of the pump, multiple pump and signal wavelengths, and polarization dependence require more complicated math or numerical schemes.

7.4.1 Simple model

In the following we present two different approaches to predict the performance of a Raman amplifier: one based on counting photons, and one based on the evolution of the optical power. The advantage of counting photons is that the photon number is conserved when considering the forward pumped Raman amplifier. As a consequence, an analytical solution of the coupled equations between pump and signal photons is obtainable even when the amplifier operates in depletion [3], under the assumption that the attenuation rate of the pump and signal are identical. This is not possible when evaluating the evolution of power because power is transferred to the material in the scattering process. However, evaluating optical power has the advantage of using directly measurable quantities.

Starting from the photon model, the propagation equation for the average signal photon number n_s is

$$\frac{dn_s}{dz} = \gamma_R n_p n_s + \gamma_R n_p (n_\nu + 1) - \alpha_s n_s, \tag{7.32}$$

where n_p is the number of pump photons, which equals the sum of the forward n_p^+ and backward n_p^- propagating pump photons ($n_p = n_p^+ + n_p^-$). The Raman gain coefficient is γ_R for photons, α_s is the signal attenuation and n_ν is the number of phonons. The first term on the right-hand side corresponds to gain, the second term represents spontaneous emission, whereas the third term is the intrinsic fiber attenuation. It is noted that the gain term in Eqn. (7.32), proportional to γ_R, is the difference between the stimulated Raman emission and the absorption, i.e. the anti-Stokes scattering of the signal.

The corresponding equations for the forward propagating pump photon number n_p^+, and for the backward propagating pump photon number n_p^- are

$$\pm\frac{dn_p^{\pm}}{dz} = -\gamma_R n_s n_p^{\pm} - \gamma_r(n_\nu + 1)n_p^{\pm} - \alpha_p n_p^{\pm}, \tag{7.33}$$

where $\pm$ relates to forward and backward propagating pump photons, respectively. The first two terms on the right-hand side counterbalance corresponding terms in Eqn. (7.32); the first term counterbalances the gain of signal photons, the second term counterbalances the spontaneous emission. The third term in Eqn. (7.33) is the intrinsic attenuation of the pump. In both equations, Eqn. (7.32) and Eqn. (7.33), Rayleigh scattering is omitted, and only scalar quantities are considered, corresponding to an unpolarized pump.

Assuming that the initial signal photon number at $z = 0$ is n_0, then the average signal photon number at position z, $n_s(z)$ is given by

$$n_s(z) = n_0 G(z) + \tilde{n}(z), \tag{7.34}$$

where the signal gain is $G(z)$, and the spontaneous emission is $\tilde{n}(z)$. The gain is given by

$$G(z) = \exp\left[\left|\int_0^z \gamma_r n_p - \alpha_s dz\right|\right], \tag{7.35}$$

where n_p is either n_p^+ or n_p^- or the sum of the two, and the spontaneous emission is given by

$$\tilde{n}(z) = G(z)\int_0^z \frac{\gamma_r n_p(n_\nu + 1)}{G(x)} dx. \tag{7.36}$$

The spontaneous emission is typically referred to as amplified spontaneous emission (ASE) since the spontaneously emitted power is seeded by the source term: $\gamma_R n_p(n_\nu + 1)$ in Eqn. (7.32) and succeedingly amplified through the term $\gamma_R n_p n_s$. Note that the gain in Eqn. (7.35), is independent on n_ν, that is temperature independent. However, the spontaneous emission, $\tilde{n}$, in Eqn. (7.36) is temperature dependent.

To arrive at equations based on more easily measurable quantities such as pump and signal powers Eqn. (7.32) is multiplied by $h\nu$ and assuming that the pump and stokes waves are monochromatic or in a well defined signal mode of relatively small bandwidth B_0 we arrive at

$$\frac{dP_s}{dz} = g_R P_p P_s + h\nu B_0(1 + n_\nu)g_R P_p - \alpha_s P_s, \tag{7.37}$$

where g_R is the Raman gain coefficient, and P_p and P_s are the powers of the pump and signal, respectively. P_p is the sum of the forward P_p^+ and backward propagating P_p^- pump power. g_R describes the Raman gain coefficient when using power equations whereas γ_R describes the corresponding gain coefficient when using photon number equations.

It is noted that sometimes a factor of 2 is included in the term $h\nu B_0 g_R P_p$, representing spontaneous emission, in Eqn. (7.37) to account for the fact that the signal typically occupies one state of polarization whereas the spontaneous emission is equally generated in both polarizations guided by single mode optical fibers.

The equations for the forward and the backward propagating pump powers corresponding to Eqn. (7.33) are

$$\pm\frac{dP_p^\pm}{dz} = -g_R\frac{\nu_p}{\nu_s}P_p^\pm P_s - h\nu\frac{\nu_p}{\nu_s}B_0(1+n_\nu)g_R P_p^\pm - \alpha_p P_p^\pm, \tag{7.38}$$

where the first two terms on the right-hand side represent depletion due to the signal and the ASE, respectively. The third term on the right-hand side describes the attenuation at the pump wavelength. In analog to Eqn. (7.32), Eqn. (7.37) may be solved and the signal gain, i.e. the signal output power relative to the signal input power, at the output end of a fiber of length L, is

$$G(L) = \exp\left(\int_0^L g_R P_p(z)dz - \alpha_s L\right). \tag{7.39}$$

Assuming that the pump is independent of the signal and is determined solely by the intrinsic fiber loss, i.e. the pump decays exponentially with the intrinsic fiber loss, the gain may be evaluated analytically. This situation is referred to as the undepleted pump regime, and the propagation of the pump power is described simply as $P_p(z) = P_p(z=0)\exp(-\alpha_p z)$ when the pump propagates in the positive z-direction, that is launched from the signal input end, referred to as the co-pumped or the forward pumped Raman amplifier. Alternatively, the pump propagates according to $P_p(z) = P_p(z=0)\exp[-\alpha_p(L-z)]$ when the pump propagates in the negative z-direction or opposite to the signal, referred to as the counter-pumped, or the backward pumped Raman amplifier, where $P_p(z=0)$ is the launched pump power.

The signal gain in the co- and counter-pumped Raman amplifiers is identical, and given by

$$G = \exp\left\{g_R L_{\text{eff}}^p P_p(z=0)\right\}\exp\left\{-\alpha_s L\right\} \tag{7.40}$$

where L_{eff}^p is the effective fiber length for the Raman amplifier, in the following referred to as the Raman effective length, given by

$$L_{\text{eff}}^p = \frac{1-\exp(-\alpha_p L)}{\alpha_p}. \tag{7.41}$$

The Raman effective length may be pictured as the length of fiber over which the same Raman interaction would be achieved if the fiber were lossless. The pump power only provides a significant contribution to the gain over a length comparable to the Raman effective length of fiber. In a typical Raman amplifier providing gain at 1555 nm, based on a conventional transmission fiber, the loss at the pump wavelength is 0.25 dB/km, and therefore the Raman effective length is ≈ 17 km.

The Raman gain is often quoted in terms of the On-Off Raman gain which is the ratio of the signal output power with the pump on to the signal output power with the pump off. The On-Off Raman gain is easily calculated in the undepleted pump regime by omitting the factor $\exp(-\alpha_s L)$ in Eqn. (7.40).

Equation. (7.40) shows that the gain in decibels increases linearly with the pump power in Watt. For a typical dispersion-shifted transmission fiber (with $A_{\text{eff}}^{ps} \approx 75$ μm^2), an unpolarized pump and an optical fiber much longer than the Raman effective length, the On-Off Raman gain is approximately 55 dB/W.

The undepleted pump approximation is very powerful and may be used to estimate the levels of gain and spontaneous emission. The criterion for its validity is that the loss of the pump should be dominated solely by the intrinsic loss, see Eqn. (7.38). Neglecting Rayleigh scattering, this results in the condition [74]

$$g_R \frac{\nu_p}{\nu_s} P_s(z) + 2B_0 h\nu_p P_p(s)(1+n_\nu) \int_{\tilde{\nu}}^{\infty} g_R d\tilde{\nu} \ll \alpha_p. \tag{7.42}$$

In a typical dispersion-shifted fiber with $\alpha_p \approx 0.057\ \text{km}^{-1} (= 0.25\ \text{dB/km})$, $g_R \approx 0.7\ (\text{Wkm})^{-1}$, the undepleted pump approximation is valid for $P_s \ll 80$ mW, when depletion due to ASE is neglected .

Eqn. (7.38) and Eqn. (7.37) are scalar, implying that the state of polarization of the pump and signal does not change during propagation, and the cross-polarization coupling is omitted. However, since the Raman amplification typically happens over many kilometers, and since the pump and signal are separated in wavelength by up to hundreds of nanometers, the state of polarization of the pump and signal change independently and the resulting Raman gain is an average taken over all states of polarizations of the pump and signal. Under these circumstances, an explicit treatment of the individual states of polarization of the pump and signals is not required.

7.4.2 Fiber-dependence

Figure 7.12 displays the Raman gain coefficient for three different fiber types, all measured using unpolarized pump light at the wavelength 1453 nm. The pure silica fiber, labeled Pure-silica, which has a fibercore of silica and a cladding with a lower refractive index, displays the lowest gain coefficient of approximately 0.5 $(\text{Wkm})^{-1}$

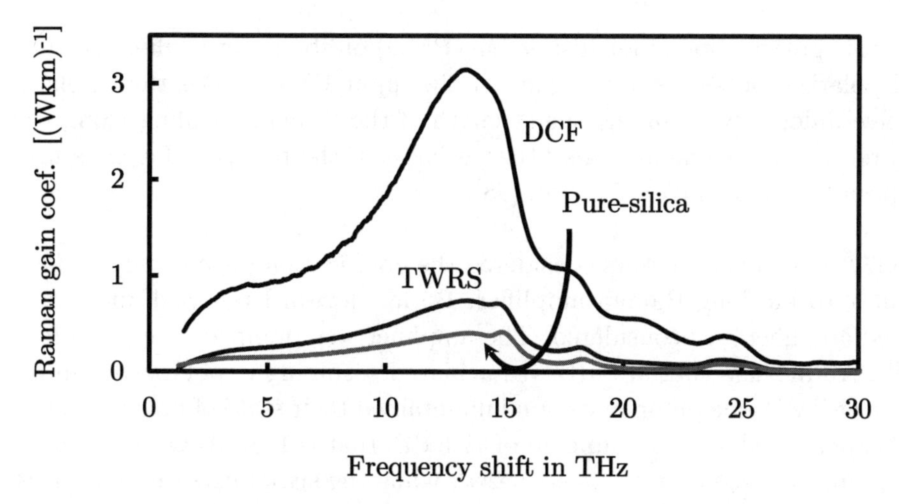

Figure 7.12: The Raman gain coefficient versus frequency shift between pump and signal for different fiber types. In the dispersion shifted fiber, DCF, the germanium content is higher and the effective area less than in the high capacity transmission fiber, the TrueWave reduced slope fiber, TWRS. The Pure-silica fiber has no germanium in the core.

at its peak. The dispersion compensating fiber, labeled DCF, which has a small core area, a silica cladding and a core of silica glass heavily doped with germanium, displays the largest Raman gain coefficient of approximately 3 $(\text{Wkm})^{-1}$ at its peak. The higher gain coefficient when compared to the silica core fiber is caused by the germanium content and the effective core area [68]. In typical high capacity transmission fibers, in Figure 7.12 exemplified by a so-called TWRS fiber, which is an OFS True-Wave-Reduced-Slope fiber, reduced slope refers to the slope of the group velocity dispersion. The TWRS fiber has a gain coefficient close to 0.7 $(\text{Wkm})^{-1}$ at its peak. This fiber has a lower concentration of germanium compared to the dispersion compensation fiber, and an effective area of approximately $75{\cdot}10^{-12}$ m^2.

The composite spectrum of any germanosilicate spectrum, with moderate fractional germanium concentration [68], may be predicted, and scaled according to the operating wavelength.

7.4.3 Polarization dependence

The Raman gain spectrum of silica depends on the relative orientation of the pump and signal polarization states. The peak gain when they are co-polarized is approximately 10 times higher than when they are orthogonally polarized. This may cause a polarization dependent gain (PDG), which may lead to transmission impairments, for instance amplitude fluctuations if the relative polarization of pump and signal vary randomly. However, the small amount of PDG produced in Raman amplifiers consisting of long lengths of fibers is typically not as large as in bulk samples because of intrinsic polarization mode dispersion (PMD) of fibers. Even when the pump and signal polarization states are aligned at the input PMD causes their polarizations to evolve differently, changing the strength of the Raman coupling along the fiber. This produces an averaging effect that is larger if the pump and signals propagate in opposite directions [76], [77] and [78].

In [79] Lin and co-workers considered the polarization dependence of the Raman gain in a 10 km long Raman amplifier. Both a forward pumped and a backward pumped amplifier was considered. The amplifier was pumped using a pump power of 1 W. An average gain of 8 dB was achieved assuming unpolarized pump. In the absence of PMD, the pump and signal maintained their state of polarization and the signal experienced a maximum gain of 17.6 dB, that is 19.6 dB On-Off Raman gain, if the pump and signal were co-polarized, while the signal experienced a loss of 1.7 dB, when the pump and signal were orthogonal polarized. The latter corresponds to an On-Off Raman gain of 0.3 dB.

The results in [79] are explained by considering the propagation of the pump and signal. Due to PMD the state of polarization of the pump, as well as the signal, changes as the pump and signal propagate. This causes an averaging of the Raman gain. In the forward pumped as well as the backward pumped Raman amplifier, the PMD causes the PDG to vanish because of the averaging of the Raman gain. This averaging is weaker in the forward pumped amplifier as compared to the backward pumped amplifier. This is because the pump and signal propagate along with each other in the forward pumped amplifier as opposed to the counter pumped amplifier where one time segment of the signal experiences many different time segments of the pump, leading to a strong averaging. As a consequence there is a significant benefit in using a backward pumped amplifier configuration. However, in the experiment in [79] a PMD of 0.01 ps/$\sqrt{\text{km}}$ was necessary to strongly reduce the impact of PDG even in a forward pumped amplifier, and only 0.001 ps/$\sqrt{\text{km}}$ in the counter pumped case.

The desire to increase the PMD causes consideration of a trade-off since a significant fiber PMD in itself directly may impair transmission. Thus, other methods may be considered to reduce PDG. This includes simple measures to depolarize the pump light, such as polarization-multiplexing outputs of two independent pump lasers at the same wavelength. Also, single pumps with sufficient spectral bandwidth can be depolarized using fiber depolarizers formed from a few meters of polarization maintaining fiber. With such measures, PDG impairments in Raman fiber amplifiers can be eliminated [80].

In most practical amplifiers the pump (or multiple pumps) is depolarized, whereas the signals, on the other hand, are typically polarized. The mutual Raman coupling among pumps and the Raman coupling between pump(s) and signals is well described in such cases by Eqn. (7.37) and Eqn. (7.38). However, the Raman coupling among individual signal channels may lead to numerical errors, particularly if their relative polarization states are preserved throughout the fiber amplifier. This error is small, however, when fiber PMD is present, since the relative polarization states of the signals evolve over short distances, which average their Raman interactions. Moreover, the largest Raman interaction occurs between signals with a large frequency separation, up to 100 nm, which are also the signals whose polarization states diverge in the shortest distance along the fiber [81], [78], [82].

7.5 Summary

- In the Raman process light is scattering against phonons, i.e. vibrational and/or rotational modes of the atomic network of the material that the light propagates through. Raman scattering is used for example within chemistry to identify constituents of a given material, or within optical communication to amplify an optical signal. The latter is possible since the scattering may happen spontaneously or be stimulated by another beam separated from the excitation beam by the phonon energy. In glass the phonon energy is centered around 13 THz, and the phonons are called optical phonons.

- Raman spectroscopy is based on spontaneous Raman scattering whereas Raman amplification is based on stimulated Raman scattering.

- Raman scattering is a third-order nonlinear process and described through $\chi(3)_{ijkl}(\omega_s; \omega_p, -\omega_p, \omega_s)$.

- The scattering may be described using a dipole description or a harmonic oscillator more specifically a damped harmonic oscillator model,

$$h(t) = ae^{(-t/\tau_s)} \sin(t/\tau_\nu). \tag{7.31}$$

- The phonon population is temperature dependent $n_\nu = \{\exp(\hbar\omega_\nu/k_B T) - 1\}^{-1}$, thus the scattering is also temperature dependent. However, in an amplifier the amplifier gain becomes temperature independent.

- The scattering requires that the electrical field of the excitation as well as the scattered beam overlap spatially and in time. Spatially the overlap is described by

$$A_{\text{eff}}^{\text{ps}} = \frac{\int_A |R_p|^2 dA \int_A |R_s|^2 da}{\int_A |R_p|^2 |R_s|^2 da}. \tag{7.29}$$

- In an optical fiber amplifier, the propagation of pump and signal may be described by a simple but yet useful approximation assuming that the pump beam is unaffected by the scattered beam. Under this assumption the Raman gain is

$$G = \exp(g_R P_p(z=0) L_{\text{eff}}) \exp(-\alpha L). \tag{7.40}$$

- The Raman gain is polarization dependent, and the response time is in the order of hundreds of femtoseconds. Consequently, the preferred configuration of a Raman amplifier is backward pumped.

Chapter 8

Brillouin Scattering

Contents

Until this point we have focused on Raman scattering when describing scattering of light as it propagates through a material. However, in the following we direct attention to Brillouin scattering. Both scattering mechanisms deal with interaction between phonons/sound waves and photons/electrical fields. Where Raman scattering is an interaction between optical phonons and electrical fields, Brillouin scattering is an interaction between acoustical phonons and electrical fields. In a more detailed description Brillouin scattering is the diffraction of optical waves on sound waves created by electrostriction. More specifically, due to electrostriction the forward propagating optical wave creates a Bragg grating that propagates in the same direction as the launched optical wave. This causes a back reflection of an optical wave at the frequency of the launched wave only corrected by the doppler shift due to the forward propagating Bragg grating. The Bragg grating, i.e. the sound wave, is caused by vibrations of the crystal (or glass) lattice. When these vibrations are associated with optical phonons i.e. high frequency phonons, the effect is called Raman scattering, whereas acoustical phonons i.e. low frequency phonons, are associated with Brillouin scattering.

The strength of the Brillouin scattered wave depends on the direction between the incoming field and the direction to the observation point, whereas Raman scattering is directional independent. In addition, generally Brillouin scattering is much stronger than Raman scattering.

Applications of Brillouin scattering include acousto optic modulators, slowing down light, optical amplifiers, and sensors, for example to measure mechanical strain or temperature. Unfortunately, Brillouin scattering may often lead to drawbacks and limitations, for example in parametric amplifiers and fiber lasers.

In the following we provide a more detailed description of Brillouin scattering. Section 8.1 provides a physical description of an acoustic wave. In Section 8.2 the impact of electrostriction on an acoustic wave is discussed and in Section 8.3 we describe how a set of coupled wave equations between the acoustic wave and the electric field govern the problem of predicting Brillouin scattering. Section 8.4 discusses one important aspect of Brillouin scattering, which is the threshold power where a significant fraction of the launched optical power is reflected. Section 8.5 discusses the methods that have been proposed to minimize the effect of Brillouin scattering in optical fibers, and finally in Section 8.6 a few applications are highlighted.

8.1 Introduction

In Brillouin scattering as well as in Raman scattering photon energy and momentum are conserved quantities during scattering by annihilation or creation of a phonon. That is

$$\text{Energy conservation:} \quad E_{ph} = E_e - E_s \quad \Leftrightarrow \quad \Omega_{ph} = \omega_e - \omega_s, \tag{8.1}$$

where E_{ph} is the phonon energy, which is related to the phonon frequency Ω_{ph} through, $E_{ph} = \hbar\Omega_{ph}$. E_e is the energy of a photon in the incoming optical wave related to the frequency ω_e of the incoming or excitation field through $E_e = \hbar\omega_e$ and E_s the energy of a photon in the scattered optical wave related to the frequency, ω_s, of the scattered wave, also referred to as the Stokes shifted wave, through $E_s = \hbar\omega_s$. In addition, to energy conservation, also the momentum is conserved

$$\text{Momentum conservation:} \quad \mathbf{p}_{ph} = \mathbf{p}_e - \mathbf{p}_s, \tag{8.2}$$

where $\mathbf{p}_{ph}$ is the phonon momentum, $\mathbf{p}_e$ the momentum of a photon that excites the phonon and $\mathbf{p}_s$ the momentum of the scattered photon.

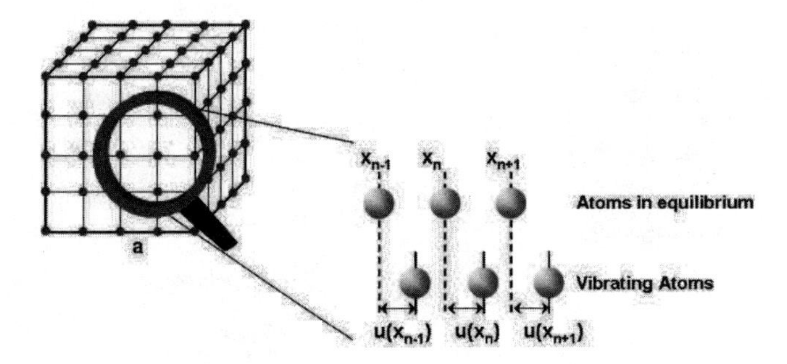

Figure 8.1: Illustration of a figurative model of a lattice consisting of similar atoms in a lattice.

There is a very important difference between Raman and Brillouin scattering. In the case of Raman scattering which is associated with scattering against an optical phonon there is a negligible angular dependence, that is, the strength of the scattered wave is the same in all directions, whereas when light is scattered against an acoustic phonon there is a strong angular dependence. More specifically, if θ is the angle between the incoming optical wave and the scattered wave then in the case of Brillouin scattering, the scattering varies according to $\sin(\theta/2)$, that is, the scattering exists in all directions except for $\theta = 0 + p2\pi$, but varies strongly with θ. We return to this in Section 8.3.1. It is important to note that for example an optical fiber only guides forward and backward propagating waves, consequently, because of the angular dependence mentioned above, the Brillouin scattering only exists in the backward propagating direction.

8.1.1 Acoustic wave

To understand the interaction between an acoustic wave and an electric field, we consider a monoatomic crystal with atoms of mass m and lattice constant a as shown in Figure 8.1, and initially we ignore any coupling to an electric field. The equilibrium position of atom number n is x_n. The displacement of atom number n from equilibrium is described by the function $u_n = u(x_n)$. We assume that the atoms oscillate back and forth parallel to the direction of wave propagation; in Figure 8.1 the wave propagation is from left to right. This is referred to as a longitudinal wave since the atom displacement is parallel to the wavevector of the optical wave that excites the acoustic wave. Figure 8.1 shows the amount of stretch from equilibrium for the spring between atoms n and $n+1$ which is given by $u_{n+1} - u_n$. The two bonds on either side of atom n produce two forces. Further, we assume that a single coupling constant β is sufficient, due to the symmetry of the crystal. Therefore, the total force on atom n is $F_{\text{tot}} = \beta[u_{n+1} - u_n] - \beta[u_n - u_{n-1}]$. From Newton's law this is rewritten as

$$m\frac{d^2u(x_n)}{dt^2} = \beta[u_{n+1} - u_n] - \beta[u_n - u_{n-1}] = \beta[u_{n+1} + u_{n-1} - 2u_n], \tag{8.3}$$

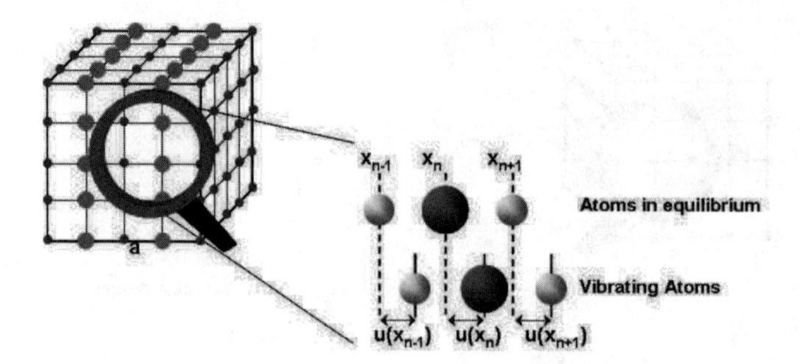

Figure 8.2: A figurative model of a lattice consisting of two types of atoms, just as in the monatomic case if we use the lattice constant a instead of the interatomic spacing a.

where m is the mass of an individual atom. Assuming that each atom may move as a spring, i.e. a harmonic motion, with a phase shift (q_a) between the sinusoidal motion of two neighboring atoms we assume a solution

$$u_{k,n} = u^0 \exp\left(-i(\omega_{ph} t + q_{ph} n a)\right), \tag{8.4}$$

where q_{ph} is the wavenumber of the accustic wave. Inserting this solution into Eqn(8.3) we get

$$m(-\omega_{ph}^2) = \beta[\exp\left(-i q_{ph} a\right) + \exp\left(i q_{ph} a\right) - 2]. \tag{8.5}$$

Using that $[\exp\left(-iq_{ph}a\right) + \exp\left(iq_{ph}a\right) - 2] = [\exp\left(-iq_{ph}a/2\right) - \exp\left(ikq_{ph}a/2\right)]^2$, we find the dispersion relation

$$\omega_{ph}^2 = \frac{4\beta}{m}[\sin(q_{ph}a/2)]^2. \tag{8.6}$$

It is noted that Eqn.(8.6) allows both positive and negative traveling waves by keeping ω_{ph} positive while changing the sign of q_{ph}. In addition to a good approximation ω_{ph} is also linear in q_{ph} for small values of q_{ph}. Consequently, at low wave numbers, the speed of sound ($= d\omega_{ph}/dq_{ph}$) is approximately $a\sqrt{\beta/m}$.

8.1.2 Optical and acoustical phonons

The description of an acoustic wave based on a chain of similar atoms is far from sufficient. Considering a lattice made of planes of different atoms as illustrated in Figure 8.2 is closer to realistic examples. Relative to the chain of similar atoms, the periodicity a is now the distance between nearest identical planes. Now we have to write two equations for the displacement corresponding to the two masses M_1 and M_2. The equations of motion are

$$\begin{aligned} M_1 \frac{d^2 u_s(x_s)}{dt^2} &= \beta[(v_s - u_s)] - \beta[(u_s - v_{s-1})] \\ &= \beta[v_s + v_{s-1} - 2u_s], \end{aligned} \tag{8.7}$$

and

$$M_2 \frac{d^2 v_s(x_s)}{dt^2} = \beta[(u_{s+1} - v_s)] - \beta[(v_s - u_s)] \qquad (8.8)$$
$$= \beta[u_{s+1} + u_s - 2v_s],$$

where s is the number of a particular unit cell. The acoustic wave, which may be a sum of multiple partial waves, is characterized by a specific frequency ω_{ph}, that is $\exp[i(sq_{ph}a - \omega_{ph}t)]$, however with different amplitudes in each plane as defined by u and v. That is, we assume solutions

$$u_s = u^0 \exp[i(sq_{ph}a - \omega_{ph}t)] \qquad (8.9)$$
$$v_s = v^0 \exp[i(sq_{ph}a - \omega_{ph}t)],$$

where u^0 and v^0 are the amplitudes of vibration of atom with mass M_1 and M_2, respectively, both in unit cell s. Inserting Eqn. (8.9) into Eqn. (8.7) and Eqn. (8.8) one find the set of coupled equations

$$M_1 u^0 \omega_{ph}^2 + \beta(v^0 + v^0 \exp(-iq_{ph}a) - 2u^0) = 0 \qquad (8.10)$$
$$M_2 v^0 \omega_{ph}^2 + \beta(u^0 \exp(iq_{ph}a) + u^0 - 2v^0) = 0,$$

which has two non-simple solutions in ω_{ph} given by

$$\omega_{ph}^2 = \beta(1/M_1 + 1/M_2) \pm \beta\sqrt{(1/M_1 + 1/M_2)^2 - 4\sin^2(q_{ph}a)/(M_1 M_2)}. \qquad (8.11)$$

These solutions are shown in Figure 8.3. It is noted that in the case when $M_1 = M_2 = m$ Eqn. (8.11) reduces to $\omega_{ph}^2 = \frac{2\beta}{m}(1 \pm |\cos(q_{ph}a)|)$. For $|q_{ph}a| < \pi/2$ the solution with the minus sign is recognized from Eqn. (8.6), i.e. a string of similar atoms.

Figure 8.3 illustrates that the allowed frequencies of propagation are split into an upper branch, the optical branch, and a lower branch, the acoustical branch. The acoustical branch is similar to the dispersion relation for a monatomic lattice, but the optical branch represents a completely different form of wave motion. In addition, the figure shows that there is a band of frequencies between the optical and acoustical branches that cannot propagate. The width of this forbidden band depends on the difference of the masses M_1 and M_2. If the two masses are equal, the two branches join (become degenerate) at $q_{ph} = \pi/2a$. The vertical axis is the frequency of the phonon, equivalent to its energy, while the horizontal axis is the wave number. The upper boundary of the horizontal axis is that of the first Brillouin zone, which extent from $q_{ph} = -\pi/a$ to $q_{ph} = \pi/a$.

8.1.3 Transverse and longitudinal acoustic waves shear waves

In the above discussion of a string of similar atoms or two types of atoms, we have been dealing with so-called compressional waves or longitudinal waves, i.e. waves of

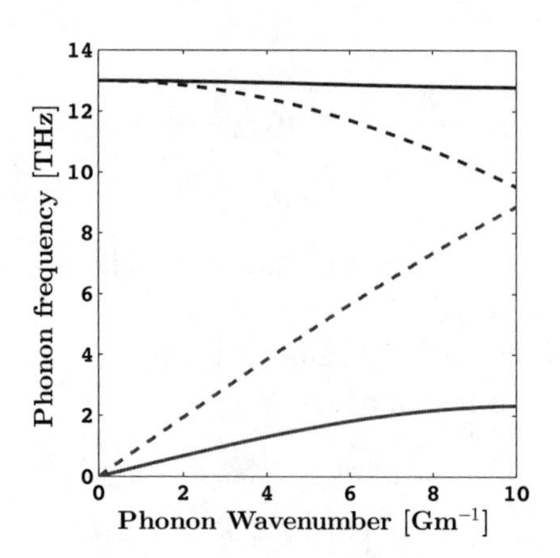

Figure 8.3: The phonon frequency as a function of the phonon wave number. In the figure the lattice spacing, a, equals 150 pm. The dashed line is calculated for masses $M_1 = M_2 = 1.66 \cdot 10^{-27}$ kg, i.e. one atomic mass unit, whereas in the solid line one of the two masses is increased by a factor of 30. The spring constant (coupling coefficient) is arbitrarily defined equal to an angular frequency $\omega_{ph} = 2\pi \cdot 13$ THz at wavenumber $q_{ph} = 0$.

alternating pressure deviations from the equilibrium pressure, causing local regions of compression and repulsion. Such waves typically exist in gases, solids, and liquids. However, an acoustic wave may also be transmitted as transverse waves of alternating shear stress at a right angle to the direction of propagation. The shear wave moves as a shear or transverse wave, so motion is perpendicular to the direction of wave propagation. Transverse waves typically propagate in solids.

The acoustic field, be it compressional waves and/or shear waves, is accurately described by fairly complicated equations as for example compressional and shear waves propagate at different speeds and may couple to each other at interfaces. In general, the mathematical problem involves prediction of a vectorial field. However, often the problem may be reduced to a much simpler task involving solution of a scalar wave equation. This is for example the case when the medium cannot support shear waves or when the excitation is small; then the acoustic field can be fully characterized by a single scalar wave function. See [83] Section 2.2 for the complete set of conditions that must be satisfied. The acoustic field is still a vector field with a velocity vector proportional to the gradient of the acoustic field. Also, certain types of losses can still be accounted for within this approximation.

In the following we treat the acoustic wave as a scalar wave.

8.2 Electrostriction

Without further details we now extend the above description of simple lattice structures to a continuous material meaning that we introduce a density wave, and the equation of motion that describes this wave is adopted from the previous section as

$$\frac{\partial^2 u}{\partial t^2} = v^2 \nabla^2 u, \tag{8.12}$$

where v^2 is the velocity of the wave, where we have used that the second-order derivative may be approximated using $\frac{d^2u}{dx^2} = \frac{u_{n+1}-2u_n+u_{n-1}}{\Delta x^2}$. This equation of motion for the acoustic wave has the general solution

$$u = a\exp\left(-i(\omega_{ph}t + \mathbf{q}_{ph}\cdot\mathbf{r})\right) + b\exp\left(-i(\omega_{ph}t - \mathbf{q}_{ph}\cdot\mathbf{r})\right), \tag{8.13}$$

where $\omega_{ph} = v|\mathbf{q}_{ph}|$ is the dispersion relation that relates the frequency ω_{ph} of the acoustic wave to the wave number $|\mathbf{q}_{ph}|$. In Section 8.3 we return to a discussion of this dispersion relation and we show that Brillouin scattering is strongest in the direction opposite to the direction of the excitation beam, and in this case $|q|_{ph} \approx 2k_e$ where k_e is the wavenumber of the excitation beam.

It is noted that where $u(x_n)$ in Section 8.1 simply described the displacement of individual atoms from their equilibrium position, $u(x)$ is now related to a change in the density of the material. However, the true material density is described as a constant contribution in addition to the varying density. For discussion of scattering of light against a propagating material density variation (a sound wave) it is not necessary to include the mean density of the material [1]. A mean density simply vanishes in Eqn. (8.12).

In addition, in the above, for example Eqn. (8.12), we have described the propagation of an acoustic wave without any coupling to the electric field, hence the description is not sufficient to describe Brillouin scattering. Inclusion of the coupling to an electric field is the main content of this and the succeeding sections. In addition, in Section 8.1 we have also described the acoustic wave through a displacement of individual atoms from their equilibrium position in units of [m]. However, the usual starting point when describing Brillouin scattering is the mutual interaction between electrical fields (photons) and an acoustic wave i.e. the density wave (phonons). Consequently the equation of motion Eqn. (8.12), is replaced by an equation in variation in density, $\Delta\rho$ in units of [kg/m^3],

$$\frac{\partial^2 \Delta\rho}{\partial t^2} = v^2 \nabla^2 \Delta\rho. \tag{8.14}$$

Later we use the fact that the variation in density is proportional to a change in dielectric constant $\Delta\varepsilon_r$ i.e.

$$\Delta\varepsilon_r = \frac{\partial\varepsilon_r}{\partial\rho}\Delta\rho = \gamma_e\frac{\Delta\rho}{\rho_0}, \tag{8.15}$$

where $\gamma_e = \rho_0 \left.\frac{\partial \varepsilon_r}{\partial \rho}\right|_\rho$ is referred to as the electrostrictive constant, which is dimensionless, and ρ_0 the mean material density.

The coupling between the electric field and the change in density, i.e. the change in dielectric constant through Eqn. (8.15), is described through electrostriction, which is the process by which materials become compressed in the presence of an electric field. The force that acts on the dielectric and appears as a source term in the acoustic wave equation is given by [33]

$$\mathbf{F} = \frac{1}{2}\varepsilon_0 \gamma_e \nabla(\mathbf{E} \cdot \mathbf{E}), \tag{8.16}$$

where ε_0 is the vacuum permittivity, that is, the force is the gradient of the dot product of the electrical field vector with itself. Inserting this force into the equation of motion of the density variations leads to

$$\frac{\partial^2 \Delta\rho}{\partial t^2} - v^2 \nabla^2 \Delta\rho = \nabla \cdot \mathbf{F}, \tag{8.17}$$

where it is noted that in the equation it is the divergence of the force as opposed to the force that appears, even though the starting point for this discussion was in fact based on Hooke's law relating the sum of forces to the displacement from equilibrium. This is explained by the fact the $\Delta\rho$ is related to the material density.

8.2.1 Attenuation

As a matter of fact the propagation of acoustic waves is more complicated than the above description, for example governed by Eqn. (8.14), which treats the acoustic wave as a simple scalar field by ignoring for example shear waves and heat dissipation [1]. However, one effect we have to include is attenuation of the acoustic wave. In the following, we simply introduce an acoustic damping constant Γ_B into the acoustic wave equation.

$$\frac{\partial^2 \Delta\rho}{\partial t^2} - \Gamma_B \nabla^2 \frac{\partial \Delta\rho}{\partial t} = v^2 \nabla^2 \Delta\rho. \tag{8.18}$$

In addition to describing damping, the second term on the left-hand side in Eqn. (8.18) also describes broadening of the linewidth of the Brillouin scattering.

Example 8.1. *Dispersion relation*
To investigate the impact of the attenuation term in the equation of motion for the acoustic wave described by Eqn. (8.18), we wish to determine the dispersion relation of the acoustic wave. Our starting point for this is to assume a plane wave for the acoustic wave

$$\Delta\rho = \frac{1}{2}\left(A(r_a) e^{i(\mathbf{q}_{ph}\cdot\mathbf{r} - \omega_{ph} t)} + c.c\right). \tag{8.19}$$

The second-order derivative of the change in density with time equals

$$\frac{\partial^2 \Delta\rho}{\partial t^2} = \left(\frac{\partial^2 A}{\partial t^2} - i2\omega_{ph}\frac{\partial A}{\partial t} - \omega_{ph}^2 A\right) e^{i(\mathbf{q}_{ph}\cdot\mathbf{r}-\omega_{ph}t)}. \quad (8.20)$$

The term in Eqn. (8.18) that includes the attenuation gives

$$\Gamma_B \nabla^2 \frac{\partial \Delta\rho}{\partial t} = \Gamma_B e^{i(\mathbf{q}_{ph}\cdot\mathbf{r}-\omega_{ph}t)} \quad (8.21)$$
$$\left(\nabla^2\frac{\partial A}{\partial t} + 2i\mathbf{q}_{ph}\cdot\nabla\frac{\partial A}{\partial t} - |\mathbf{q}_{ph}|^2\frac{\partial A}{\partial t} - i\omega_{ph}(\nabla^2 A + 2i\mathbf{q}_{ph}\cdot\nabla A - |\mathbf{q}_{ph}|^2 A)\right),$$

where $|\mathbf{q}_{ph}|^2$ is the sum of the squared coordinates of $\mathbf{q}_{ph}$. Finally the Laplacian of the change in density is

$$v^2\nabla^2\Delta\rho = v^2\left(\nabla^2 A + 2i(\nabla A\cdot\mathbf{q}_{ph}) - |\mathbf{q}_{ph}|^2 A\right) e^{i(\mathbf{q}_{ph}\cdot\mathbf{r}-\omega_{ph}t)}. \quad (8.22)$$

Inserting Eqn (8.20) through Eqn. (8.22) into Eqn. (8.18) and ignoring all time derivatives and all second-order derivatives in space, corresponding to stead-state and applying the slowly varying envelope approximation, leads to

$$-\omega_{ph}^2 A - i\Gamma_B\omega_{ph}|\mathbf{q}_{ph}|^2 A + v^2|\mathbf{q}_{ph}|^2 A - 2v^2 i(\nabla A\cdot\mathbf{q}_{ph}) - 2\Gamma_B\omega_{ph}\mathbf{q}_{ph}\cdot\nabla A. \quad (8.23)$$

It is customary to assume that the phonon only propagates over a very short distance and consequently drop terms with first order spatial derivative. Consequently we find

$$-\omega_{ph}^2 A - i\Gamma_B\omega_{ph}|\mathbf{q}_{ph}|^2 A + v^2|\mathbf{q}_{ph}|^2 A = 0, \quad (8.24)$$

which is fulfilled for

$$|\mathbf{q}_{ph}|^2 = \omega_{ph}^2\frac{1}{v^2 - i\Gamma_B\omega_{ph}}. \quad (8.25)$$

It is noted that $|\mathbf{q}_{ph}|$ in fact is a complex number. Using the approx. $(1-x)^{-1} \approx 1+x$, valid for $|x| < 1$ and $(1-x)^{1/2} \approx 1 + x/2$, we find

$$|\mathbf{q}_{ph}| \approx \frac{\omega_{ph}}{v} + i\frac{1}{2}\frac{\Gamma_B\omega_{ph}^2}{v^3}. \quad (8.26)$$

Inserting this into the assumed solution to the acoustic wave equation shows that the sound wave is attenuated with an absorption described through the imaginary part of $|\mathbf{q}_{ph}|$ that is $\Gamma_B\omega_{ph}^2/v^3$.

—— ■ ——

8.2.2 Response time

The response time is related to the phonon lifetime $T_B \approx 10$ ns [3]. The dynamic aspects may be even more important than in the case of Raman due to the long response time. The quasi CW is valid only for pulse widths of 100 ns or more. The coupled equations Eqn. (8.16) and Eqn. (8.17) together with the relevant propagation equations in Chapter 5 need to be solved. One example where the response time of Brillouin scattering impacts the propagation is when considering pump pulses much wider than the response time. An example was demonstrated by Bar-Joseph in 1985 [84] and elaborated to Raman amplifiers by Ott and coworkers in 2009 [85].

8.3 Coupled wave equations

Generally an acoustic wave is as mentioned earlier related to the pressure i.e. described by a vector field. However, for simplicity, we assume that it can be modeled as a scalar field. This approximation is valid for example when considering only small acoustic fields. In addition, in the following we consider only plane waves, however with arbitrary direction. That is

$$\text{the acoustic wave is:} \qquad \Delta\rho = \frac{1}{2}\left(A(r_a)e^{i(\mathbf{q}_{ph}\cdot\mathbf{r}-\omega_{ph}t)} + c.c\right), \tag{8.27a}$$

$$\text{the excitation beam is:} \qquad E_e = \frac{1}{2}\left(E_e^0(r_e)e^{i(\mathbf{k}_e\cdot\mathbf{r}-\omega_e t)} + c.c\right), \tag{8.27b}$$

$$\text{and the scattered field is:} \qquad E_s = \frac{1}{2}\left(E_s^0(r_s)e^{i(\mathbf{k}_s\cdot\mathbf{r}-\omega_s t)} + c.c\right), \tag{8.27c}$$

where r_i is the algebraic distance measured along the respective distance of propagation $r_i = \mathbf{k}_i \cdot \mathbf{r}/|\mathbf{k}_i|$ and for the acoustic wave $r_{ph} = \mathbf{q}_{ph} \cdot \mathbf{r}/|\mathbf{q}_{ph}|$ [86].

8.3.1 Acoustic wave equation

Under the assumption of plane waves we can now evaluate the acoustic wave. Using the example from above in Section 8.2.1, using $\omega_{ph} = \omega_2 - \omega_1$ and $|\mathbf{q}_{ph}| = \omega_{ph}/v$ we may rewrite Eqn. (8.17) as

$$\frac{\partial^2 A}{\partial t^2} - 2i\omega_{ph}\frac{\partial A}{\partial t} - (\omega_{ph}^2)A - v^2\left(-|\mathbf{q}_{ph}|^2 A + 2i|\mathbf{q}_{ph}|\nabla A + \nabla^2 A\right) = \nabla \cdot \mathbf{F}. \tag{8.28}$$

Assuming a slowly varying amplitude in space

$$2\,|\mathbf{q}_{ph}|\,|\nabla A| \gg \left|\nabla^2 A\right|, \tag{8.29}$$

and steady state in time

$$\frac{\partial^2 A}{\partial t^2} = \frac{\partial A}{\partial t} = 0, \tag{8.30}$$

we find

$$(v^2|\mathbf{q}_{ph}|^2 - \omega_{ph}^2)A - v^2 2i\,|\mathbf{q}_{ph}|\,\nabla A = \nabla \cdot \mathbf{F}. \tag{8.31}$$

Assuming that the acoustic wave is highly damped, we include a term

$$\frac{dA}{dz} = \frac{1}{2}\frac{\omega_{ph}|q_{ph}\mathbf{q}_{ph}|}{v^2}\Gamma_B A, \tag{8.32}$$

in agreement with Eqn. (8.21). Then A needs to satisfy

$$\left((v^2\,|\mathbf{q}_{ph}|^2 - \omega_{ph}^2) - i\omega_{ph}\Gamma_B|\mathbf{q}_{ph}|^2\right)A = \nabla \cdot \mathbf{F}. \tag{8.33}$$

That is

$$A = \frac{\nabla \cdot \mathbf{F}}{(v^2\,|\mathbf{q}_{ph}|^2 - \omega_{ph}^2) - i\omega_{ph}\Gamma_B|\mathbf{q}_{ph}|^2}. \tag{8.34}$$

Inserting the force from Eqn. (8.16)

$$A = \frac{\frac{1}{2}\varepsilon_0\gamma_e\nabla \cdot \nabla(\mathbf{E}\cdot\mathbf{E})}{(v^2\,|\mathbf{q}_{ph}|^2 - \omega_{ph}^2) - i\omega_{ph}\Gamma_B|\mathbf{q}_{ph}|^2}. \tag{8.35}$$

Regarding the term described by the dot product of the electrical field with itself, we note that the change in refractive index is defined on the basis of static electric fields and a fixed temperature. When dealing with fields at optical frequencies, $\mathbf{E}\cdot\mathbf{E}$ is replaced by $\langle\mathbf{E}\cdot\mathbf{E}\rangle$ where the brackets denote the time average over one optical cycle. Note that there may be contributions not only from the static component but also from the hypersonic component (even though the average of the hypersonic component equals zero)[1].

Considering a case where $\mathbf{E}$ consists of one frequency only, then the hypersonic component is at twice the optical frequency. More specifically, assuming an electric field $\mathbf{E} = \frac{1}{2}(\mathbf{E^0}e^{-i\omega t} + c.c.)$ then $\langle\mathbf{E}\cdot\mathbf{E}\rangle = \langle 1/4(\mathbf{E^0}\cdot\mathbf{E^0}e^{-2i\omega t} + (\mathbf{E^0})^*\cdot(\mathbf{E^0})^*e^{2i\omega t} + 2\mathbf{E^0}\cdot(\mathbf{E^0})^*)\rangle \cong \frac{1}{2}\langle\mathbf{E^0}\cdot(\mathbf{E^0})^*\rangle$. In the remaining we assume that the only contribution to the electrostriction comes from $\frac{1}{2}\langle\mathbf{E^0}\cdot(\mathbf{E^0})^*\rangle$, and therefore we ignore the hypersonic contributions.

Example 8.2. *Direction*
Let us consider the source term for the acoustic wave, i.e. the term proportional to $\nabla\cdot\nabla(\mathbf{E}\cdot\mathbf{E})$ as induced by the fields E_e and E_s noted in the beginning of this section. The relevant source term at $\omega_{ph} = \omega_e - \omega_s$ originates from $E_eE_s^*$, that is

$$\nabla\cdot\nabla(\mathbf{E}\cdot\mathbf{E}) \propto -|\mathbf{k}_e - \mathbf{k}_s|^2 = -|\mathbf{q}_{ph}|^2. \tag{8.36}$$

Figure 8.4 illustrates the three wavevectors of relevance; the wavevectors of the electric fields E_e and E_s, that is: $\mathbf{k}_e$ and $\mathbf{k}_s$ and the resulting wavevector of the acoustic wave $\mathbf{q}_{ph}$.

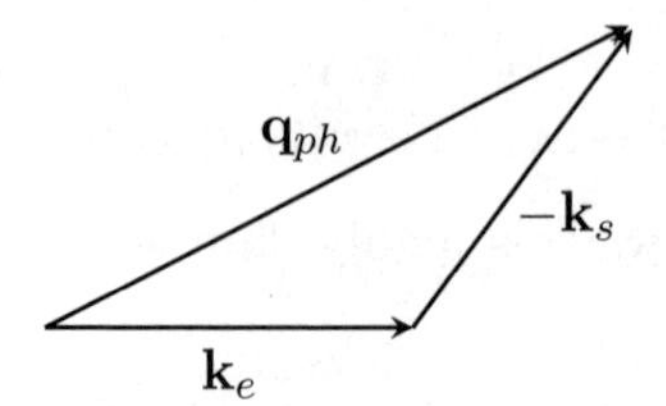

Figure 8.4: $\mathbf{k}_e$ is the wavevector of the excitation beam, $\mathbf{k}_s$ is the wavevector of the scattered beam, and $\mathbf{q}_{ph}$ is the wavevector of the acoustic wave. These are related through the conservation of momentum, and using the coupled equations for the acoustic wave and the electric field one can see that the Brillouin scattering is strongest in the backward direction.

Since the frequency of the phonon is very low compared to optical frequencies i.e. the frequency of the scattered wave is very close to that of the incident wave, $|\mathbf{k}_e| \approx |\mathbf{k}_s|$, one finds

$$|\mathbf{q}_{ph}| = 2|\mathbf{k}_e| \sin\left(\frac{\theta}{2}\right). \tag{8.37}$$

Thus, the source terms of the acoustic wave is strongest in the direction of $\theta = \pi$, that is in the backward direction, which is the direction opposite to the launched beam.

In addition, since $\omega_{ph} = v|\mathbf{q}_{ph}|$ we get $\omega_{ph} = 2v|\mathbf{k}_e| \sin\left(\frac{\theta}{2}\right)$, that is, the frequency shift varies with angle θ and is largest in the backward direction.

—— ■ ——

8.3.2 Wave equations for electric field

The induced polarization of relevance to the study of Brillouin scattering is

$$\mathbf{P}^0_{\omega_s} = \varepsilon_0 \left(\chi^{(1)} + K\chi^{(3)}\mathbf{E}^0_e(\mathbf{E}^0_e)^*\right)\mathbf{E}^0_s, \tag{8.38}$$

where the relevant susceptibility is $\chi^{(3)}_{ijkl}(-\omega_s, \omega_e, -\omega_e, \omega_s)$. When dealing with the refractive index, we recall that the relative permittivity may be written as $\varepsilon_r = 1 + \chi^{(1)} + K\chi^{(3)}\mathbf{E}^0_e(\mathbf{E}^0_e)^*$.

With this we can derive the nonlinear propagation equation for the two fields at ω_e and ω_{ph} respectively. To do this we need to evaluate the induced polarization at the two frequencies due to the propagating acoustic wave.

First we need to relate the change in density $\Delta\rho$ to the relative permittivity. Expressing the total mass per unit volume ρ as the sum of the average mass density

ρ and a small variation $\Delta\rho$, i.e. ($\rho = \rho_0 + \Delta\rho$), we can write the relative permittivity as a function of the density by using a simple Taylor expansion

$$\varepsilon_r(\rho) = \varepsilon_r(\rho_0) + \left.\frac{\partial\varepsilon_r}{\partial\rho}\right|_{\rho_0}(\rho - \rho_0). \tag{8.39}$$

One notes that the variation in density is a wave at frequency ω_{ph}, consequently ε_r consists of a DC term ($= \varepsilon_r(\rho_0)$) and a term that oscillates with frequency ω_{ph} ($= \left.\frac{\partial\varepsilon_r}{\partial\rho}\right|_{\rho_0}(\rho - \rho_0)$), which is set up by the electric field components $\mathbf{E}_s^0$ and $(\mathbf{E}_e^0)^*$.

From this, using the above definition of the relative permittivity, we get

$$(\mathbf{P}_{\omega_s}^0)^{(\mathrm{NL})} = \varepsilon_0(\varepsilon_r(\rho_0) - 1)\mathbf{E}_s^0 + \varepsilon_0 \left.\frac{\partial\varepsilon_r}{\partial\rho}\right|_{\rho_0}\Delta\rho\mathbf{E}_e^0, \tag{8.40}$$

where the first term on the right-hand side ($\varepsilon_0(\varepsilon_r(\rho_0) - 1)\mathbf{E}_s^0$) is recognized as the linear induced polarization, and the second term as the nonlinear induced polarization, i.e.

$$(\mathbf{P}_{\omega_s}^0)^{(\mathrm{NL})} = \varepsilon_0 \left.\frac{\partial\varepsilon_r}{\partial\rho}\right|_{\rho_0}\Delta\rho\mathbf{E}_e^0, \tag{8.41}$$

which may also be expressed in terms of the electrostrictive constant γ_e defined through $\Delta\varepsilon_r = \gamma_e\frac{\Delta\rho}{\rho_0}$ for $\Delta\varepsilon_r$, see Eqn. (8.15).

When comparing to the description using susceptibilities, we find using Eqn. (8.38) that the Brilliouin susceptibility is defined through the relation

$$K\chi_{ijkl}^{(3)}(-\omega_s, \omega_e, -\omega_e, \omega_s)\mathbf{E}_e^0(\mathbf{E}_e^0)^*\mathbf{E}_s^0 = \left.\frac{\partial\varepsilon_r}{\partial\rho}\right|_{\rho_0}\Delta\rho\mathbf{E}_e^0. \tag{8.42}$$

Just like Raman scattering, Brillouin scattering may be either stimulated or spontaneous. If the process is stimulated it requires the existence of a component of the electrical field at frequency at ω_s in the counter-propagating direction through the scattering medium, whereas if the process is spontaneous it only requires the presence of an excitation field at the input. In all cases the scattering depends on the population of phonons just as in the case of Raman scattering, that means that the process is temperature dependent.

Example 8.3. *Spontaneous emission*
As discussed in the introduction to this chapter, the Brillouin scattered light is strongest in the backward direction. We now consider how the induced polarization acts as a source term in the wave equation for the spontaneously scattered field. As previously, we start with a scalar plane acoustic wave and a scalar plane electric field wave, given by

$$\text{the acoustic wave:} \qquad \Delta\rho = \frac{1}{2}\left(A(r_a)e^{i(\mathbf{q}_{ph}\cdot\mathbf{r} - \omega_{ph}t)} + c.c\right), \tag{8.43}$$

$$\text{the excitation beam:} \qquad E_e = \frac{1}{2}\left(E_e^0(r_e)e^{i(\mathbf{k}_e\cdot\mathbf{r} - \omega_e t)} + c.c\right). \tag{8.44}$$

From this we evaluate the source term

$$\Delta\rho E_e = \frac{1}{4}\left(A(r_a)E_e^0(r_e)e^{i((\mathbf{q}_{ph}+\mathbf{k}_e)\cdot\mathbf{r}-(\omega_{ph}+\omega_e)t)} + A(r_a)^* E_e^0(r_e)e^{i(\mathbf{k}_e-\mathbf{q}_{ph})\cdot\mathbf{r}-(\omega_e-\omega_{ph})t)} + c.c\right). \tag{8.45}$$

From this it appears that the scattered frequency has two components, one at frequency $\omega_{ph} + \omega_e$ and one at $\omega_e - \omega_{ph}$. In addition, the corresponding wavevectors are $\mathbf{q}_a + \mathbf{k}_e$ and $\mathbf{k}_e - \mathbf{q}_{ph}$. The first corresponds to taking energy from the phonons, also referred to as anti-Stokes scattering, whereas the latter corresponds to giving energy to the phonon, referred to as Stokes scattering similar to the Raman case. The Stokes is much more likely than anti-Stokes scattering.

—— ■ ——

Example 8.4. *Brillouin Gain*
In this example we are interested in evaluating the Brillouin gain in an optical fiber. As opposed to the example above this means that we may replace $\mathbf{r}$ by $\pm z$. With respect to the density variations we then consider the acoustic wave

$$\Delta\rho = \frac{1}{2}\left(A(r_a)e^{i(q_{ph}z-\omega_{ph}t)} + c.c\right). \tag{8.46}$$

The direction of the wave is included i.e. since the scattered wave propagates in the opposite direction as the incoming wave, i.e. $\mathbf{k}_s$ is in opposite direction of $\mathbf{k}_e$, we use

$$\tilde{E} = E_e + E_s = \frac{1}{2}\left((E_e^0 e^{i(k_e z-\omega_e t)} + c.c.) + (E_s^0 e^{i(-k_s z-\omega_s t)} + c.c.)\right). \tag{8.47}$$

Consequently

$$\tilde{E}\tilde{E} = \frac{1}{4}\left(|E_e^0|^2 + |E_s^0|^2 + (E_e^0)^2 e^{2i(k_e z-\omega_e t)} + (E_s^0)^2 e^{2i(-k_s z-\omega_s t)} + 2E_e^0 E_s^0 e^{i(k_e z-\omega_e t)}e^{i(-k_s z-\omega_s t)} + 2E_e^0 e^{i(k_e z-\omega_e t)}(E_s^0)^* e^{-i(-k_s z-\omega_s t)} + c.c.\right). \tag{8.48}$$

From this we identify the term operating at the frequency $\omega_a = \omega_e - \omega_s$, that is

$$2E_e^0(E_s^0)^* e^{i((k_e+k_s)z-(\omega_e-\omega_s)t)}. \tag{8.49}$$

From Eqn. (8.35) the amplitude to the density variation is then

$$A = \frac{\varepsilon_0\gamma_e E_e^0(E_s^0)^*(-q_{ph}^2)}{(v^2|\mathbf{q}_{ph}|^2 - \omega_{ph}^2) - i\omega_{ph}\Gamma_B|\mathbf{q}_{ph}|^2}. \tag{8.50}$$

Secondly, from the electric fields we also obtain an expression for the induced polarization

$$\begin{aligned} P^{(\mathrm{NL})} &= \frac{\varepsilon_0 \gamma_e}{\rho_0} \Delta\rho (E_e + E_s) \qquad (8.51)\\ &= \frac{\varepsilon_0 \gamma_e}{\rho_0} \Delta\rho \frac{1}{4} \Big(A E_e^0 e^{i((q_{ph}+k_e)z-(\Omega+\omega_e)t)} + A E_s^0 e^{i((q_{ph}-k_s)z-(\Omega+\omega_s)t)} \\ &+ A^* E_e^0 e^{i((k_e-q_{ph})z-(\omega_e-\Omega)t)} + A^* E_s^0 e^{i(-(k_s+q_{ph})z-(\omega_s-\Omega)t)} + c.c. \Big). \end{aligned}$$

It is emphasized that to arrive at this amplitude, it is assumed that the amplitudes of the electric field have no changes as a function of distance. With the amplitude now known we have an expression for the acoustic wave by inserting Eqn. (8.51) into Eqn. (8.46) remembering that $q_{ph} = k_s + k_e$ and with $\omega_s = \omega_e - \Omega$ we identify $A^* E_e^0 e^{i((k_e-q_{ph})z-(\omega_e-\Omega)t)}$ as matching the scattered beam, i.e. compare $e^{i(-k_s z-\omega_s t)}$, and $A E_s^0 e^{i((q_{ph}-k_s)z-(\Omega+\omega_s)t)}$ as matching the incoming beam, i.e. compare $e^{i(k_e z-\omega_e t)}$.

From Chapter 5 or equivalently Eqn. (7.24) in [2]

$$i\left(\frac{\partial}{\partial z} - v_g^{-1}\frac{\partial}{\partial t}\right) E_s^0 = -\frac{\omega_\sigma^2 \mu_0}{2k_s} (P^0)^{(\mathrm{NL})} \exp(ik_s z), \qquad (8.52)$$

where $(P^0)^{(\mathrm{NL})}$ is the amplitude of the induced polarisation at ω_s. Using the result from Eqn. (8.51) $(P^0)^{(\mathrm{NL})} = \frac{1}{2}\frac{\varepsilon_0\gamma_e}{\rho_0} A^* E_e^0 e^{i((k_e-q_{ph})z)}$, where the reader is reminded that $P^{(\mathrm{NL})} = \frac{1}{2}\left((P^0)^{(\mathrm{NL})} + c.c.\right)$ gives

$$i\left(\frac{\partial}{\partial z} - v_g^{-1}\frac{\partial}{\partial t}\right) E_s^0 = \frac{\omega_2}{4cn}\frac{\gamma_e}{\rho_0} A^* E_e^0, \qquad (8.53)$$

where we have used that $c^2 = 1/(\varepsilon_0\mu_0)$, and where

$$A = \frac{\varepsilon_0 \gamma_e E_e^0 (E_s^0)^* (-q_{ph}^2)}{(v^2 |\mathbf{q}_{ph}|^2 - \omega_{ph}^2) - i\omega_{ph}\Gamma_B |q_{ph}|^2}. \qquad (8.54)$$

Likewise we find for the field at ω_e

$$i\left(\frac{\partial}{\partial z} + v_g^{-1}\frac{\partial}{\partial t}\right) E_e^0 = \frac{\omega_2}{4cn}\frac{\gamma_e}{\rho_0} A E_s^0. \qquad (8.55)$$

With this, assuming q_{ph} is real and considering steady-state, we get conditions

$$\begin{aligned} i\frac{\partial E_s^0}{\partial z} &= \frac{\omega_2}{4cn}\frac{\gamma_e}{\rho_0} \frac{\varepsilon_0\gamma_e(-q_{ph}^2)}{\left((v^2 q_{ph}^2 - \omega_{ph}^2) + i\omega_{ph}\Gamma_B |q_{ph}|^2\right)} |E_e^0|^2 E_s^0, \qquad (8.56)\\ i\frac{\partial E_e^0}{\partial z} &= \frac{\omega_2}{4cn}\frac{\gamma_e}{\rho_0} \frac{\varepsilon_0\gamma_e(-q_{ph}^2)}{\left((v^2 q_{ph}^2 - \omega_{ph}^2) - i\omega_{ph}\Gamma_B |q_{ph}|^2\right)} |E_s^0|^2 E_e^0. \end{aligned}$$

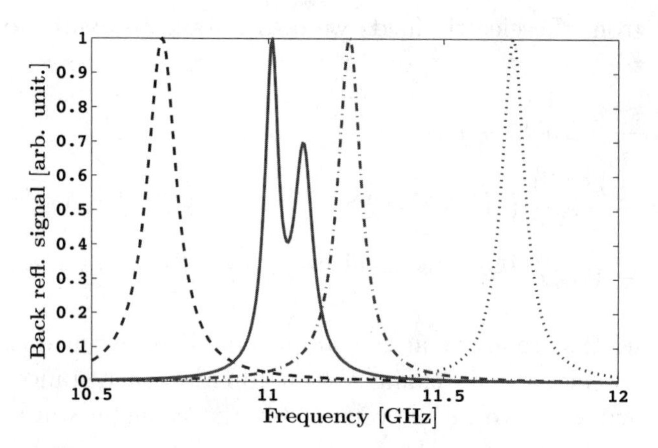

Figure 8.5: Brillouin gain spectra (spontaneous emission) as a function of phonon frequency for four types of optical fibers. Dashed-dotted: is a silica core fiber, Solid: a depressed cladding fiber, Dashed: a dispersion shifted fiber, and Dotted: an aluminum doped fiber. Adapted from [88] and [89].

Rewriting the above differential equations in electric field to equations in intensity defined through: $I = \frac{1}{2}cn\varepsilon_0|E^0|^2$ we obtain an equation for the amplification of the intensity at the frequency ω_s

$$\frac{dI_s}{dz} = -gI_sIe, \tag{8.57}$$

where the gain coefficient is approximated by

$$g_B = \frac{16\omega_2^2\gamma_e^2}{c^3n^2\rho_0 v\Gamma_B q_{ph}^2} \frac{(\Gamma_B q_{ph}^2/2)^2}{((vq_{ph} - \omega_{ph})^2 + (\Gamma_B|q_{ph}|^2/2)^2)}, \tag{8.58}$$

where we have used that $\omega_{ph} = q_{ph}v \approx 2k_2v = 2\omega_2 nv/c$.

If we consider values from silica [87], that is the material density $\rho_0 = 2.21 \cdot 10^3$ kg/m^{-3}, the acoustic velocity $v = 5.96 \cdot 10^3$ m/s, and an electrostrictive coefficient $\gamma_e = 0.286$, and a spontaneous linewidth $\Delta\nu_B = \Gamma_B q^2 = 38.44$ MHz at 1.0 μm, then we find a peak Brillouin gain coefficient of $g_B = 1.5 \cdot 10^{-10}$ m/W.

Figure 8.5 shows examples of Brillouin gain spectra from typical communication fibers. The peak frequency of the Brillouin spectrum corresponding to the different fibers varies from 10.7 to 11.8 GHz. The bandwidth of the Brillouin spectra of each fiber is very narrow (tens of MHz) which may be comparable to the width of an excitation laser. Consequently one needs to consider the spectral shape of the excitation source as well as the signal source when predicting Brillouin scattering.

Provided that the laser spectrum consists of a single mode broadened by stochastic fluctuations, the average effect on Brillouin scattering of the finite spectral width of the laser is found by convolving the laser spectrum $g(\nu_L)$ with the Brillouin frequency spectrum $f(\nu)$. Assuming that the laser line is a Lorentzian profile of width

$\Delta\nu_L$, the convolution $f(\nu) * g(\nu_L)$ is a Lorentzian of width $\Delta\nu_L + \Delta\nu_B$ and thus the gain coefficient is multiplied by

$$\frac{\Delta\nu_B}{\Delta\nu_L + \Delta\nu_B}. \tag{8.59}$$

It is noted that because of the narrow width of the Brillouin spectrum (tens of MHz) it is easy to create a case where the gain coefficient simplifies to

$$\frac{\Delta\nu_B}{\Delta\nu_L}. \tag{8.60}$$

The dependence on the laser spectral bandwidth is also true for example in the case of Raman scattering. However, it is noted that contrary to the case of Brillouin scattering, the bandwidth of the Raman scattering (tens of THz) is typically much broader than the laser linewidth and consequently Eqn. (8.59) reduces to a factor of one; the reader is reminded that in the case of Raman scattering $\Delta\nu_B$ needs to be replaced by $\Delta\nu_R$.

The above reduction Eqn. (8.59) may not be accurate in all cases as for example if the spectrum is an information carrying signal where the modes in the laser signal have a specific time-correlation. In this case this time dependence may need to be included.

—— ■ ——

8.4 Threshold

In many applications of optical fibers Brilliouin scattering needs to be considered, since electrostriction in glass is more dominant than for example both the Kerr effect and the Raman scattering, at least when working with continuous wave optical fields.

As discussed previously, the consequence of Brilliouin scattering is that light is shifted in frequency and backreflected. This represents a significant limitation for example when designing fiberlasers, fiber optics parametric amplifiers or even when an optical pump is launched to a fiber amplifier. Thus Brillouin scattering is undesirable and it makes sense to introduce a threshold for Brillouin scattering. There are several ways one may define a threshold, below we highlight two.

In Smith [90] a threshold was defined as

$$(P_p^0)_{\text{th}} = \frac{21 A_{\text{eff}}}{g_B L_{\text{eff}}}. \tag{8.61}$$

This definition was used in 1982 when D. Cotter reported measurements of a threshold in a 13.6 km long fiber [87].

Another definition of a threshold is when the Rayleigh scattered power equals the backscattered Brillouin power. The Rayleigh scattered power equals

$$P_r^R = P_p^0 \frac{\alpha S_B(1 - \exp(-2\alpha L))}{2\alpha}, \tag{8.62}$$

where the recapture cross-section is defined as in Eqn. (7.7). Assuming that backscattered Brillouin power is undepleted and furthermore predicting the backscattered power as if the power $h\nu d\nu$ that is one photon per unit bandwidth were present at the output end of the fiber and finally, neglecting any temperature effect, then

$$P_r^{\text{SBS}} = h\nu\Delta\nu \exp(g_B P_p^0 L_{\text{eff}}). \tag{8.63}$$

The approximation of undepleted power is valid since the threshold is low. Evaluating the input power where these are identical leads to

$$P_p^0 \frac{\alpha S_B(1 - \exp(-2\alpha L))}{2\alpha} = h\nu\Delta\nu \exp(g_B P_p^0 L_{\text{eff}}), \tag{8.64}$$

from which the threshold power $(P_p^0)_{\text{th}}$ may be found. Figure 8.6, right illustrates the threshold predited using this method in a 10 km long optical fiber.

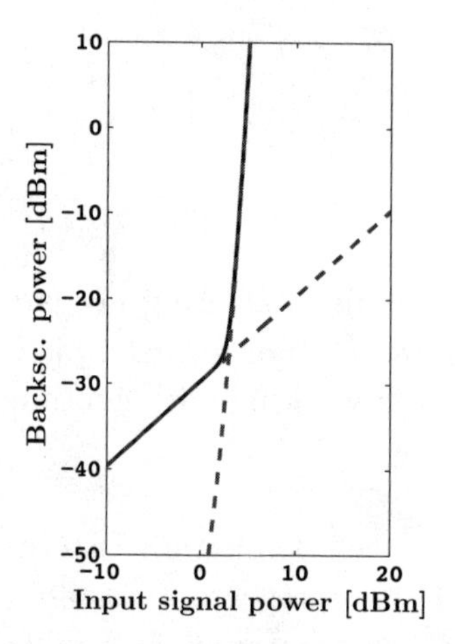

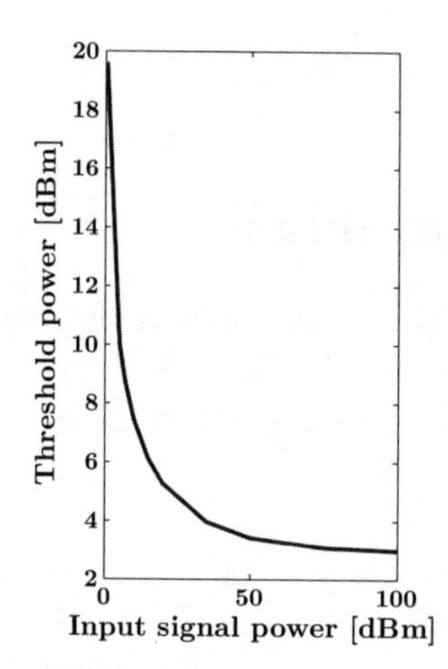

Figure 8.6: Left Predicted backreflected and scattered power from a 10 km long fiber as a function of input signal power. **Right** Predicted threshold power as a function of the length, i.e. the cross-point of the two curves in the left figure calculated as a function of length. In both figures a loss of 0.2 dB/km, a gaincoefficient g_B/A_{eff} of $5{\cdot}10^{-11}/75{\cdot}10^{-12}$ $(\text{Wm})^{-1}$ wavelength of 1555 nm and a bandwidth $d\nu$ equal to 15 MHz have been used.

8.5 Reduced SBS fibers

In many applications Brillioin scattering represents a limitation, for example in high power fiber lasers, where Brillouin scattering works against achieving high power levels with a narrow laser line width, and in fiber optical parametric amplifiers where Brillouin scattering prevents the use of narrow linewidth high pump power to be launched together with a signal.

Consequently there has been a strong desire to find methods to reduce Brillouin scattering and to minimize the impact of Brillouin scattering. Such methods include fiber design, i.e. doping with suitable materials, varying certain parameters along the fiber length for example stress or temperature and phase modulation, or dithering of the optical signal. In the following a brief description of the results obtained, using different codopant materials or by modifying the refractive index profile of an optical fiber is given.

In analog to an optical field a sound wave may be guided by an optical fiber simply because the sound wave experiences an acoustic refractive index profile defined through the difference in sound velocity in the core relative to the cladding. The acoustic refractive index n_a is defined as the ratio of the speed of sound in the cladding to that of the core: $n_{ph} = v_{cl}/v_c$ where v_c is the speed of velocity in the core.

To a first-order approximation the optical field and the longitudinal acoustic field are governed by similar scalar wave equations for the fundamental mode [91]

$$\frac{d^2 f_i}{dr^2} + \frac{1}{r}\frac{df_i}{dr} + k_i^2(n_i^2(r) - n_{i,\text{eff}}^2)f_i = 0 \quad i = 0, ph, \tag{8.65}$$

$n_{ph} = v_{\text{clad}}/v_L$, where v_L is the velocity of the longitudinal acoustic wave. For a silica host fiber doped with germanium, the acoustic velocity as a function of the germanium concentration w_{GeO_2} equals $5944(1 - 7.2 \cdot 10^{-3}\ w_{GeO_2})$ m/s. Note that $w = 0$, is identical to silica. In such fibers the optical refractive index of the core as a function of germanium doping is: $n_c = n(1 + 1.0 \cdot 10^{-3} w_{GeO_2})$, whereas the refractive index of the acoustic wave is: $n_{ph} = 1 + 7.2 \cdot 10^{-3} w_{GeO_2}$.

When calculating the threshold power, see Section 8.4, the threshold scales with the reciprocal overlap between the electric and acoustic fields [91]. Consequently one may design fibers that guide both the acoustic and the optical fields but where the overlap between the optical and the acoustic mode is minimized, even for a fixed optical mode.

In [92] Xin Chen and coworkers have explored the possibility of using the fact that optical fibers may be designed with a small difference between the core and cladding refractive index to obtain a larger core diameter compared to that of conventional fibers, while at the same time minimizing the Brillouin threshold. Without

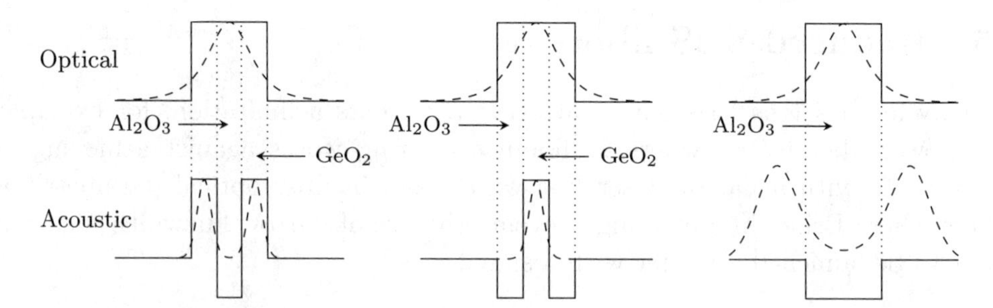

Figure 8.7: Three different examples of fiber design where aluminium and germanium are used as dopants to maintain a fixed optical mode while modifying the acoustic mode slightly. From [91].

other dopant materials than germanium they demonstrated an increase in Brillouin the threshold of 2 to 5 dB.

Doping the core with aluminium instead of germanium has been suggested by Nakanishi in 2006 [93]. Germanium doping decreases the sound speed in the core relative to the sound speed in the silica cladding, whereas aluminium doping increases the sound speed relative to the cladding. With the acoustic refractive index defined as noted with Eqn. (8.65), as the ratio of the sound speed in the cladding relative to that of the core, it is possible to design fibers where the acoustic mode is significantly different from the optical mode, and consequently, the overlap between the two modes is minimized, which improves the Brillouin threshold. Examples are illustrated in Figure 8.7.

In 2010 L. Grüner-Nielsen and coworkers demonstrated a design of a highly nonlinear optical fiber with improved Brillouin threshold. Compared against a standard germanium doped single mode fiber, a 4.3 dB reduction in the Brillouin gain coefficient was reported. [89]. However, it was pointed out that for such a fiber a better suited figure of merit is n_2/g_B. When evaluating this figure of merit an improvement of 4.8 dB was achieved. In addition, it was pointed out that an additional improvement of 4 dB may be obtainable by applying a linear strain. As a result of the aluminum doping a Brillouin threshold of 1.6 W (defined as the power level where the reflected power equals 1 % of the input power) in a 134 m long fiber was reported.

Other dopant materials may be used, and Table 8.1 provides an overview.

Table 8.1: Trends in optical and acoustic refractive index for various co-dopants [91].

	GeO_2	P_2O_3	TiO_2	B_2O_3	F_2	Al_2O_3
Optical refractive index	↑	↑	↑	↓	↓	↑
Acoustic refractive index	↑	↑	↑	↑	↑	↓

8.6 Applications

The first experimental demonstration of Brillouin scattering in optical fibers was reported in 1972 by E. Ippen and R. Stolen [94]. Since then there has been a significant interest in Brillouin scattering focused on undesired effects as well as applications. In Section 8.4 we have briefly mentioned some of the undesired effects and below we highlight some applications of Brillouin scattering

Sensors In Section 8.5 it was briefly discussed how Brillouin scattering could be reduced by applying strain or temperature gradient along the fiber length. Where a mechanical sensor measuring strain or stress is based on the fact the the electrostriction depends on stress and strain, a temperature sensor is based on using the fact the the phonon population is temperature dependent, see [95].

Fiber amplifiers and laser Due to the high efficiency an obvious application is to make amplifiers. However, in this context a significant drawback is the narrow spectral width of the Brillouin gain coefficient. However, such an amplifier may be very attractive to make a fiber laser. Several fiber lasers, CW as well as pulsed, have been demonstrated.

Acousto optic modulators Brillouin scattering can be used to diffract and shift the frequency of light. Such modulators are used in lasers for Q-switching, in telecommunications for signal modulation, and in spectroscopy for frequency control.

Slow light because of the narrow linewidth of the Brillouin gain spectrum the Brillouin induced refractive index changes significantly with frequency. Consequently this results in large derivative in the wavevector with frequency, and hence the potential of altering the group velocity of light [96].

The above list of applications is far from complete, but such a complete list would be is outside the scope of this book.

8.7 Summary

- The Brillouin scattering process is analog to the Raman scattering process, in the sense that light is scattering against phonons. However, in the case of Brillouin scattering the phonos are acoustic and they may be described as a sound wave, i.e. propagating density variations or electrostriction. In an optical fiber Brillouin scattering may be pictured as a bragg grating, similar to a mirror, set up by the propagating wave. The created scattered wave is then backward propagating. Like Raman scattering Brillouin scattering also causes a Stokes as well as an anti-Stokes wave. In glass the Stokes scattered wave is centered around 10 GHz from the excitation wave and the width and the Brillouin spectrum is close to 50 MHz.

- The strength of the Brillouin scattering depends on the linewidth of the excitation laser. The broader the excitation laser linewidth the lower the Brillouin scattering. Consequently Brillouin scattering is typically no problem in communication systems.

- The strength of the scattering depends on temperature, strain and material. All three parameters have been used to suppress Brillouin scattering in optical fibers

- Brillouin is like Raman scattering a third-order nonlinear process, and is described through $\chi(3)_{ijkl}(\omega_s; \omega_p, -\omega_p, \omega_s)$. In the time domain Brillouin scattering may be described using a damped harmonic oscillator.

- Brillouin scattering may be described as electrostriction that is changes in the material density caused by an applied electric field. This causes a density wave i.e. an acoustic wave. Consequently the sound velocity is an important parameter. In analog to an optical refractive index, an acoustic refractive index is described as the speed of the acoustic wave in a material relative to the acoustic speed i vacuum.

- In analog to an optical waveguide, where the optical refractive index is used create a waveguide, the acoustic refractive index may be used to create a waveguide for the acoustic wave - or suppress guiding of an acoustic wave. The latter has been used in optical fibers to suppress Brillouin scattering.

- Application of Brillouin scattering include amplification, slow light, modulators for light and sensing.

Chapter 9

Optical Kerr effect

Contents

In the previous Chapters 7 and 8 we discussed nonlinear interaction in an electrical field having two frequencies, a pump, and a signal, more specifically we focused on continuous waves, Raman and Brillouin scattering which are both phase insensitive processes. In the following we focus on the so-called nonlinear refractive index which in this chapter is mainly a change in refractive index caused by the electrical field itself, i.e. a nonlinear interaction with the electrical field itself. This significantly impacts pulse propagation.

In general, the term Kerr effect is used when the refractive index that a propagating beam experiences, is modified by the electric field itself. This includes the so-called DC Kerr effect and the optical Kerr effect, sometimes also called the AC Kerr effect. In the DC Kerr effect, a static electric field is used to modify the refractive index of the beam that propagates through the modulator. In the optical Kerr effect, an optical beam is modifying the refractive index either for the beam itself or for a beam at another wavelength, the latter is more correctly termed cross phase modulation. Where the optical Kerr effect is described using the susceptibility

$\chi^{(3)}(-\omega;\omega,-\omega,\omega)$, cross phase modulation is described using $\chi^{(3)}(-\omega;\omega_1,-\omega_1,\omega)$, where the beam at ω_1 modifies the refractive index for the beam at frequency ω.

In this chapter we only consider the optical Kerr effect. However, before doing so, it is useful to remind ourselves about the linear refractive index.

From Chapter 1, the linear induced polarization is

$$(\mathbf{P}^0)^{(1)} = \varepsilon_0\chi^{(1)}(-\omega;\omega)\mathbf{E}^0. \tag{9.1}$$

This is the induced polarization that is used to describe linear propagation, and for example may be used to describe birefringence.

For simplicity we now consider a linearly polarized electric field. For this field it is sufficient to describe the relative permittivity as the scalar quantity

$$\varepsilon_r = 1 + \chi^{(1)}(-\omega;\omega), \tag{9.2}$$

and the linear refractive index n_0 equals

$$n_0 = \sqrt{1 + \mathrm{Re}[\chi^{(1)}(-\omega;\omega)]}. \tag{9.3}$$

In this chapter focus is directed toward the optical Kerr effect. This is described through the third-order induced polarization, which in this case equals

$$(P_i^0)^{(3)} = \frac{3}{4}\varepsilon_0\chi_{ijkl}^{(3)}(-\omega;\omega,-\omega,\omega)E_j^0(E_k^0)^*E_l^0, \tag{9.4}$$

where E_i, $i = (j,k,l)$ is the amplitude of the electrical field at the signal frequency ω. In general the components of the induced polarization are coupled to similar and other components of the electrical field vector through the susceptibility-tensor. Often the vector aspect is avoided simply by assuming that all fields have one, and only one, non-zero component, for example, by assuming that the electric field is linearly polarized along the x-axis and that it remains polarized along the x-axis during propagation. Alternatively, a scalar quantity for nonlinear susceptibility, sometimes referred to as the effective susceptibility, may be introduced.

Effective susceptibility

If the amplitude of the electric field is written as

$$\mathbf{E} = \frac{1}{2}\sum_j \left(\mathbf{e}_j E_j^0 e^{-i(\omega_j t - \beta_j z)} + c.c.\right), \tag{9.5}$$

where the amplitude E_j^0 is a complex scalar and $\mathbf{e}_j$ is a unit vector in the direction of polarization of the electric field component, then a scalar susceptibility may be defined through the projection of the induced polarization on the electric field vector i.e. through the dot product

$$\chi_{\text{eff}}^{(n)} = \mathbf{e}_{\omega_\sigma}^* \cdot \left(\chi^{(n)}(-\omega_\sigma;\omega_1,\omega_2,\cdots,\omega_n)\mathbf{e}_1\mathbf{e}_2\cdots\mathbf{e}_n\right). \tag{9.6}$$

where $\mathbf{e}^*_{\omega_\sigma}$ corresponds to the negative frequency $-\omega_\sigma$ which appear as the output argument in the susceptibility

The induced polarization vector of order n is then expressed as

$$(\mathbf{P}^0)^{(n)} = \varepsilon_0 \chi_{\text{eff}}^{(n)} K E_1^0 E_2^0 \cdots E_n^0 \mathbf{e}_\sigma. \tag{9.7}$$

In the case when an electrical field consisting of only one frequency at the input is considered, $\mathbf{E} = \frac{1}{2}\left(\mathbf{e}E^0 e^{-i(\omega t - \beta z)} + c.c.\right)$. From this the induced polarization vector may be expressed as

$$\mathbf{P}^0 = \varepsilon_0 \left(\chi_{\text{eff}}^{(1)} + \frac{3}{4}\chi_{\text{eff}}^{(3)} |E^0|^2 \right) E^0 \mathbf{e}, \tag{9.8}$$

where $\chi_{\text{eff}}^{(1)}$ and $\chi_{\text{eff}}^{(3)}$ are evaluated according to Eqn. (9.6). Applying this, then the relative permittivity equals

$$\varepsilon_r = 1 + \chi_{\text{eff}}^{(1)} + \frac{3}{4}\chi_{\text{eff}}^{(3)} |E^0|^2. \tag{9.9}$$

Since the refractive index n is related to the relative permittivity through $n = \sqrt{\varepsilon_r}$ the refractive index approximates

$$n \approx n_0 \left(1 + \frac{3}{8} \frac{\chi_{\text{eff}}^{(3)} |E^0|^2}{n_0^2} \right), \tag{9.10}$$

where the linear refractive index $n_0 = \sqrt{1 + \chi_{\text{eff}}^{(1)}}$ has been inserted. From this the nonlinear refractive index, n_2, also referred to as the intensity dependent refractive index or the Kerr refractive index is defined from $n = n_0 + n_2|E^0|^2$ as

$$n_2 = \frac{3}{8} \frac{\chi_{\text{eff}}^{(3)}}{n_0}, \tag{9.11}$$

resulting in an intensity dependent change of the propagation constant $\Delta\beta$ which may be written as

$$\Delta\beta = \omega n_2 |E^0|^2 / c = \frac{\omega}{c n_0} \frac{3}{8} \chi_{\text{eff}}^{(3)} |E^0|^2. \tag{9.12}$$

In the above there is no waveguide, and the change in propagation constant is expressed through the electric field. However, by comparing against Eqn. (5.66), when considering propagation through a waveguide, the change in the propagation constant, $\Delta\beta$, is

$$\Delta\beta = \frac{\omega n_2^I}{c} \frac{P}{A_{\text{eff}}}, \tag{9.13}$$

where n_2^I indicate that the intensity refractive index needs to be multiplied with the intensity of the electric field rather than simply the electric field to give the refractive index. It is emphasized at this point that the refractive index, as already indicated, may be given either as a refractive index which is multiplied with the electrical field squared, that is terms like $n_2|E^0|^2$, or as a term multiplied by the power of the propagating light, that is like $n_2^I P/A_{\text{eff}}$. To differentiate between the two we denote the latter with superscript I. From Eqn. (5.77), the nonlinear propagation is

$$-i\frac{\partial A}{\partial z} = -\frac{1}{2}\beta_2\frac{\partial^2 A}{\partial t^2} + \frac{\omega n_2^I}{c}\frac{P}{A_{\text{eff}}}A + i\frac{\alpha}{2}A. \tag{9.14}$$

It is noted that Eqn. (9.14) is the scalar propagation equation, where A is the envelope of the optical signal, and the first term on the right-hand side represents group velocity dispersion through β_2 which is the second-order derivative of the propagation constant β with respect to frequency; evaluated at the carrier frequency. Note, the group velocity dispersion that is typically provided by the fiber-manufacturer, D, is related to β_2 through $\beta_2 = -\lambda^2 D/(2\pi c)$. The second term describes the impact of the intensity dependent refractive index, which in silica is $n_2^I = 3.2 \cdot 10^{-20}$ W/m. P is the power of the pulse, which is a function of time and position, and A_{eff} is the effective area of the fiber. Typically A_{eff} is around $75 \cdot 10^{-12}$ m^2 in conventional transmission optical fibers. As opposed to Eqn. (5.77), a term on the right-hand side, which represents the loss or gain, α, is included.

In Eqn. (9.14) several effects are omitted, for example Raman and Brillouin scattering within the pulse, i.e. short wavelengths within the pulse amplify longer wavelengths in the pulse. The following section focuses on propagation of pulses described by Eqn. (9.14), in Section 9.2 the focus is directed toward solitons, while Section 9.4.2 describes supercontinuum generation, and Section 9.4.2 describes briefly the full vectorial propagation equation. For completeness, the optical Kerr effect may also be used to describe other nonlinear phenomena such as self-focusing.

9.1 Short pulse propagation

For envelope equations we express the field in terms of an envelope by factoring the electric field $E(r, z, t)$ into a carrier wave at a chosen angular frequency ω_σ and an amplitude $E^0(r, z, t)$

$$E(r, z, t) = \frac{1}{2}\left(E^0(r, z, t)\exp^{i(\beta_\sigma z - \omega_\sigma t)} + c.c.\right). \tag{9.15}$$

Furthermore, as described in Chapter 5, the amplitude of the electric field is separated into a function that describes the radial dependence $R(r)$ of the amplitude and a function that describes the longitudinal dependence $A(z, t)$ of the electrical field. Then from Chapter 5 we adopt directly

$$\frac{\partial A}{\partial z} + \beta_1\frac{\partial A}{\partial t} + i\frac{1}{2}\beta_2\frac{\partial^2 A}{\partial t^2} - i\Delta\beta A = 0, \tag{9.16}$$

and also from Chapter 5

$$\Delta\beta = \frac{k^2}{2\beta_0}\frac{\int_A \Delta\varepsilon |R|^2 da}{\int_A |R|^2 da}, \tag{9.17}$$

where $\Delta\varepsilon$ includes the $\chi^{(3)}$ nonlinearity. Since we consider pulse propagation, i.e. a carrier wave multiplied by an envelope, we focus on the impact due to Raman scattering, denoted intra-pulse Raman scattering, as well as the effects of electronic origin that is the Kerr effect. This is described through $\Delta\varepsilon = \chi^{(3)}(-\omega_j;\omega_j,-\omega_j,\omega_j)\mathbf{E}^0(\mathbf{E}^0)^*\mathbf{E}^0$ which we describe using $\Delta\varepsilon = (\chi_R^{(3)} + \chi_K^{(3)})\mathbf{E}^0(\mathbf{E}^0)^*\mathbf{E}^0$ where $\chi_R^{(3)}$ and $\chi_K^{(3)}$ represent the Raman susceptibility and the Kerr susceptibility, respectively.

Often the radial dependence is separated from the nonlinearity by assuming that the entire transverse electric field interacts with one and only one material, for example that the entire electric field propagates in the core of an optical fiber only. Including a factor of $(\varepsilon_0 n_0 c/2)$ in $(\chi_R^{(3)} + \chi_K^{(3)})$ allows us to write $\Delta\beta$ as a function of power P, while keeping it as a function of the susceptibility - in analog with Eqn. (9.12), that is

$$\Delta\beta = \frac{\omega}{2cn_{\text{eff}}(\omega)A_{\text{eff}}(\omega)}\left[\chi_R^{(3)}(\omega) + \chi_K^{(3)}(\omega)\right]P, \tag{9.18}$$

where n_{eff} is the effective refractive index, as defined in Appendix F. For completeness, it is noted that the change in the propagation constant is in fact more complicated since all spectral components may interact mutually with each other, especially when considering very short pulses.

To be able to see the impact of the frequency dependence of the effective refractive index and the effective area of the electric field, it is customary to introduce the auxiliary function $H(\omega)$ as

$$H(\omega) = \frac{1}{n_{\text{eff}}(\omega)A_{\text{eff}}(\omega)}, \tag{9.19}$$

and a simple Taylor approximation of $H(\omega)$ is

$$H(\omega)\approx H(\omega_0) + \left.\frac{dH(\omega)}{d\omega}\right|_{\omega_0}(\omega-\omega_0) = H(\omega_0)\left(1 + \frac{\left.\frac{dH(\omega)}{d\omega}\right|_{\omega_0}}{H(\omega_0)}(\omega-\omega_0)\right). \tag{9.20}$$

Using that $d(\ln(y))/dx = (1/y)dy/dx$ we may rewrite the Taylor approximation of $H(\omega)$ as

$$H(\omega) = H(\omega_0)\left(1 + \left.\left(\frac{d}{d\omega}\ln[H(\omega)]\right)\right|_{\omega_0}(\omega-\omega_0)\right). \tag{9.21}$$

With this we arrive at

$$\begin{aligned}\Delta\beta &= \frac{\omega_0}{2c}\left(\frac{\omega-\omega_0}{\omega_0}+1\right)H(\omega)(\chi_R^{(3)}+\chi_K^{(3)})P \qquad (9.22)\\ &\approx \frac{\omega_0}{2c}\left(\frac{\omega-\omega_0}{\omega_0}+1\right)H(\omega_0)\left(1+\left.\left(\frac{d}{d\omega}\ln\left[H(\omega)\right]\right)\right|_{\omega_0}(\omega-\omega_0)\right)(\chi_R^{(3)}+\chi_K^{(3)})P.\end{aligned}$$

To first-order in $(\omega - \omega_0)$ we may write $\Delta\beta$ as

$$\Delta\beta \approx H(\omega_0)\frac{\omega_0}{2c}\left(1 + (\omega - \omega_0)\left[\frac{1}{\omega_0} + \frac{d}{d\omega}\ln\left[H(\omega)\right]\Big|_{\omega_0}\right]\right)(\chi_R^{(3)} + \chi_K^{(3)})P. \quad (9.23)$$

The first term in the bracket is responsible for self phase modulation four-wave mixing and intra Raman scattering, i.e. self frequency shift whereas the second term in the square brackets is associated with self steepening and optical shock formation. The term in the square brackets is often written as τ_{shock}, i.e.

$$\tau_{\text{shock}} = \left[\frac{1}{\omega_0} + \frac{d}{d\omega}\ln\left[H(\omega)\right]\Big|_{\omega_0}\right]. \quad (9.24)$$

It is customary to describe pulse propagation in the time domain, and consequently the product defined in the last term on the left-hand side of Eqn. (9.23) becomes a convolution, compare with the BO approximation in Chapter 2. The modification to the shock term due to the frequency dependence of the effective area is typically larger than the modification due to the frequency dependence of the effective refractive index.

From this the term originating from $\Delta\beta$ equals

$$\frac{H(\omega_0)\omega_0}{2c}\left(1 + i\tau_{\text{shock}}\frac{d}{dt}\right)\int R^{(3)}(t')|A(t - t')|^2 dt', \quad (9.25)$$

where $R(t')$ is the third order nonlinear response i.e. from Chapter 2 $R^{(3)}(t') = \sigma\delta(t') + h_r(t')$. By inserting this response into Eqn. (9.16) we get

$$\frac{\partial A}{\partial z} + \beta_1\frac{\partial A}{\partial t} + i\frac{1}{2}\beta_2\frac{\partial^2 A}{\partial t^2} = i\frac{H(\omega_0)\omega_0}{2c}\left(1 + i\tau_{shock}\frac{d}{dt}\right)\int R^{(3)}(t')|A(t - t')|^2 dt' A(t). \quad (9.26)$$

It is important to note at this point that $R^{(3)}(t)$ is normalized such that when integrated it equals one. Consequently the nonlinear strength needs to be included in the function H. Inserting the response function $R^{(3)}(t) = (1 - f_R)\delta(t) + f_R h_R(t)$ the last term is rewritten as

$$\begin{aligned} &\frac{H(\omega_0)\omega_0}{2c}\left(1 + i\tau_{\text{shock}}\frac{\partial}{\partial t}\right)\left[A(z,t)\int R^{(3)}(t')|A(z, t - t')|^2 dt'\right] \quad (9.27) \\ &= \frac{H(\omega_0)\omega_0}{2c}\left(1 + i\tau_{\text{shock}}\frac{\partial}{\partial t}\right) \\ &\times \left[(1 - f_r)A(z,t)|A(z,t)|^2 + f_r A(z,t)\int h(t')|A(z, t - t')|^2 dt'\right]. \end{aligned}$$

This expression may be evaluated numerically for example by utilizing that a convolution in the time domain equals a product in the frequency domain. Alternatively, we may make another Taylor expansion, this time of $|A(z,t-t')|^2$ around $t'=0$. By applying the chain rule then $|A(z,t-t')|^2$ is approximated by

$$|A(z,t-t')|^2 \approx |A(z,t)|^2 + t'\frac{\partial}{\partial t}\left[|A(z,t)|^2\right], \tag{9.28}$$

by inserting Eqn. (9.28) into Eqn. (9.27) and integrating over t' we arrive at an equation from which we can identify the normalized function H. For instance one may compare against Eqn. (9.26) that is [3]

$$\frac{\partial A}{\partial z} + \beta_1\frac{\partial A}{\partial t} + i\frac{1}{2}\beta_2\frac{\partial^2 A}{\partial t^2} = i\gamma\left(|A|^2A + \frac{i}{\omega_0}\frac{\partial}{\partial t}(|A|^2A) - T_R A\frac{\partial(|A|^2)}{\partial t}\right), \tag{9.29}$$

where the second-order time derivative of the convolution integral has been ignored and where T_R is defined through the slope of the Raman gain spectrum

$$T_R = f_R\int th_R(t)dt = f_R\left.\frac{dH(\omega)}{d\omega}\right|_{\omega=0} = f_R\left.\frac{d\mathrm{Im}[H(\omega)]}{d\omega}\right|_{\omega=0}. \tag{9.30}$$

To arrive at the last equation, we have used: $F(\omega) = \int_{-\infty}^{\infty} f(t)e^{-i\omega t}dt$ and consequently $F(\omega=0) = \int_{-\infty}^{\infty} f(t)dt$ and that the real part of $H(\omega)$ is symmetric. The first term on the right-hand side of Eqn. (9.29) is responsible for self phase modulation (SPM), the second term is responsible for self-steepening and shock formation and the last term is responsible for self frequency shift.

The propagtion equation is often solved using the so-called **Split Step Fourier method**. In this method the propagation is divided into a linear part, i.e. group velocity dispersion and loss and the nonlinear part which include the nonlinearity, Raman effect etc. The propagation is then solved in small steps where in each step, first the linear part is solved and the output from this used as an input to the nonlinear part. A detailed description is found in Agrawal [3].

9.2 Propagation of short pulses

The following section is divided into three subsections. In the first subsection, linear propagation, the effect of group velocity dispersion, is briefly discussed. In the second subsection, the group velocity dispersion is assumed to equal zero, and only the impact of the intensity dependent refractive index is discussed. In the third subsection, the combined effects of the group velocity dispersion and the nonlinear refractive index are discussed. Finally, system demonstrations and the impact due to loss and amplification are discussed in the two final subsections.

9.2.1 Linear propagation

In this regime, the nonlinear refractive index is ignored, and only effect of group velocity dispersion is described. This may be a valid approximation if the power is low enough that nonlinearities are irrelevant, or if the impact from group velocity dispersion is much larger than the impact from the nonlinear response.

From Eqn. (9.14), the nonlinear propagation then reduces to

$$-i\frac{\partial A(t)}{\partial z} = -\frac{1}{2}\beta_2\frac{\partial^2 A(t)}{\partial t^2}. \tag{9.31}$$

This equation may be solved by changing to the Fourier domain, where the corresponding differential equation is

$$-i\frac{\partial A(\omega)}{\partial z} = \frac{1}{2}\beta_2\omega^2 A(\omega), \tag{9.32}$$

this has the solution

$$A(\omega, z) = A(\omega, z=0)\exp\left(\frac{1}{2}\beta_2\omega^2\right), \tag{9.33}$$

where $A(\omega)$ is the Fourier transform of $A(t)$. In typical fibers the group velocity dispersion is a few to several [ps/(nm·km)]. This corresponds to $|\beta_2|$ in the order of tens of [ps^2/km].

9.2.2 Self-phase modulation (SPM)

When the group velocity is close to or equal to zero or when its impact is much smaller than the nonlinearity, only the intensity dependent refractive index needs to be considered. Consequently, the propagation equation is of the form

$$-i\frac{\partial A(t)}{\partial z} = \frac{\omega n_2^I}{c}\frac{P}{A_{\text{eff}}}A(t). \tag{9.34}$$

This has the solution

$$A(t, z) = A(t, z=0)\exp\left\{i\frac{\omega n_2^I}{c}\frac{P}{A_{\text{eff}}}z\right\}. \tag{9.35}$$

Assuming that the initial pulse is a Gaussian pulse, $A(t) = \exp(-t^2/T_0^2)$, then the output is simply the input multiplied by the exponential in Eqn. (9.35). The pure imaginary argument to the exponential function is referred to as a chirp. From this, the intensity of the pulse in the time domain is unchanged whereas the pulse in the frequency domain is changed and broadened, i.e. new frequencies have been generated.

The consequence of the nonlinear refractive index is accumulation of a nonlinear phase shift. This phase shift results in spectral broadening of the pulse: however, a positive group velocity dispersion may lead to compression of the pulse in the time domain. Thus, by carefully counterbalancing the nonlinear phase shift against the group velocity dispersion, a balance between the two effects is achieved, resulting in a pulse that propagates without changing shape. If the balance between dispersion and nonlinearity is static, the pulse is referred to as a fundamental soliton, whereas if the balance is dynamic, the pulse shape changes in a periodic manner along the propagation distance, referred to as a higher order soliton.

9.2.3 Solitons

The nonlinear propagation equation that governs the propagation of a pulse with envelope $A(t, z)$ is given in Eqn. (9.14). Omitting loss, the nonlinear propagation equation has solutions that exist when $\beta_2 < 0$ and has the form

$$A(t, z = 0) = a_0 \sqrt{P_{\text{sol}}}\,\text{sech}(bt). \tag{9.36}$$

These solutions are referred to as solitons. P_{sol} is the peak power required to get a fundamental soliton [3]. a_0 is the amplitude relative to the fundamental soliton amplitude. If a_0 in Eqn. (9.36) equals one Eqn.(9.35) is a fundamental soliton, if a_0 is an integer higher than one, then Eqn. (9.36) is still a solution to the nonlinear propagation equation. However, the balance between the dispersion and the nonlinearity is now dynamic, implying that the pulse shape changes in a periodic manner along the propagation distance. This is referred to as a higher order soliton, and more specifically when $a_0 = 2$ a second-order soliton, $a_0 = 3$ a third-order soliton etc. If a_0 is a non-integer larger than 0.5 the pulse adjust itself, amplitude and width, to become a soliton either with a static balance between dispersion and nonlinearity, as the fundamental soliton, or with a dynamic balance between dispersion and nonlinearity as a higher order solition as a higher order soliton.

The fundamental soliton is a pulse that propagates without any changes in its shape during propagation. This only exist under ideal conditions, i.e. no loss, no birefringence etc., further more, the soliton solution is based on a critical balance between the group velocity dispersion and the optical Kerr effect. This is illustrated in Figure 9.1.

The peak power required to get a fundamental soliton increases with increasing group velocity dispersion and shortened pulse width according to

$$P_{\text{sol}} = \frac{\ln^2(\sqrt{2}+1)}{\pi^2} \frac{A_{\text{eff}}}{c n_2^I} \frac{\lambda^3 D}{\Delta\tau^2}, \tag{9.37}$$

where $\Delta\tau$ is the pulse width at half maximum intensity [3]. It is noted that the lower the dispersion, the lower the peak power is required, which raises a fundamental challenge because it causes a low signal-to-noise ratio. In a fiber with a group velocity dispersion of 1 ps/(nm·km) and an effective area of $A_{\text{eff}} = 75 \cdot 10^{-12}$ m^2, 93 mW of peak power is required to transmit a 5 ps wide pulse at 1555 nm. In [97], [98] it was proposed to solve the problem of a low signal power by alternating the dispersion between two large values - this is referred to as a dispersion managed soliton.

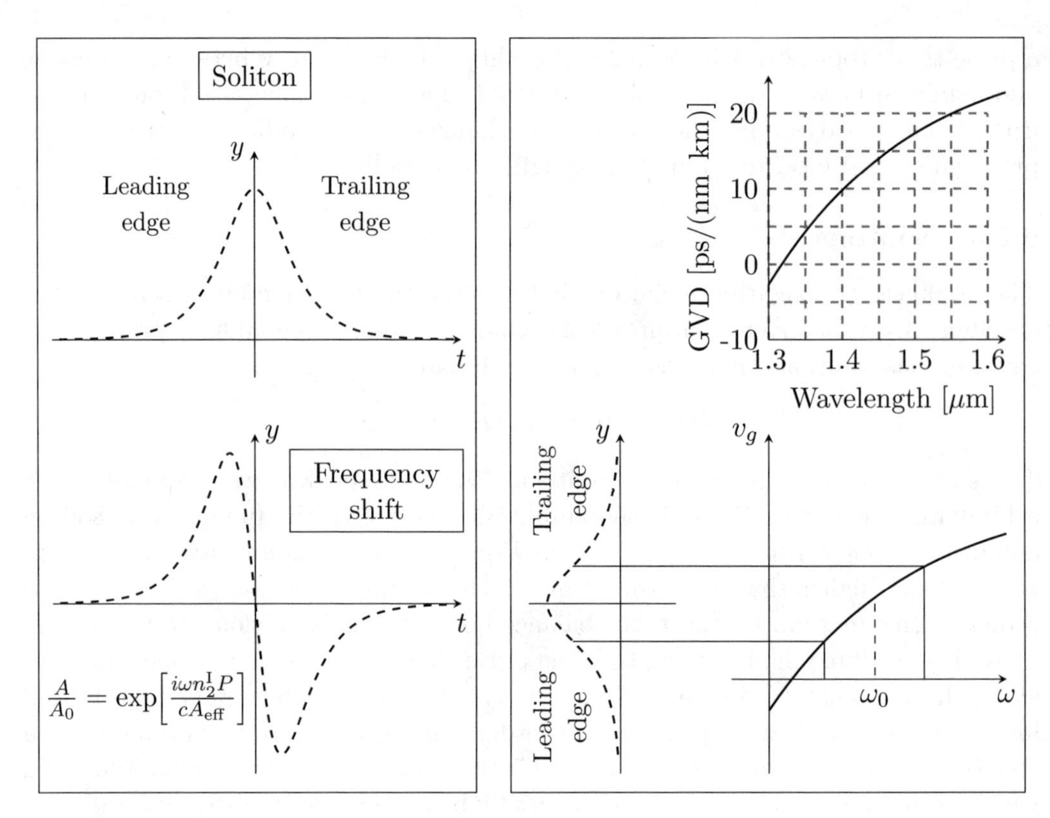

Figure 9.1: The nonlinear phaseshift results in a reduction of the spectral width at the leading edge of the pulse while the spectral width is increased at the trailing edge of the pulse. When the sign of the group velocity dispersion is negative, also called anomalous dispersion, the pulse is narrowed.

A pulse that does not have the correct hyperbolic secant shape at launch or an amplitude different from an integer times the fundamental soliton amplitude as described below Eqn. (9.36), asymptotically reshapes itself into a soliton. The excess power from this process radiates as a dispersive wave.

9.3 Pulse characterization

The pulse width is defined through the parameter b in Eqn. (9.36). The full-width at half maximum intensity pulse width is

$$\Delta\tau = 2\ln(\sqrt{2}+1)/b, \tag{9.38}$$

in the time domain. In the frequency domain, the full width at half maximum intensity pulse width equals

$$\Delta\nu = 2\ln(\sqrt{2}+1)\frac{b}{\pi^2}. \tag{9.39}$$

From the pulse width in the time and frequency domain, the time-bandwidth product equals: $\Delta\tau\Delta\nu \approx 0.315$ for Gauss pulse. It is noted that the time bandwidth product evaluated above represents a minimum, the so-called transform limited time-bandwidth product, see Example 9.3. The time bandwidth product of a pulse is larger than the minimum if for example a pulse has a chirp.

Often the above quantities, $\Delta\tau$ and $\Delta\nu$, are used to characterize pulses since they are measurable quantities. However, in some cases, for example when pulses are heavily distorted, it may make more sense to use quantities as the mean time $< t >$ and mean frequency $< \nu >$, or mean angular frequency $< \omega >= 2\pi < \nu >$, to describe the location of the center of a pulse. To describe the width of a pulse, the root mean square time T_{RMS} (a similar quantity exists in the frequency domain) may be used, and the skewness γ_{sk} (in time and/or frequency domain) may be used to describe the symmetry of the pulse relative to the mean. These quantities are based on mean, second, and third moment, similar to quantities used in analysis of probability density distributions.

The mean time of an amplitude envelope pulse $A(t)$ in the time domain is

$$\langle t \rangle = \frac{1}{E_s} \int_{-\infty}^{\infty} AtA^* dt, \tag{9.40}$$

where $E_s = \int_{-\infty}^{\infty} AA^* dt$ ensures that the envelope is normalized. The second-order time moment $\langle t^2 \rangle$ is

$$\langle t^2 \rangle = \frac{1}{E_s} \int_{-\infty}^{\infty} At^2 A^* dt. \tag{9.41}$$

Knowing the mean time and the second moment, the root mean square time is

$$T_{\text{RMS}} = \sqrt{\langle t^2 \rangle - \langle t \rangle^2}. \tag{9.42}$$

Note that T_{RMS}^2 is equivalent to the variance of a probability density function. Typically the mean time $\langle t \rangle$ equals zero, or may be used to define an artificial zero, and the root mean square width of the pulse is simply defined from Eqn. (9.41). For a soliton $A(t) = \text{sech}(bt)$, the RMS width in the time domain is $T_{\text{RMS}}^2 = (\pi^2/12)(1/b^2)$, see Example 9.2.

By replacing t with frequency ω and $A(t)$ with the pulse in the frequency domain $A(\omega)$, the mean and RMS width in the frequency domain Ω_{RMS} is evaluated. For a soliton, $\Omega_{\text{RMS}} = b^2/3$, see Example 9.2. This results in a RMS time bandwidth product of $T_{\text{RMS}}\Omega_{\text{RMS}} = \pi/6$.

In addition to the pulse width in the time and frequency domain, it is often useful to quantify if a pulse has become asymmetric. This may for example happen in pulse propagation because of a change in group velocity dispersion with frequency or due to depletion in amplifiers. The skewness, γ_{sk}, of a pulse may be evaluated as

$$\gamma_{sk} = \frac{\langle t^3 \rangle - 3\langle t \rangle T_{\mathrm{RMS}}^2 - \langle t \rangle^3}{T_{\mathrm{RMS}}^3}. \tag{9.43}$$

Note, γ_{sk} is dimensionless. If $\gamma_{sk} > 0$ the right trail is longer than the left trail, and the pulse is referred to as positive skew, or right-skewed, see Fig. 9.2 (a) in Example 9.1. On the contrary a pulse is negative skew, or left-skewed, if the left trail is longer than the right trail, $\gamma_{sk} < 0$, see Fig. 9.2 (c) in Example 9.1.

Example 9.1. *Positive and negative skewed pulses*
We consider a Gaussian envelope signal as a reference, i.e.

$$A(t) = \exp(-(t/T_0)^2), \tag{9.44}$$

and let us compare this against

$$A_s(t) = \begin{cases} \exp(-(t/T_1)^2) & t < 0 \\ \exp(-(t/T_2)^2) & t > 0. \end{cases} \tag{9.45}$$

These two functions, $A(t)$ and $A_s(t)$, are relatively easy to consider since they have analytical solutions.

The Gaussian pulse has $E_s = T_0\pi/2$. The time mean and third-order time moment both equal zero, i.e. $\langle t \rangle = \gamma_{sk} = 0$. From Eqn. (9.41), the second-order moment equals

$$\langle t^2 \rangle = \frac{\Gamma(3/2)}{2\sqrt{\pi}} T_0^2, \tag{9.46}$$

where Γ is the gamma function defined as $\Gamma(t) = \int_0^\infty x^{t-1} e^{-x} dx$.

The skewed pulse $A_s(t)$ has $E_s = \frac{1}{2}\sqrt{\frac{\pi}{2}}(T_1 + T_2)$. This pulse has mean time

$$\langle t \rangle = \frac{\Gamma(1)}{\sqrt{2\pi}} \frac{T_2^2 - T_1^2}{T_1 + T_2}. \tag{9.47}$$

The second-order moment is

$$\langle t^2 \rangle = \frac{\Gamma(3/2)}{2\sqrt{\pi}} \frac{T_1^3 + T_2^3}{T_1 + T_2}, \tag{9.48}$$

and the third-order moment is

$$\langle t^3 \rangle = \frac{\Gamma(2)}{2\sqrt{2\pi}} \frac{T_2^4 - T_1^4}{T_1 + T_2}. \tag{9.49}$$

Figure 9.2 illustrates a Gaussian pulse, in sub figure 9.2 (a), a positive skewed pulse, in Figure 9.2 (c), and a negative skewed pulse and in Figure 9.2 (b) a non-skewed,

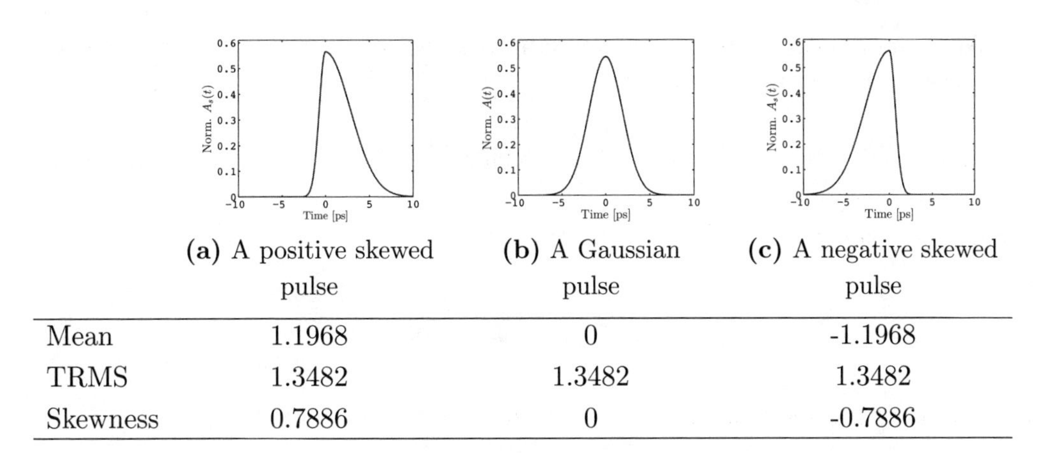

(a) A positive skewed pulse

(b) A Gaussian pulse

(c) A negative skewed pulse

Mean	1.1968	0	-1.1968
TRMS	1.3482	1.3482	1.3482
Skewness	0.7886	0	-0.7886

Figure 9.2: A symmetric and two skewed pulses. **(a)** A positive skewed pulse $A_s(t)$ from Eqn. (9.45) by using $T_1 = 1$ ps and $T_2 = 4$ ps, **(b)** A Gaussian pulse $A(t)$ from Eqn. (9.44) using $T_0 = 2.69628$ ps and **(c)** A negative skewed pulse $A_s(t)$ from Eqn. (9.45) by using $T_1 = 4$ ps and $T_2 = 1$ ps.

or symmetric, pulse. More specifically, the symmetric pulse is a Gaussian pulse obtained from Eqn. (9.44) by using $T_0 = 2.69628$ ps whereas the positive skewed pulse in Figure 9.2 is obtained by Eqn. (9.2) by using $T_1 = 1$ ps and $T_2 = 4$ ps. Finally the negative skewed pulse is obtained also from Eqn. (9.2) but now by using $T_1 = 4$ ps and $T_2 = 1$ ps.

— ■ —

Example 9.2. *Soliton*

We evaluate the RMS time-bandwidth product of a soliton i.e. sech(bt). To start with we note the the pulse is symmetric, i.e. $\langle t \rangle = 0$. We then evaluate the normalization parameter, E_s,

$$E_s = \int_{-\infty}^{\infty} \text{sech}^2(bt)dt = \tanh(t)/b = 2/b. \tag{9.50}$$

Using this result, we can now evaluate $\langle t^2 \rangle$

$$\langle t^2 \rangle = \frac{1}{E_s} \int t^2 \text{sech}^2(bt)dt = \frac{b}{2} \frac{1}{b^3} \frac{\pi^2}{6} = \frac{\pi^2}{12} \frac{1}{b^2}, \tag{9.51}$$

where we have used $\int_{-\infty}^{\infty} \frac{x^2}{cosh^2(x)} dx = \pi^2/6$ [99]. As expected, it may be noted that as a soliton gets wider, relative to $b = 1$, in the time domain i.e. b less than one, $\langle t^2 \rangle$ increases.

We then move to the frequency domain. First, we note $\langle\omega\rangle = 0$. We then calculate $\langle\omega^2\rangle$. However, we first evaluate the normalization parameter

$$E_s = \int_{-\infty}^{\infty} (\frac{\pi}{b})^2 \operatorname{sech}^2(\frac{\pi\omega}{2b}) = 4\frac{\pi}{b}. \tag{9.52}$$

Using this result, we can evaluate $\langle\omega^2\rangle$

$$\langle\omega^2\rangle = \frac{1}{E_s}\int \omega^2 (\frac{\pi}{b})^2 \operatorname{sech}^2\left(\frac{\pi\omega}{2b}\right) d\omega = \frac{b^2}{3}. \tag{9.53}$$

From this we get the time-bandwidth product $\sqrt{\langle t^2\rangle\langle\omega^2\rangle} = \pi/6$.

—— ■ ——

Example 9.3. *The minimum time-bandwidth product*
Assuming the time and frequency axes are defined such that $\langle t\rangle = 0$ and $\langle\omega\rangle = 0$ [100], which is always possible, then it is possible to show that a function can not be jointly localized in time and frequency arbitrarily well, either one has poor frequency localization or poor time localization.
Using

$$\Delta_t^2 = \frac{1}{E}\int_{-\infty}^{\infty} t^2|f(t)|^2 dt \quad \text{and} \quad \Delta_\omega^2 = \frac{1}{2\pi E}\int_{-\infty}^{\infty} \omega^2|F(\omega)|^2 d\omega. \tag{9.54}$$

If $\sqrt{|t|}f(t) \rightarrow 0$ as $|t| \rightarrow \infty$, then

$$\Delta_t\Delta_\omega \geq \frac{\pi}{2}. \tag{9.55}$$

It is noted that the equality only holds if $f(t)$ is Gaussian, that is $f(t) = Ce^{-\alpha t^2}$, where C and α are constants.

To show this, we start by applying the Schwartz inequality to the product $\Delta_t^2\Delta_\omega^2$. In the time domain this product is expressed as

$$\Delta_t^2\Delta_\omega^2 = \left(\frac{1}{E}\int_{-\infty}^{\infty} t^2|f(t)|^2 dt\right)\left(\frac{1}{2\pi E}\int_{-\infty}^{\infty}\left|\frac{d^2 f}{dt^2}\right| dt\right). \tag{9.56}$$

From the Schwartz inequality we may evaluate the product of the integrals on the right-hand side

$$\int_{-\infty}^{\infty} t^2|f(t)|^2 dt \int_{-\infty}^{\infty}\left|\frac{df}{dt}\right|^2 dt \geq \left|\int_{-\infty}^{\infty} tf(t)\frac{df}{dt}dt\right|^2. \tag{9.57}$$

When evaluating the right-hand side it is noted that

$$\int_{-\infty}^{\infty} tf\frac{df}{dt}dt = \int_{-\infty}^{\infty} t\frac{d(f^2/2)}{dt}dt = \left[t\frac{f^2}{2}\right]_{-\infty}^{\infty} - \int_{-\infty}^{\infty}\frac{f^2}{2}dt. \tag{9.58}$$

Since we have required that $\sqrt{|t|}f(t) \to 0$ we now find that

$$\int_{-\infty}^{\infty} tf\frac{df}{dt}dt = -\frac{E}{2}, \tag{9.59}$$

where we have used $\int_{-\infty}^{\infty} f^2 dt = E$. Inserting gives the simple result: $\Delta_t \Delta_\omega \geq \pi/2$.

——— ■ ———

As mentioned, the balance between the dispersion and the nonlinearity may be dynamic for a_0 larger than one, with the consequence that the pulse shape changes in a periodic manner along the propagation distance. The periodic distance equals the soliton period, which equals

$$z_0 = \left(\frac{\tau^2}{4\ln^2(\sqrt{2}+1)} \right) \frac{1}{D} \frac{\pi^2 c}{\lambda^2}. \tag{9.60}$$

Pulses that do not have the correct secant hyperbolic shape at launch or an amplitude different from an integer times the fundamental soliton power, asymptotically shape themselves into a soliton, while excess power is dispersed from the initial pulse. Figure 9.3 illustrates how a square wave adjusts itself to a soliton. The figure also illustrates how power from the initial pulse is dispersed from the pulse while a soliton is formed.

Figure 9.3 is calculated assuming propagating through an optical fiber with an effective are of $50 \cdot 10^{-12}$ m^2, $n_2 = 3.2 \cdot 10^{-20}$ m/W, and a $\beta_2 = -0.05$ ps^2/km wavelength of 1.55 μm. The figure shows square shape pulses that try to adjust themselves to form solitons. In the left figure the pulse energy is sufficient and a soliton is quickly formed. In the right figure the pulse energy is not sufficient to form a soliton after propagation through 20 km.

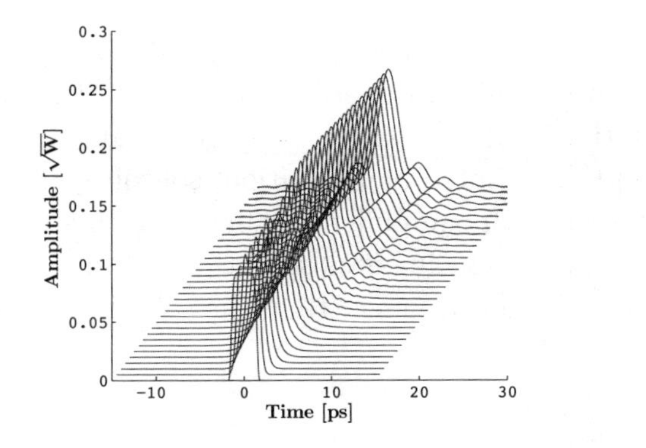

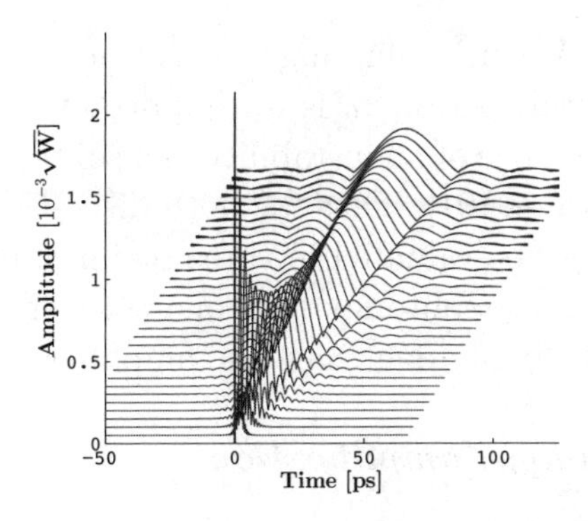

Figure 9.3: Left A square shape pulse 3 ps and amplitude 0.1 $\sqrt{W}$ propagated 25 km. **Right** A square shape pulse 0.3 ps wide peak power $2 \cdot 10^{-3}\sqrt{W}$ propagated 20 km.

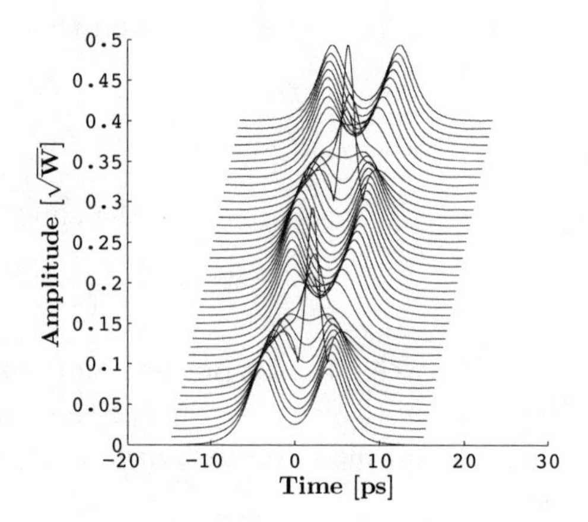

Figure 9.4: Collision of two first-order solitons, with a peakpower of 8.6 mW, full-width half max of 2.6 ps, and propagating 2000 km.

Two solitons that are close to each other in time may attract each other, if they are sufficiently close and if they have the same phase. This is shown in Figure 9.4. One hand-waving explanation for this phenomena is that light is attracted toward high refractive indices.

The opposite case, repulsion of two solitons, occurs if the two solitons have opposite phases.

9.3.1 Impact due to loss/amplification

Due to intrinsic fiber attenuation, a true soliton does strictly speaking not exist in an optical fiber, see Figure 9.5. To counteract the attenuation, optical amplification is used. This may be done either at discrete points along the transmission, referred to as lumped amplification, or the loss may be counterbalanced over long distances using distributed amplification.

When evaluating the soliton stability, the amplifier spacing z_a relative to the soliton period z_0 is an important measure. Elgin a coworkers showed in 1993 the existence of an instability of a soliton if the amplifier spacing is equal to eight soliton periods[101]. Thus for both discrete and distributed amplification, soliton stability may be evaluated in three regimes, one when the amplifier spacing is much shorter than the soliton period, one when the amplifier spacing is close to the soliton period, and finally one when the amplifier spacing is much longer than the soliton period.

Lumped amplification

Amplifier spacing much shorter that the soliton period, $z_a \ll z_0$. The soliton is not capable of adjusting its width to its peak power. As a consequence the soliton propagates as a pulse that does not change pulse width but only

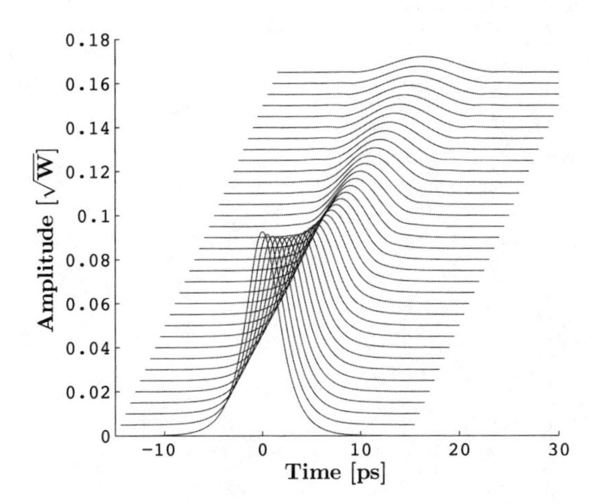

Figure 9.5: The propagation of a fundamental soliton under the influence of loss of 0.2 dB/km, using a pulse with the same parameters as in Figure 9.4 and propagation in 100 km fiber.

amplitude. The launched soliton power thus needs to be higher than the required soliton power, to ensure that the path average power equals the soliton power [102]. The adjusted soliton power relative to the fundamental soliton power is then

$$P_0/P_{\text{sol}} = z_a/L_{\text{eff}}. \tag{9.61}$$

These solitons are referred to as **path average solitons**.

Amplifier spacing similar to the soliton period, $z_a \approx z_0$. In this case the soliton starts to respond to the attenuation by adjusting its width to the amplitude. However, the soliton is not capable of adjusting its width with a rate that matches the amplitude loss rate. As a consequence there is a strong radiation of power from the soliton. This radiation exhibits a maximum when the amplifier spacing equals eight soliton periods ($z_a = 8z_0$) [101].

Amplifier spacing much longer than the soliton period, $z_a \gg z_0$. This final regime is highly relevant when transmitting ultra short pulses and the soliton period therefore approaches a few kilometers. In this case the rate of change in power with distance is slow relative to the soliton period, and as a consequence the soliton is capable of adiabatically adjusting its width to the power. This is also referred to as **adiabatic transmission**.

The stability of a soliton being transmitted and amplified may be evaluated by comparing a soliton after propagation with its expected shape, i.e. a soliton. This may be done by calculating the radiated energy $W = \int_{-\infty}^{\infty} |q - q_e|^2 dt$ where q is the actual pulse and q_e the expected pulse [103]. Figure 9.6 illustrates the radiated energy as a function of amplifier spacing relative to the soliton period. In addition,

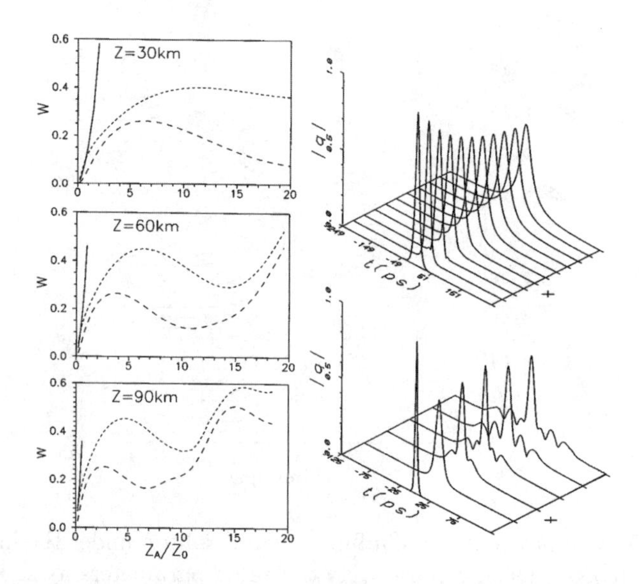

Figure 9.6: Left radiated energy from a soliton as a function of the ratio between the amplification and the soliton period, z_a/z_0. The radiated energy is shown after one, two and three amplifiers. For the dashed curve an expected pulse similar to a soliton with arbitrary amplitude a, and width b is assumed: $A(t) = a_0 \operatorname{sech}(bt)$. For the dotted curve a soliton with phase $-4\eta^2 x$ is assumed $A(t) = 2\eta \operatorname{sech}(2\eta t) \exp(-4i\eta^2 x)$ included. Both dashed and dotted curve is based on numerical solutions to the nonlinear wave equation. The solid line is a theoretical prediction assuming a soliton solution to the propagation equation. **Right** Traces of a 10 ps (top) and 2.5 ps (bottom) pulse propagated through 10, (corresponding to 300 km) and 5 amplifier sections (corresponding to 150 km), respectively. For the 10 ps pulse $z_a/z_0 = 0.75$ and for the 2.5 ps pulse $z_a/z_0 = 12$. From [103].

two specific examples, a 10 ps pulse and a 2.5 ps pulse, illustrate the radiation of energy.

The figure shows that discrete amplification leads to very unstable soliton propagation, only very rapid amplification $z_a \ll z_0$ results in stable soliton propagation. The adiabatic amplification regime is seen however, the point in z_a/z_0 resulting in minimum radiated energy shifts as more and more amplification stages are considered.

Distributed amplification

Amplification of solitons through a distributed amplifier may also be divided into the three regimes as discussed for lumped amplification

Amplifier spacing much shorter than than the soliton period, $z_a << z_0$. As in the case of lumped amplification, there is also a stable regime for distributed amplification when $z_a << z_0$. However, there is not a close form

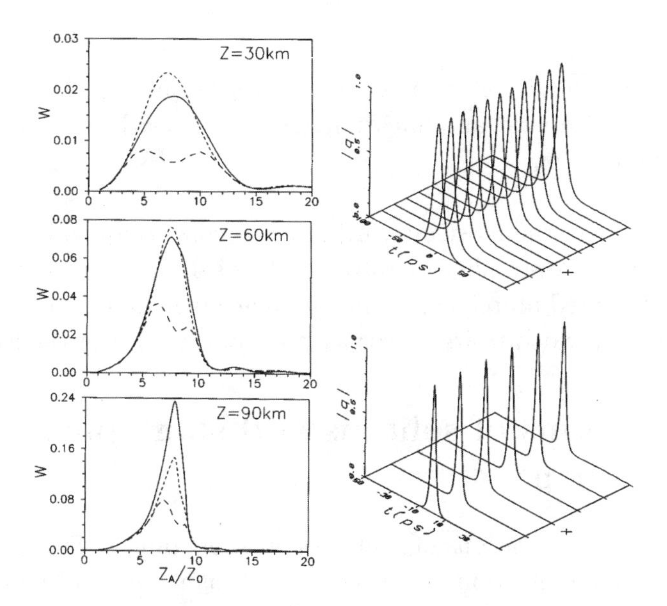

Figure 9.7: Left radiated energy from a soliton as a function of z_a/z_0. The radiated energy is shown after one, two and three amplifiers. As in Figure 9.6 the dashed curve compares an actual pulse to an expected pulse: $A(t) = a\,\mathrm{sech}(bt)$. The dotted curve compares to a soliton: $A(t) = 2\eta\,\mathrm{sech}(2\eta t)\exp(-4i\eta^2 x)$. The solid line shows theoretical predictions. **Right** Traces of a 10 ps **(top)** and 2.5 ps **(Bottom)** pulse propagated through 10, (corresponding to 300 km) and 5 amplifier sections (corresponding to 150 km), respectively. for the 10 ps pulse $z_a/z_0 = 0.75$ for the 2.5 *ps* pulse $z_a/z_0 = 12$. From [103].

expression for the initial amplitude in this case, since the average power depends on the amplifier design.

Amplifier spacing comparable to the soliton period, $z_a \approx z_0$. In analog to lumped amplification, the soliton propagation becomes instable, and a clear peak in the radiated energy from the pulse is seen for $z_a = 8z_0$, Figure, 9.7 left, middle.

Amplifier spacing much larger than the soliton period, $z_a >> z_0$. This is the most interesting regime when comparing distributed to lumped amplification. When applying distributed amplification the solitons are even more capable of adjusting their pulse width to the current pulse power than in the similar regime for lumped amplification. Consequently, when using distributed amplification transmission there exists a regime for transmission that does not exist in lumped amplification. This regime is even more interesting as shorter and shorter pulses are transmitted. In fact this regime may also be used to compress solitons by applying gain to a soliton.

As in the case of discrete amplification, the soliton stability may be predicted by evaluating the radiated energy. In analog to Figure 9.6, Figure 9.7 illustrates the soliton stability when using distributed amplification.

From Figure 9.7 the three regimes are now very clear, and by considering propagated pulses the adiabatic amplification regime is clearly accessible when using distributed amplification.

In summary, one of the most interesting noise contributions to soliton propagation originates from the stability or robustness of solitons against loss and amplification. Here distributed amplification has a great advantage, since the soliton may be able to adjust its width to its power as it propagates along the fiber.

9.4 Applications of solitons and short pulse propagation

There are numerous applications of solitons and short pulses. Two important examples are optical communication and sources such as super continuum sources and fiber lasers, where the latter field has a manifold of applications including molding, bio chemistry, material processing etc.

The applications are determined by the availability of nonlinear materials as novel materials and waveguides. Recent material systems include gas-filled fibers, silicon waveguides and multimode or higher order mode fibers.

In the following we highlight applications of solitons within optical communication and we briefly mention super continuum generation.

9.4.1 Optical communication system demonstrations

In 1988, Mollenauer and Smith published experimental results showing the transmission of solitons through 4000 km of fiber, where loss was compensated solely by Raman gain [104]. The setup was a loop with a length of one span being 41.7 km, see Figure 9.8. The fiber had a loss of 0.22 dB/km at the wavelength 1600 nm. The signal wavelength was close to 1600 nm and the pump wavelength was 1497 nm. About 300 mW of pump power was used to make the fiber loop transparent. The transmission distance was limited by an effective pulse broadening exceeding a factor of $\sqrt{2}$.

Even though solitary pulses in a transmission system may be pictured as pulses that propagate without changing their shape, this is in practice not the case since noise is added to the pulses during propagation, for example when the pulses are amplified and in addition, pulse distortion may also occur because a true soliton only exists when there is no loss. In a real system, the attenuation or amplification of a soliton may lead to instability. This is discussed below. The accumulation of noise causes amplitude noise i.e. degradation of the signal-to-noise ratio similar to a linear system, and jitter in the arrival time of the solitons, often denoted the Gordon-Haus

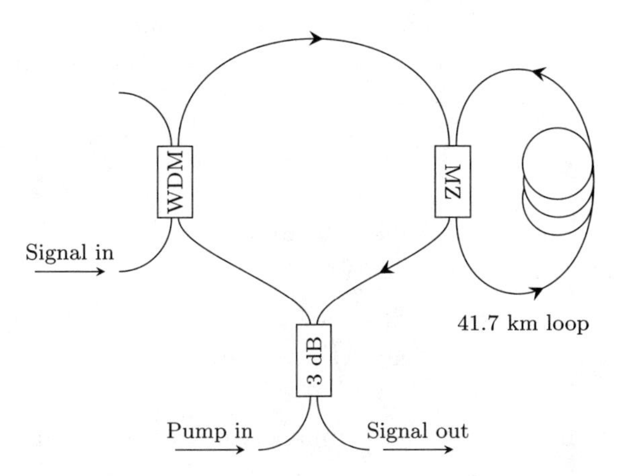

Figure 9.8: The March Zender (MZ) allows pump light to be efficiently coupled into the loop, while at the same time about 95% of the signal light recirculates around the loop [104]. The arrows mark the route of the signal.

jitter. A detailed analysis of these noise phenomena is however, outside the scope of this book.

9.4.2 Supercontinuum generation

Light emitted over a wide smooth continues bandwidth, is referred to as super continuum. Such light has many applications and is consequently of high importance. One way to generate supercontinuum is by using short pulse propagation in optical fibers. Because of nonlinear effects such as Raman scattering, self phase modulation and four-wave mixing, see Chapter 10, the spectrum of the pulses are broadened over hundreds of nanometers.

In the remaining of this subsection we only provide a brief introduction to the basics and application of supercontinuum generation. A comprehensive description of the topic is outside the scope of this book. For this readers are referred to [64].

The treatment of solitons in this chapter has been based on simple scalar equations and only the impact due to group velocity dispersion, loss, and gain has been discussed. In addition, the pulses are also assumed to contain many cycles of the carrier wave, since the slowly varying envelope approximation is used. Finally, Raman scattering among spectral components within the pulse itself is excluded. These approximations may hold if the fiber used has a small, but non-vanishing group velocity dispersion, and if in addition, the pulses are not sub-picosecond.

If on the contrary the pulses are rather short, then it is not a valid approximation to ignore higher order group velocity dispersion terms (i.e. β_3 and/or higher order derivatives of the propagation constant with frequency) nor intra pulse Raman scattering. As discussed in Section 9.1, these effects need to be included, and consequently short pulse propagation is modeled by the differential equation

$$\frac{\partial A}{\partial z} + i\frac{\beta_2}{2}\frac{\partial^2 A}{\partial t^2} - \frac{\beta_3}{6}\frac{\partial^3 A}{\partial t^3} = i\gamma|A|^2 A + i\gamma\left(\frac{i}{\omega_0}\frac{\partial}{\partial t}(|A|^2 A) - T_R A\frac{\partial |A|^2}{\partial t}\right), \quad (9.62)$$

where, β_3 is the derivative of β_2 at the carrier frequency, ω_0, and T_R is the first moment of the Raman response function

$$T_R = \int_{-\infty}^{\infty} tR(t)dt, \quad (9.63)$$

which is related to the slope of the Raman gain in the vicinity of the carrier frequency.

Supercontiuum generation is possible using pulsed sources, femto- or pico-second pulsed light. However, supercontinuum generation has been demonstrated even with continuous waves.

Of special interest for supercontinuum generation is the dispersion zero of the fiber used. In this relation photonic crystal fibers have attracted much attention, since the dispersion may be tailored and may have two zero crossings.

Currently, supercontinuum generation is a subject of strong interest. Applications of supercontinuum sources for characterization of optical components covering a wide bandwidth, but also sources for optical coherence tomography are being investigated.

9.5 Summary

The governing equation when describing short pulse propagation is

$$\frac{\partial A}{\partial z} + \beta_1 \frac{\partial A}{\partial t} + i\frac{1}{2}\beta_2 \frac{\partial^2 A}{\partial t^2} - i\Delta\beta A = 0, \tag{9.16}$$

where $A(z,t)$ is the envelope of the pulse, β the propagation constant that is β_1 the first-order derivative of beta with frequency evaluated at the propagation frequency, β_2 the second-order derivative of the propagation constant with frequency also evaluated at the propagation frequency. β_1 relates to the group velocity whereas β_2 relates to the group velocity dispersion and is responsible for linear pulse broadening. The term with $\Delta\beta$ is in its simplest form related to the intensity dependent refractive index but may include more complicated terms such as Raman scattering, or frequency dependence of the spot size of the propagating beam, see Section 9.1.

A particular solution is a soliton where the linear pulse broadening caused by group velocity dispersion is balanced by the effect of the intensity dependent refractive index. A soliton has a hyperbolic secant shape

$$A(t, z = 0) = a_0 \sqrt{P_{\text{sol}}}\, \text{sech}(bt), \tag{9.36}$$

where a_0 is the soliton amplitude, which in normalized units needs to be an integer number and b relates to the temporal width of the soliton. For a fundamental soliton the two parameters are related through

$$P_{\text{sol}} = \frac{\ln^2(\sqrt{2}+1)}{\pi^2} \frac{A_{\text{eff}}}{cn_2} \frac{\lambda^3 D}{\tau^2}. \tag{9.37}$$

As pulse widths get shorter and shorter more and more effects need to be included into the governing equation. Because of a complicated interplay between numerous nonlinear effects, a subset of pulses are formed and the spectral content of the output reaches an enormous bandwidth. This is referred to as supercontinuum generation.

It is equally important to characterize pulses in the frequency and time domains. One of the important quantities is the pulse within the time as well as in the frequency domain. The width may be either based on measured quantities as the full width at half maximum intensity or more derived quantities like the RMS width, see Section 9.3. The product of the two quantities is called the time bandwidth product. If a pulse is undistorted and has no chirp, then the time-bandwidth product takes its minimum value. For a Gaussian pulse the minimum time-bandwidth product, based on RMS pulse widths is $\pi/2$, see Example 9.3

Chapter 10

Four-wave mixing

Contents

Until this point we have described nonlinear pulse propagation, more specifically, an envelope function multiplied on a carrier frequency. This has been done using a quasi monochromatic wave in Chapter 9. In Chapters 7 and 8 we have discussed mutual interactions among two frequencies in Raman and Brillouin scattering. In this chapter we focus on mutual interaction among three or four frequencies, in centrosymmetric materials, i.e. materials where $\chi^{(2)}$ equals zero. This is often referred to as four-wave mixing or degenerate four-wave mixing when two out of the four frequencies are identical. More specifically we turn our attention toward nonlinear interactions that occur because of the intensity dependent refractive index. When the effects of Raman and Brillouin scattering together with intrinsic loss are ignored in this nonlinear interaction, the material in which the nonlinear interaction happens, only plays a passive role in the sense that no energy is transferred to or from the material. Consequently, the nonlinear interaction that we discuss in this Chapter is referred to as a parametric process.

Four-wave mixing is important in many fields of optics and has many applications, including fiber optical parametric amplifiers (FOPAs), wavelength converters and optical waveform sampling. In addition, the potential of generation and processing of quantum states makes four-wave mixing of particular interest. Finally, four-wave mixing among channels in a communication system is also of great importance. The reason for this is that the power per channel is increasing, which leads to challenges due to four-wave mixing among information carrying channels.

10.1 Physical description

To picture what happens in four-wave mixing it is useful to imagine light as being photons. During propagation two photons at a specific frequency change frequency – one photon loses energy, i.e. shifts to lower frequency (longer wavelength), while the other photon gains energy, i.e. shifts to higher frequency (shorter wavelength). Consequently, the energy of the system is conserved and the material only acts as a catalyst, hence the name parametric process.

If the photons initially are launched at frequency ω_p, and one photon loses energy by shifting to the frequency ω_s, ($\omega_s < \omega_p$) it has lost the energy corresponding to $\omega_p - \omega_s$. The other photon then gains the corresponding energy and shifts to $\omega_p + (\omega_p - \omega_s)$. This latter frequency is denoted the idler frequency, ω_i.

In the above the two photons were initially at the same frequency, this does not have to be the case. In a more general picture the two photons may be at different frequencies, the key is that one photon loses as much energy as the other gains.

In the above descriptions, either three or four frequencies were interacting with each other. In addition, the power level at one or two of the considered frequencies were significantly larger than the power level at the remaining frequencies. In summary, considering three or four frequencies, each with a low or high power level, gives a large number of different configurations of four-wave mixing to consider.

Three frequencies - degenerate four-wave mixing Two configurations are typically considered, one denoted as modulation interaction (or single pump) the other as inverse modulation interaction (or dual pump) [105], see Figure 10.1. In the single pump configuration, energy conservation requires $2\omega_p = \omega_i + \omega_s$. In the dual pump configuration, energy conservation requires $2\omega_s = \omega_{p1} + \omega_{p2}$. In the simplest configuration there is one strong pump and a signal at launch while the idler is absent. This configuration is used in the design of a phase insensitive parametric amplifier or a wavelength converter. If the idler is present at launch, then the configuration is a phase sensitive amplifier. In the other configuration, the so-called inverse modulation interaction, or dual pump, two pumps are used for example to create a flat gain spectrum for a signal located in between the two pumps [106] [107] or to achieve a polarization independent gain.

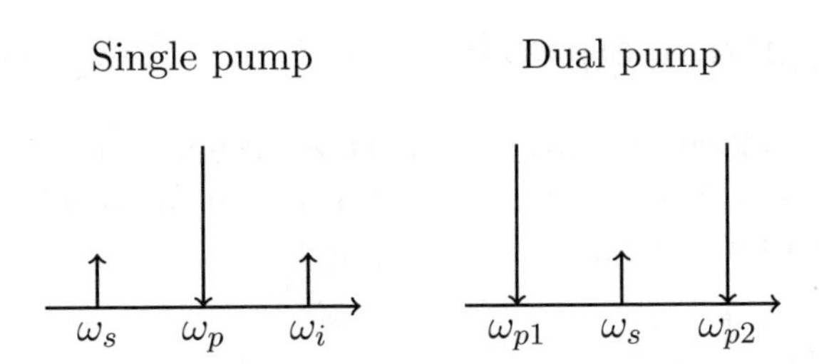

Figure 10.1: Three wave interaction; **Left** modulation interaction or single pump configuration and **right** inverse modulation interaction or dual pump configuration. Note, the length of the arrows indicate the power at the respective frequency. In addition, the direction of the arrow show what waves lose energy and what waves gain energy, upward arrow indicates gain whereas downward arrow indicates loss.

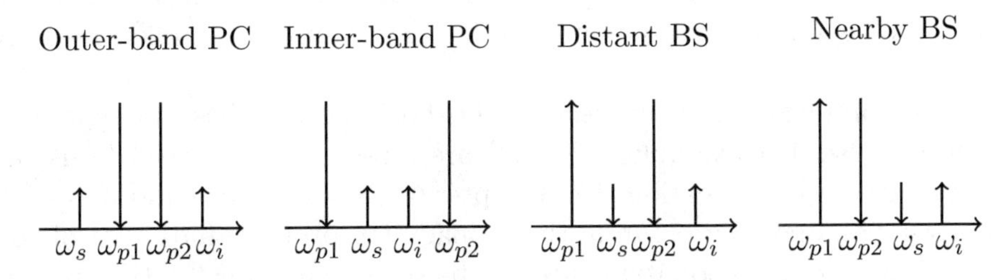

Figure 10.2: Outer-band and Inner-band phase conjugation, distant and nearby Bragg scattering.

Four frequencies Often treated scenarios are divided into two groups: phase conjugation (PC), where two different pump photons are annihilated and two different sidebands are created, and Bragg scattering (BS), where a sideband and a pump photon are annihilated and a different sideband and a pump photon is created. In both PC configurations, energy conservation requires $\omega_{p1} + \omega_{p2} = \omega_i + \omega_s$. In both BS configurations, energy conservation requires $\omega_{p1} + \omega_i = \omega_{p2} + \omega_s$. In inner-band PC, two high power pumps surround two weaker signals. It is titled phase conjugation since the generated idler is proportional to the complex conjugate of the amplified signal. In inverse phase conjugation or outer-band PC, two strong pumps are surrounded by two weaker signals. Bragg scattering is divided into nearby and distant frequency Bragg scattering - In nearby BS power is transferred between signal and idler and between the pumps - In distant BS power is transferred between pump and signal and pump and idler.

For simplicity we focus in the following on the degenerate case i.e. the case where only three frequencies interact with each other. This could be one strong pump and two weaker frequencies, a signal and an idler, as in Figure 10.1, the left-hand side illustration.

10.2 Propagation equations—three frequencies

Since we have constrained ourselves to consider degenerate four-wave mixing, the electric field consists of three frequencies, the pump frequency ω_p, the signal frequency ω_s and the idler frequency ω_i

$$\begin{aligned}\mathbf{E}(z,t) =& \frac{1}{2}\left[\mathbf{e}_p E_p^0(z,t) R_p(r) \exp\{i(\beta_p z - \omega_p t)\}\right. \\ &+ \mathbf{e}_s E_s^0(z,t) R_s(r) \exp\{i(\beta_s z - \omega_s t)\} \\ &\left.+ \mathbf{e}_i E_i^0(z,t) R_i(r) \exp\{i(\beta_i z - \omega_i t)\} + \text{c.c.}\right],\end{aligned} \tag{10.1}$$

where $\mathbf{e}_j$, $j = s, p, i$ is a unit polarization vector corresponding to the signal, pump, and idler respectively. The subindex j is used in the following for signal, pump, and idler, respectively. β_j is the propagation constant as determined from the waveguide eigenvalue problem, also corresponding to the signal, pump, and idler, respectively.

To predict propagation of the electrical field one has to choose an approach that serves the purpose. For example if $E_j(z,t)$ are pulses one may need to use a split step Fourier method, as mentioned in Chapter 9. On the other hand if $E_j(z,t)$ are continuous waves, one may separate the propagation problem into a set of coupled equations. These coupled equations need to be solved numerically. Under some circumstances as for example when neglecting loss, depletion, and Raman scattering, the problem may be solved analytically [108] [109].

In the following we focus on the description of CW or quasi CW cases, i.e. we derive a set of coupled propagation equations for the pump signal and idler. To obtain such a set of coupled propagation equations we identify the induced polarization at the respective frequencies, i.e. ω_s, ω_i, and ω_p.

From Chapter 3, the μth coordinate of the induced third-order polarization may be written in terms of the third-order susceptibility tensor $\chi^{(3)}_{\mu jkl}$ as

$$(P_\mu^0)^{(3)} = \varepsilon_0 \sum_{jkl} K \chi^{(3)}_{\mu jkl} E_j^0 E_k^0 E_l^0. \tag{10.2}$$

Assuming that the electric field is written as in Eqn. (10.1), the induced polarization contains many different frequencies. Let us start by identifying the contributions oscillating at the signal frequency ω_s.

In accordance with the introductory picture discussed in Section 10.1, one contribution to the induced polarization at ω_s originates from the photon at ω_p losing the energy corresponding to the frequency shift $\omega_p - \omega_i$, see also the left-hand side illustration in Figure 10.1. Consequently, the photon ends up having the frequency $\omega_s = \omega_p - (\omega_i - \omega_p)$. Note, this equation also represents the requirement of energy conservation, $2\omega_p = \omega_s + \omega_i$. In terms of a description using the electric field, this interaction is described through the term

$$\begin{aligned} &\mathbf{e}_p E_p^0(z,t) R_p(r) \exp\left(i(\beta_p z - \omega_p t)\right) \\ &\times \mathbf{e}_i^* (E_i^0)^*(z,t) R_i^*(r) \exp\left(-i(\beta_i z - \omega_i t)\right) \mathbf{e}_p E_p^0(z,t) R_p(r) \exp\left(i(\beta_p z - \omega_p t)\right). \end{aligned} \tag{10.3}$$

To find the induced polarization we need to count the number of permutations of the input frequencies, i.e. permutations of $(\omega_p, -\omega_i, \omega_p)$. In this case there are three permutations. In addition, the process is a third-order nonlinear process, thus the prefactor also includes a factor of 2^{-3}. Furthermore, none of the input frequencies are zero and none of the output frequencies are zero. Finally, we are interested in the amplitude of the induced polarization oscillating at frequency ω_s and with complex notation for the electric field we need to include a factor of 2 in the prefactor. Thus, in summary the prefactor equals $K = 3(1/2^3)2 = 3/4$ as seen in table 3.2.

In addition to the case described above there are three other frequency combinations that all contribute to the induced polarization at ω_s. These are listed symbolically in Eqn. (10.4), which also includes the number of permutations possible for each case.

$$\begin{aligned} \omega_s &= \omega_s - (\omega_s - \omega_s) \quad , 3 \text{ permutations}, \quad K = 3/4 \\ \omega_s &= \omega_s - (\omega_i - \omega_i) \quad , 6 \text{ permutations}, \quad K = 3/2 \\ \omega_s &= \omega_s - (\omega_p - \omega_p) \quad , 6 \text{ permutations}, \quad K = 3/2. \end{aligned} \tag{10.4}$$

The first process, in Eqn. (10.4) corresponds to self-phase modulation, while the last two corresponds to cross-phase modulation. We can now write the amplitude of the induced polarization at ω_s

$$\begin{aligned} (P_\mu^0)^{(3)} = &\frac{3}{4}\varepsilon_0 (R_p)^2 (R_i)^* \sum_{jkl} \chi^{(3)}_{\mu jkl}(-\omega_s; \omega_p, \omega_p, -\omega_i)(E_p^0)_j (E_p^0)_k (E_i^0)^*_l \\ &+ \frac{6}{4}\varepsilon_0 |R_p|^2 R_s \sum_{jkl} \chi^{(3)}_{\mu jkl}(-\omega_s; \omega_p, -\omega_p, \omega_s)(E_p^0)_j (E_p^0))^*_k (E_s^0)_l \\ &+ \frac{6}{4}\varepsilon_0 |R_i|^2 R_s \sum_{jkl} \chi^{(3)}_{\mu jkl}(-\omega_s; \omega_i, -\omega_i, \omega_s)(E_i^0)_j (E_i^0)^*_k (E_s^0)_l \\ &+ \frac{3}{4}\varepsilon_0 |R_s|^2 R_s \sum_{jkl} \chi^{(3)}_{\mu jkl}(-\omega_s; \omega_s, -\omega_s, \omega_s)(E_s^0)_j (E_s^0)^*_k (E_s^0)_l, \end{aligned} \tag{10.5}$$

with a complex conjugate at $-\omega_s$. That is, the induced polarization at the frequency ω_s is

$$P_\mu^{(3)}(z,t) = \frac{1}{2}(P_\mu^0)^{(3)} \left(e^{-i(\omega_s t - \beta_s z)} + c.c. \right). \tag{10.6}$$

By inserting $\mathbf{E}_s(z,t) = \mathbf{e}_s E_s(z) R_s(r) \exp\{i(\beta_s z - \omega_s t)\}$, and assuming that the state of polarization for none of the frequency components changes during propagation, and neglecting $\partial^2 \mathbf{E_s}/\partial z^2$, which correspond to the slowly varying envelope approximation, we obtain a wave equation similar to Eqn. (5.48) for each vector component of the electric field at the signal frequency

$$\frac{\partial (E_s^0 R_s)_\mu}{\partial z} = i\frac{\mu_0 \omega_s^2}{2\beta_s}(P_\mu^0)^{(3)} e^{-i\beta_s z}. \tag{10.7}$$

To continue we assume that all components of the electric field are linearly polarized for example along the x-axis, and remain in this state of polarization. Consequently, we may skip the sub-index to the electric field and the induced polarization. In addition, we assume that the nonlinearity is dispersion free, by which we mean that the susceptibility is frequency independent, for example: $\chi^{(3)}_{\mu jkl}(-\omega_s;\omega_p,-\omega_p,\omega_s) = \chi^{(3)}_{\mu jkl}(-\omega_s;\omega_i,-\omega_i,\omega_s)$. Consequently, we may omit the frequency arguments to the susceptibility. In summary, we may simply write the tensor as a constant, we simply use $\chi^{(3)}$, and Eqn. (10.7) may be written as

$$\begin{aligned} \frac{\partial E_s^0}{\partial z} R_s = i\frac{\omega_s^2}{2c^2\beta_s}\frac{3}{4}\chi^{(3)} \Big[&(R_p E_p^0)^2 (R_i E_i^0)^* e^{iz(2\beta_p - \beta_i - \beta_s)} \\ &+ \left\{ 2|R_p|^2|E_p^0|^2 + 2|R_i|^2|E_i^0|^2 + |R_s|^2|E_s^0|^2 \right\} E_s^0 R_s \Big]. \end{aligned} \tag{10.8}$$

Extending the assumption of dispersion free susceptibility to include all frequencies, so for example $\chi^{(3)}_{\mu jkl}(-\omega_s;\omega_i,-\omega_i,\omega_s) = \chi^{(3)}_{\mu jkl}(-\omega_i;\omega_s,-\omega_s,\omega_i)$, i.e. assuming Kleinman symmetry we may write a propagation equation for the electric field at the idler frequency by swapping index 's' by 'i', i.e.

$$\begin{aligned} \frac{\partial E_i^0}{\partial z} R_i = i\frac{\omega_i^2}{2c^2\beta_i}\frac{3}{4}\chi^{(3)} \Big[&(R_p E_p^0)^2 (R_s E_s^0)^* e^{iz(2\beta_p - \beta_s - \beta_i)} \\ &+ \left\{ 2|R_s|^2|E_s^0|^2 + 2|R_p|^2|E_p^0|^2 + |R_i|^2|E_i^0|^2 \right\} E_i^0 R_i \Big]. \end{aligned} \tag{10.9}$$

For the pump one has to remember that the cross-modulation term, i.e. the first term in Eqn. (10.8) and Eqn. (10.9) is slightly different. More specifically, at the frequency of the pump, the cross modulation term is caused by ω_s, ω_i and ω_p, more specifically ($\omega_p = \omega_s + (\omega_i - \omega_p)$). This term has 6 permutations consequently

$$\begin{aligned} \frac{\partial E_p^0}{\partial z} R_p = i\frac{\omega_p^2}{2c^2\beta_p}\frac{3}{4}\chi^{(3)} \Big[&2(R_s E_s^0)(R_i E_i^0)(R_p E_p^0)^* e^{iz(\beta_s + \beta_i - 2\beta_p)} \\ &+ \left\{ 2|R_s|^2|E_s^0|^2 + 2|R_i|^2|E_i^0|^2 + |R_p|^2|E_p^0|^2 \right\} E_p^0 R_p \Big]. \end{aligned} \tag{10.10}$$

The above set of coupled equations, Eqn. (10.8) through Eqn. (10.10) may be solved numerically and enable one to predict for example the gain of an amplifier or the conversion efficiency of a wavelength converter. It is important to note that even if for example only a signal and a pump field is present at the input, the above equations predict four-wave mixing accurately since the first term on the right-hand side of the propagation equations acts as a source term. From Eqn. (10.8) to Eqn. (10.10), it is obvious that the difference in propagation constant between pump, signal, and idler ($\beta_s + \beta_i - 2\beta_p$) is an important parameter. In the following we refer to this parameter as the linear phase mismatch $\Delta\beta = \beta_s + \beta_i - 2\beta_p$. For $\chi^{(3)}$ real, the equations Eqn. (10.8) through Eqn. (10.10), show that the first term on the right-hand side is responsible for energy transfer while the last three terms cause a phase change.

Often it provides significant insight to solve propagation equations in power, rather than electric field. To derive power equations we introduce the power into Eqn. (10.8) through Eqn. (10.10). The power is $P_j = \int_A I_j da$, where I_j is the intensity of the pump, signal, and idler, corresponding to $j = p, s, i$. The intensity is $I_j = (1/2)\varepsilon_0 c n_j (E_j^0 R_j)(E_j^0 R_j)^*$, where n_j is the refractive index. In a waveguide we approximate the refractive index by the effective refractive index, see Appendix E. It is also noted that the transverse distribution of the electric field, i.e. R_j for a guided mode in a waveguide is a real function.

The rate of change of the power with propagation distance is then

$$\frac{dP_j}{dz} = \frac{1}{2}\varepsilon_0 c n_j 2Re \int_A \left[(E_j^0 R_j)^* \frac{d(E_j^0 R_j)}{dz} \right] da. \tag{10.11}$$

From Eqn. (10.11) we find propagation equations for the signal, pump, and idler power. Using Eqn. (10.8) through Eqn. (10.10) we find

$$\begin{aligned}\frac{dP_s}{dz} = \varepsilon_0 c n \frac{\omega_s^2}{2c^2\beta_s}\frac{3}{4} Re \int_A \Big[(E_s^0 R_s)^* i\chi^{(3)} \Big[(R_p E_p^0)^2 (R_i E_i^0)^* e^{-i\Delta\beta z} \\ + \left\{ 2|R_p|^2|E_p^0|^2 + 2|R_i|^2|E_i^0|^2 + |R_s|^2|E_s^0|^2 \right\} E_s^0 R_s \Big]\Big] da.\end{aligned} \tag{10.12}$$

By replacing the electric field of the pump, signal, and idler, i.e. E_j , $j \in (i, s, p)$, by the square root power multiplied by a phase $\phi_j(z)$, more specifically

$$E_j^0 = \frac{\sqrt{P_j}}{\sqrt{(1/2) n_j \varepsilon_0 c}} \frac{e^{i\phi_j}}{\sqrt{\int_A |R_j|^2 da}}, \qquad j \in (i, s, p), \tag{10.13}$$

and by assuming that both the susceptibility and the transverse distribution of the electric field are real, i.e. $\chi^{(3)}$ and R are real this may be rewritten as

$$\frac{dP_s}{dz} = \frac{3\omega_s}{2c^2\varepsilon_0} \frac{P_p\sqrt{P_i P_s}}{n_p\sqrt{n_i n_s}} \int_A \frac{\chi^{(3)} R_p^2 R_i^* R_s^* \sin(\theta)}{\int_A |R_p|^2 da \sqrt{\int_A |R_i|^2 da \int_A |R_s|^2 da}} da, \tag{10.14}$$

where we have used $\beta_s = \omega_s n_s / c$, where n_s is the effective refractive index of the waveguide at the wavelength of the signal, and the angle $\theta = \Delta\beta z - 2\phi_p + \phi_s + \phi_i$ has been introduced. For a guided mode supported by a waveguide it is a valid approximation, since a guided mode is a solution to an eigenmode problem, having real eigen-values i.e. real propagation constants, and real eigenvectors, i.e real transverse electric field vectors, as solutions.

Similar equations are found for the idler and pump power

$$\frac{dP_i}{dz} = \frac{3\omega_i}{2c^2\varepsilon_0} \frac{P_p\sqrt{P_iP_s}}{n_p\sqrt{n_in_s}} \int_A \frac{\chi^{(3)} R_p^2 R_i^* R_s^* \sin(\theta)}{\int_A |R_p|^2 da \sqrt{\int_A |R_i|^2 da \int_A |R_s|^2 da}} da, \tag{10.15a}$$

$$\frac{dP_p}{dz} = -2\frac{3\omega_p}{2c^2\varepsilon_0} \frac{P_p\sqrt{P_iP_s}}{n_p\sqrt{n_in_s}} \int_A \frac{\chi^{(3)} (R_p^*)^2 R_i R_s \sin(\theta)}{\int_A |R_p|^2 da \sqrt{\int_A |R_i|^2 da \int_A |R_s|^2 da}} da. \tag{10.15b}$$

It is noted that the three coupled equations Eqn. (10.14) through Eqn. (10.15b) have a common integral in the following denoted Γ

$$\Gamma = \int_A \frac{\chi^{(3)} (R_p^*)^2 R_i R_s}{n_p \int_A |R_p|^2 da \sqrt{n_i \int_A |R_i|^2 da \, n_s \int_A |R_s|^2 da}} da. \tag{10.16}$$

If we redefine $\chi^{(3)}$ as $\tilde{\chi}_j^{(3)} = \chi^{(3)} \frac{3\omega_j}{4c^2\varepsilon_0}$, $j \in (i, p, s)$ and approximate the frequencies as being identical $\omega_s = \omega_p = \omega_i$ we may rewrite the power equations as

$$\frac{dP_s}{dz} = 2P_p\sqrt{P_iP_s} \sin(\theta)\tilde{\Gamma}_s \tag{10.17a}$$

$$\frac{dP_i}{dz} = 2P_p\sqrt{P_iP_s} \sin(\theta)\tilde{\Gamma}_i \tag{10.17b}$$

$$\frac{dP_p}{dz} = -4P_p\sqrt{P_iP_s} \sin(\theta)\tilde{\Gamma}_p, \tag{10.17c}$$

where $\tilde{\Gamma}_j$ is

$$\tilde{\Gamma}_j = \int_A \frac{\tilde{\chi}_j^{(3)} (R_p^*)^2 R_i R_s}{n_p \int_A |R_p|^2 da \sqrt{n_i \int_A |R_i|^2 da \, n_s \int_A |R_s|^2 da}} da \quad , \quad j \in (i, s, p). \tag{10.18}$$

Note that if θ equals $\pi/2$ the signal and idler increases with a maximum rate, and energy is removed from the pump; this is the **phase sensitive amplifier**. However, if θ equals $-\pi/2$ the signal and idler lose power. This is the **phase sensitive attenuator**. In addition, if for example no idler is launched, Eqn. (10.17a) is not able to describe amplification. However, this case is the **phase insensitive amplifier** where the idler grows in the absence of an idler at the input. To describe this using Eqn. (10.17a) one needs to include so-called vacuum fluctuations at the input. We return to this in Section 10.3.

Since θ is a function of position, Eqn 10.17 is not sufficient to describe evolution of power as a function of position. We need a propagation equation for θ. From Eqn. (10.13) a governing equation for the phase of the signal, pump, and idler is

$$\frac{d\phi_j}{dz} = \frac{i(1/2)n_j\varepsilon_0 c \int_A |R_j|^2 da}{2P_j} \left(E_j^0 \frac{d(E_j^0)^*}{dz} - (E_j^0)^* \frac{dE_j^0}{dz} \right) \quad , \quad j \in (s, p, i). \tag{10.19}$$

Replacing the rate equations for the electric field in Eqn. (10.19) using Eqn. (10.8) through Eqn. (10.10) we find rate equations for the phases of pump, signal, and idler. For example for the signal we find

$$\begin{aligned}\frac{d\phi_s}{dz} =& \frac{3\varepsilon_0\omega_s}{16P_s}\int_A \chi^{(3)}\left[Re\left[(E_s^0R_s)((E_p^0R_p)^*)^2(E_i^0R_i)e^{i\Delta\beta z}\right]\right. \\ &+ \left.\left[2|E_p^0R_p|^2 + 2|E_i^0R_i|^2 + |E_s^0R_s|^2\right]|E_s^0R_s|^2\right]da.\end{aligned} \tag{10.20}$$

Replacing the electric fields by their corresponding powers using Eqn. (10.13) we find

$$\frac{d\phi_s}{dz} = \left[P_p\sqrt{\frac{P_i}{P_s}}\cos(\theta)\tilde{\Gamma}_s + \int_A \tilde{\chi}_s^{(3)}(2P_p\tilde{r}_p + 2P_i\tilde{r}_i + P_s\tilde{r}_s)\tilde{r}_s da\right] \tag{10.21a}$$

$$\frac{d\phi_i}{dz} = \left[P_p\sqrt{\frac{P_s}{P_i}}\cos(\theta)\tilde{\Gamma}_i + \int_A \tilde{\chi}_i^{(3)}(2P_s\tilde{r}_s + 2P_p\tilde{r}_p + P_i\tilde{r}_i)\tilde{r}_i da\right] \tag{10.21b}$$

$$\frac{d\phi_p}{dz} = \left[2\sqrt{P_iP_s}\cos(\theta)\tilde{\Gamma}_p + \int_A \tilde{\chi}_p^{(3)}(2P_s\tilde{r}_s + 2P_i\tilde{r}_i + P_p\tilde{r}_p)\tilde{r}_p da\right], \tag{10.21c}$$

where $\tilde{r}_i = \frac{|R_j|^2}{n_j\int_A |R_j|^2 da}$, $j \in (i, s, p)$. Using, $\theta = \Delta\beta z - 2\phi_p + \phi_s + \phi_i$, i.e. $\frac{d\theta}{dz} = \Delta\beta - 2\frac{d\phi_p}{dz} + \frac{d\phi_s}{dz} + \frac{d\phi_i}{dz}$ we may now derive a final rate equation for θ by inserting the above rate equations for the phases, see example 10.1 below.

Example 10.1. *Fiber optical parametric amplifier (FOPA)*
Let us approximate the angular frequencies of the pump, signal, and idler by one frequency i.e. we assume that they are close enough to each other so we may consider them all identical $\omega_p = \omega_i = \omega_i = \omega_0$. Consequently, the transverse field distribution for the pump, signal, and idler is the same $R_s = R_p = R_i = R$, and the refractive index for pump, signal, and idler is the same $n_s = n_p = n_i = n$. If we further assume that $\chi^{(3)}$ is constant across the waveguide, then $\tilde{\Gamma}$ is

$$\tilde{\Gamma} = \frac{3\omega_0}{4c^2\varepsilon_0}\frac{\chi^{(3)}}{n^2}\frac{\int_A |R|^4 da}{(\int_A |R|^2 da)^2}, \tag{10.22}$$

and

$$\tilde{\chi}^{(3)}\int_A \tilde{r}^2 da = \frac{3\omega_0}{4c^2\varepsilon_0}\frac{\chi^{(3)}}{n^2}\frac{\int_A |R|^4 da}{(\int_A |R|^2 da)^2} = \tilde{\Gamma}. \tag{10.23}$$

Note, the fraction of the integrals are recognized as the reciprocal effective area $(A_{\text{eff}})^{-1}$ introduced in Chapter 9. In fiber optics it is customary to use

$$\tilde{\Gamma} = \frac{2\pi n_2^I}{\lambda A_{\text{eff}}}, \tag{10.24}$$

where n_2^I is the intensity dependent refractive index, discussed in Chapter 9.

In addition, rather than expressing propagation equations in electric fields, propagation equations are given in terms of square root of the optical power. In the case of the single pump parametric amplifier the rate equations for the electric field i.e. Eqn. (10.8) through Eqn. (10.10) may be rewritten as

$$\frac{dA_p}{dz} = i\tilde{\Gamma}\left[\left\{|A_p|^2 + 2(|A_s|^2 + |A_i|^2)\right\} A_p + 2A_s A_i A_p^* \exp(i\Delta\beta z)\right], \tag{10.25a}$$

$$\frac{dA_s}{dz} = i\tilde{\Gamma}\left[\left\{|A_s|^2 + 2(|A_i|^2 + |A_p|^2)\right\} A_s + A_i^* A_p^2 \exp(-i\Delta\beta z)\right], \tag{10.25b}$$

$$\frac{dA_i}{dz} = i\tilde{\Gamma}\left[\left\{|A_i|^2 + 2(|A_s|^2 + |A_p|^2)\right\} A_i + A_s^* A_p^2 \exp(-i\Delta\beta z)\right]. \tag{10.25c}$$

The corresponding propagation equations for power levels in the pump, signal, and idler are given by Eqn. (10.17a), only with the same Γ, given by Eqn. (10.24) for pump, signal, and idler. The equation for the the phases ϕ_j in Eqn (10.21a) also reduces significantly, and it is possible to derive a differential equation for the angle θ, which appears in the propagation equations for the power levels. The propagation equation for the angle θ is

$$\frac{d\theta}{dz} = \Delta\beta + \cos(\theta)\tilde{\Gamma}P_p\sqrt{P_s P_i}\left[-\frac{4}{P_p} + \frac{1}{P_i} + \frac{1}{P_s}\right] + \tilde{\Gamma}\left[2P_p - P_s - P_i\right]. \tag{10.26}$$

—— ■ ——

10.3 Spontaneous emission in four-wave mixing

Like Raman scattering, four-wave mixing may be used to transfer energy from one wavelength to another. This may be used for example to make an optical amplifier or a wavelength converter. However, as opposed to Raman scattering, four-wave mixing may be used to generate not only light at longer wavelength but also light at shorter wavelength. Another significant difference between Raman scattering and four-wave mixing is that Raman scattering is a light matter interaction involving an energy reservoir defined by phonons and light, whereas four-wave mixing is a parametric process where the material is only a catalyst between mutual wave interactions. Consequently, the process of four-wave mixing is free of spontaneous emission. However, the latter statement may be considered as not completely accurate. The reason is that in for example wavelength conversion where only a pump and a signal is launched, light grows based on so-called vacuum fluctuations, as we return to in section 10.3.1.

When discussing four-wave mixing, and more specifically spontaneous four-wave mixing, we need to be able to describe not only vacuum fluctuations but equally well laser light and its signal to noise ratio.

10.3.1 Input conditions

Throughout this book we have described signals as electrical fields, i.e. a carrier wave with an amplitude, in some cases a time dependent amplitude, and a phase. This description is typically referred to as a classical description. In addition, we have also used photon number, i.e. counted the number of photons in a signal, for example in our discussion of Raman scattering. In relation to the classical description it is well accepted that the signal is characterized not only by amplitude and phase, but equally important, by its signal to noise ratio. Using the photon description the signal to noise ratio is defined as the mean photon number squared relative to the variance in photon number.

From laser physics and quantum optics it is established that an electric field which has the least uncertainty in phase as well as amplitude is a so-called minimum uncertainty state; an example of such a state is a so-called Coherent state [110]. This state is characterized by having a photon number distribution which follows a Poissonian probability density distribution, where the mean and variance in photon number are identical. Since the signal to noise ratio equals the mean photon number squared relative to the variance in photon number, the signal to noise ratio for a Coherent state equals the mean photon number.

In this book we wish to describe nonlinear optics by classical electrical fields rather than using operators as in quantum optics. This has the advantage that the description of various phenomena more easily may be adapted to numerical approaches. Consequently, more comprehensive analysis may be done, leading to numerical results that agree well with experiments. This is for example the case when including different effects, as Raman scattering, pump depletion, and loss when describing phenomena based on four-wave mixing. However, to describe spontaneous emission in four-wave mixing it is important to be able to seed the process with an input that obeys quantum optics rules, i.e. an input in the absence of a classical field. This may best be explained by considering an example.

Let us consider the degenerate case of four-wave mixing where one strong pump field is launched together with a signal, as for example in a phase insensitive amplifier or a wavelength converter. After propagation through a short fiber, the idler has grown. This process is described automatically by equations Eqn. (10.8) through Eqn. (10.10). However, it is noted that the power in the idler does not grow automatically from Eqn. (10.17a). For completeness, readers interested in further details of approaches based on quantum optics are referred to for example [111] and [112].

To incorporate spontaneous emission accurately in the four-wave mixing process, it is essential that the input fields are accurately described, not the least the idler, where there is no power at launch. The condition, i.e. the absence of a classical field, is referred to as vacuum field, or zero point field. The strong pump is described classically by an amplitude and a phase. To describe noise on the pump, i.e. uncertainty in the amplitude and phase, a small uncertainty is added to the amplitude and phase. As an approach to a quantum mechanical description the pump is described as a Coherent state.

The Coherent state, i.e. the amplitude and phase of the corresponding classical signal is described using so-called quadrature signals. Note, the term signal is used here as a general term not reflecting whether it is a pump, signal or an idler in four-wave mixing, but a signal is simply an electric field, with an amplitude and a phase. More specifically if the amplitude of the classical signal is A, and the phase of the classical signal ϕ, then the quadrature signals denoted, x_0 and p_0, equal $|\alpha|\cos(\phi)$ and $|\alpha|\sin(\phi)$, respectively. The uncertainty of the Coherent state is then included by adding fluctuations δx and δp, to both of the quadratures i.e. the signal is then

$$A = x + ip = |\alpha|\cos(\phi) + \delta x + i(|\alpha|\sin(\phi) + \delta p). \tag{10.27}$$

To describe spontaneous emission, δx and δp may have Gaussian distributions with a zero mean, $\langle \delta x \rangle = \langle \delta p \rangle = 0$, but finite variances. We return to an example of a Coherent state below. In our semiclassical description we treat α and ϕ as fixed values, i.e. values without any uncertainty. In the following $|\alpha|\cos(\phi)$ is denoted x_0 and $|\alpha|\sin(\phi)$ is denoted p. The power of the electric field, proportional to the photon number is then

$$\begin{aligned} |A|^2 &= (x_0 + \delta x)^2 + (p + \delta p)^2 \\ &= x_0^2 + \delta x^2 + 2x_0\delta x + p^2 + \delta p^2 + 2p\delta p. \end{aligned} \tag{10.28}$$

Using that $\langle x_0 \delta x \rangle = \langle p \delta x \rangle = 0$, we get the average signal power, i.e. equivalently the expected photon number,

$$\langle |A|^2 \rangle = x_0^2 + \delta x^2 + p^2 + \delta p^2 = |\alpha|^2 + \delta x^2 + \delta p^2. \tag{10.29}$$

To characterize the signal, including its signal to noise ratio, we need the variance of the signal. Using the power given by Eqn. (10.28) the expected second-order moment of the power is

$$\begin{aligned} \langle |A|^4 \rangle = & (x_0^2 + p^2)^2 + (\delta x^2 + \delta p^2)^2 + 2(x_0^2 + p^2)(\delta x^2 + \delta p^2) \\ & + (2x_0\delta x)^2 + (2p\delta p)^2. \end{aligned} \tag{10.30}$$

The variance of the power is $V_{|A|^2} = \langle (|A|^2)^2 \rangle - \langle (|A|^2) \rangle^2$. Inserting Eqn. (10.29) and Eqn. (10.30) we find

$$V_{|A|^2} = \langle (2x_0\delta x)^2 + (2p\delta p)^2 \rangle. \tag{10.31}$$

- **Coherent state** In the asymptotic limit of a signal much larger than the spontaneous terms, Eqn. (10.29) show

$$\langle |A|^2 \rangle \to |\alpha|^2 \quad \text{for} \quad |\alpha|^2 \gg \delta x^2 + \delta p^2, \tag{10.32}$$

that is the expected photon number in the signal approach $|\alpha|^2$.

The Coherent state is characterized by having equal mean and variance. In the case of a Coherent state $\langle \delta x^2 \rangle = \langle \delta p^2 \rangle = 1/4$ in units of photon numbers i.e. $(1/4)\hbar\omega_0 B_0$ in physical units, where ω_0 is the frequency of the signal being considered, and B_0 the bandwidth of the signal or the detector used to measure power. With this the variance equals: $V_{|A|^2} = |\alpha|^2$, which is identical to the mean.

- **Vacuum fluctuations**[1] A vacuum state holds no photons but has an electric field that fluctuates in magnitude corresponding to the energy of half a photon. In the limit when there is no laser light launched at a given frequency i.e. $|\alpha|^2 = 0$, Eqn. (10.29) shows

$$\langle |A|^2 \rangle \rightarrow \langle \delta x^2 \rangle + \langle \delta p^2 \rangle \quad \text{for} \quad |\alpha|^2 \rightarrow 0. \tag{10.33}$$

  That is, the expected signal power approaches $1/2$ in units of photon number i.e. $1/2\hbar\omega_0 B_0$ in physical units.

In the classical picture we may calculate the energy flux, or power, of the electric field, $P = \langle |A^2| \rangle$. It is noted that this quantity is proportional to the photon number. From $\langle |A^2| \rangle$ we may also calculate the energy flux per unit photon energy flux, $P_{\text{field}}/P_{\text{ph}} = \langle |A^2| \rangle / (\hbar\omega B_0)$. The latter quantity is, for large photon numbers, completely equivalent to the number of photons in the field. But in the case of $|\alpha| \rightarrow 0$ then $P_{\text{field}}/P_{\text{ph}} \rightarrow 1/2$, thus explicitly stating the energy of the vacuum field, which can be confusing since the result leads to the conclusion that there is half a photon on average in the vacuum state, even though such a conclusion is wrong, given that the field amplitude is zero and thus holds zero photons.

10.4 Amplifiers

One of the obvious examples of the use of four-wave mixing is as a parametric amplifier. Compared to other optical amplifiers, for example the erbium doped fiber amplifier [113] and the Raman amplifier, discussed in Chapter 7, the parametric amplifier theoretically offers noise free amplification, phase sensitive amplification and an instantaneous response (when ignoring Raman scattering). The latter is important for regeneration. In addition, one of the benefits of parametric amplifiers is that their bandwidth is determined by the fiber design, more specifically, by the slope of the group velocity dispersion with respect to wavelength, and the available pump power. Their operation wavelength is determined by the wavelength of zero group velocity dispersion around which the gain is symmetrically centered. Most important in many applications, certainly if one considers an amplifier or a wavelength converter, is not only the gain or conversion efficiency but the added spontaneous

[1] The goal of this book is not that the reader understands vacuum fluctuations or zero point fields to the degree of detail needed in quantum optics. For such a level of knowledge the reader is referred to text books on quantum optics as for example [110]. The goal is rather that the reader obtains sufficient insight to model spontaneous emission in four-wave mixing.

emission. In an amplifier the added spontaneous emission is typically quantified by the noise figure. In the following we discuss the gain and noise figures of a single pump amplifier.

10.4.1 Single pump–phase insensitive amplification (PIA)

The single pump amplifier is in principle very simple, while still being able to provide a high gain and a wide bandwidth. However, to achieve a significant gain a relatively high pump power level has to be used. Unfortunately this causes the onset of Brillouin scattering. To avoid this, the spectrum of the pump may be broadened. This may be done by applying a phase modulation to the pump. Figure 10.3 shows an example of an experimental setup of a single pump parametric amplifier.

To achieve gain through four wave mixing, one has to combine pump light with signal light using a pump combiner. The combined electric field is then propagated through a highly nonlinear fiber having a group velocity dispersion that is optimized for four-wave mixing. To avoid the onset of Brillouin scattering, one may choose to broaden the spectrum as described in [109], [114]. Figure 10.3 shows an example of applying four modulation tones. Other methods include applying a temperature gradient to the fiber [115] [116], or by using strained fiber [117],[118], or by changing the fiber design [119], [120], [91], [89].

The setup in Figure 10.3 supports phase insensitive amplification. However, by combining the signal with another signal, phase sensitive amplification has been demonstrated. Several configurations of phase sensitive amplifiers have been reported for example using a nonlinear fiber interferometer [121] or as presented in [122] using a so-called copier. In the following, focus is directed toward phase insensitive amplification, but we return briefly to phase sensitive amplifiers in Section 10.4.2.

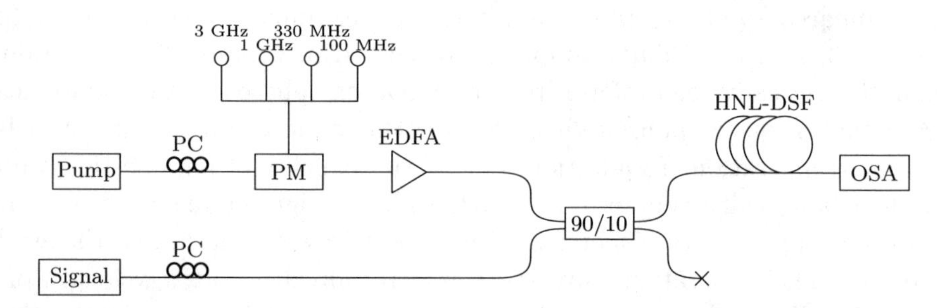

Figure 10.3: Experimental setup of phase insensitive parametric amplifier. Polarization controller (PC), phase modulator (PM), Erbium-doped fiber amplifier (EDFA), highly nonlinear dispersion shifted fiber (HNL-DSF).

Gain

The gain is defined as the signal output power relative to the signal input power. To evaluate this, one may solve the coupled equations in Section 10.2 numerically or use the analytical approach as described by Stolen et al. [108] and Hansryd et al. [109]. To find the analytical solution one may use an expansion of the propagation constant around the frequency of the pump ω_p

$$\beta(\omega) = \beta_0(\omega_p) + \beta_1(\omega - \omega_p) + \frac{1}{2}\beta_2(\omega - \omega_p)^2 + \cdots , \tag{10.34}$$

where β_1 is the first-order derivative of the propagation constant evaluated at the pump frequency, β_2 is the second-order derivative of the propagation constant evaluated at the pump frequency etc. To evaluate $\Delta\beta$ we need to find $\beta(\omega_p)$, $\beta(\omega_s)$ and $\beta(\omega_i)$. These are

$$\beta(\omega_p) = \beta_0(\omega_p) \tag{10.35a}$$

$$\beta(\omega_s) = \beta_0(\omega_p) + \beta_1(\omega_s - \omega_p) + \frac{1}{2}\beta_2(\omega_s - \omega_p)^2 + \cdots \tag{10.35b}$$

$$\beta(\omega_i) = \beta_0(\omega_p) + \beta_1(\omega_i - \omega_p) + \frac{1}{2}\beta_2(\omega_i - \omega_p)^2 + \cdots . \tag{10.35c}$$

Since $\omega_i = \omega_p + \Delta\omega$ and $\omega_s = \omega_p - \Delta\omega$ we have to second-order in frequency

$$\Delta\beta = (\beta(\omega_s) + \beta(\omega_i) - 2\beta(\omega_p)) = \beta_2(\Delta\omega)^2. \tag{10.36}$$

To find the second-order derivative of β at ω_p we use a Taylor expansion of β_2 around the zero dispersion frequency ω_0

$$\beta_2(\omega) \approx \beta_2(\omega_0) + \left.\frac{d\beta_2}{d\omega}\right|_{\omega_0} (\omega - \omega_0). \tag{10.37}$$

Since β_2 is expanded around the zero dispersion wavelength, $\beta_2(\omega_0) = 0$, and since $\omega = 2\pi c/\lambda$, i.e. $d\omega = (-2\pi c/\lambda^2)d\lambda$ we get

$$\Delta\beta = \left.\frac{d\beta_2}{d\omega}\right|_{\omega_0} (\lambda_p - \lambda_0)\frac{-2\pi c}{\lambda^2}(\lambda_s - \lambda_p)^2(\frac{-2\pi c}{\lambda^2})^2. \tag{10.38}$$

The derivative of the β_2 with frequency, may be rewritten as the derivative with wavelength using

$$\frac{d\beta_2}{d\omega} = \frac{d\beta_2}{d\lambda}\frac{d\lambda}{d\omega} = -\frac{d\beta_2}{d\lambda}\frac{\lambda^2}{2\pi c}. \tag{10.39}$$

It is customary to express the group velocity dispersion not through β_2 but rather through a dispersion parameter D defined as $\beta_2 = -\frac{\lambda^2}{2\pi c}D$. Applying this we finally get

$$\Delta\beta = -\frac{2\pi c}{\lambda_0^2}\left.\frac{dD}{d\lambda}\right|_{\lambda_0} (\lambda_p - \lambda_0)(\lambda_s - \lambda_p)^2. \tag{10.40}$$

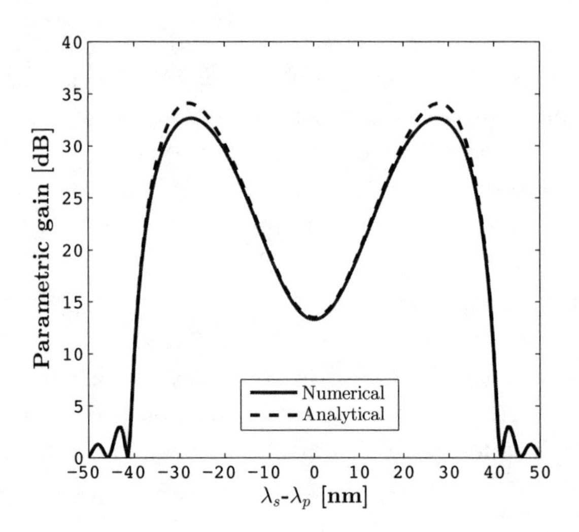

Figure 10.4: Gain versus wavelength for a single pumped phase insensitive parametric amplifier. The dashed curve is obtained from Eqn. (10.41), while the solid curve is obtained by solving the amplitude equations numerically with a signal power of 0.1 mW.

It is noted that if β is expanded to higher orders in Eqn. (10.36) only even orders in the derivative of β appear. With this the gain of the amplifier may now be expressed as a function of $\Delta\beta$. In [109] this is shown to be

$$G = 1 + \left[\frac{P_p(z=0)\gamma}{g_{\text{FWM}}} \sinh(g_{\text{FWM}} L) \right]^2, \tag{10.41}$$

where g defined through $g_{\text{FWM}}^2 = \left[(\gamma P_p(z=0))^2 - (\kappa/2)^2 \right]$, where $\kappa = \Delta\beta + 2\gamma P_p(z=0)$. It is noted that the maximum gain is obtained for perfect phase matching i.e. $\kappa = 0$. Under this condition the gain equals $G = {}^1\!/\!_4 \exp(2\gamma P_p(z=0)L)$. For comparison, in Chapter 7 the maximum on-off Raman gain in a Raman amplifier was found to be: $G = \exp(g_R P_p(z=0) L_{\text{eff}})$.

Let us consider an example of a 300 m long fiber with a nonlinear strength $\gamma = 11.0$ $(\text{Wkm})^{-1}$, a zero dispersion wavelength at 1559 nm, and a group velocity dispersion slope of: $0.03{\cdot}10^3$ s/m^3. Using furthermore a pump at the wavelength: 1560.7 nm, and a pump power of 1.4 W we predict the gain from Eqn. (10.41), see Figure 10.4.

The figure also shows the predicted gain obtained by solving the amplitude equation Eqn. (10.25a) numerically, but with a signal power of 0.1 mW. The gain is clearly depleted in this case and even though the spectral shape is unchanged, the maximum gain is reduced by more than 1 dB.

Noise figure

The noise performance of a parametric amplifier is different from many other optical amplifiers, including amplifiers based on rare earth doped fiber amplifiers and the Raman amplifier. There are several reasons for this. First of all, parametric amplification is not achieved through the creation of an energy reservoir. On the contrary, energy is simply transferred directly from the pump to the signal under the assumption that an idler is present. However, the transfer of energy may still happen spontaneously because of vacuum fluctuations at the idler wavelength. Consequently, when describing spontaneous parametric emission, sometimes also referred to as parametric fluorescence, this may be done by including vacuum fluctuations. Another important difference between for example a Raman amplifier and the parametric amplifier is the fact that the parametric amplifier is based on a phase dependent process, where the Raman amplifier only depends on power. As a consequence of this it may not be sufficient just to predict the power of the spontaneous emission, it may in some applications be more relevant to predict a constellation diagram; we return to this later in this section.

An optical amplifier is characterized by its gain, and equally important, by its noise figure. The latter is defined as how much the amplifier degrades the signal to noise ratio (SNR) from the input to the output, i.e.

$$F = \frac{(\mathrm{SNR})_{\mathrm{in}}}{(\mathrm{SNR})_{\mathrm{out}}}, \tag{10.42}$$

where $(\mathrm{SNR})_{\mathrm{in}}$ is the signal to noise ratio at the input and $(\mathrm{SNR})_{\mathrm{out}}$ is the signal to noise ratio at the output of the amplifier. The signal to noise ratio is defined as the mean number of photons $\langle n \rangle$, squared relative to the variance V_n, of the photon number

$$(\mathrm{SNR}) = \frac{\langle n \rangle^2}{V_n}. \tag{10.43}$$

It is noted that for the phase sensitive amplifier the above definition is not sufficient. The reason for this is that in the case of the phase sensitive amplifier, the phase as well as the amplitude of the signal are parameters of equal significance. In addition, it is important that the idler is well defined, meaning ideally with a fixed phase and a fixed amplitude, since uncertainty in the amplitude or phase of the idler transfers to uncertainty in the signal. Finally, the amplifier may also alter the state from the input to the output. More specifically, it has been demonstrated that the parametric amplifier may be used to squeeze light[25]. Consequently, it may be necessary to characterize the variance in both quadrature signals rather than just the variance in photon number. This may be done using so-called constellation diagrams. We show an example of a constellation diagram at the end of this section, Figure 10.6.

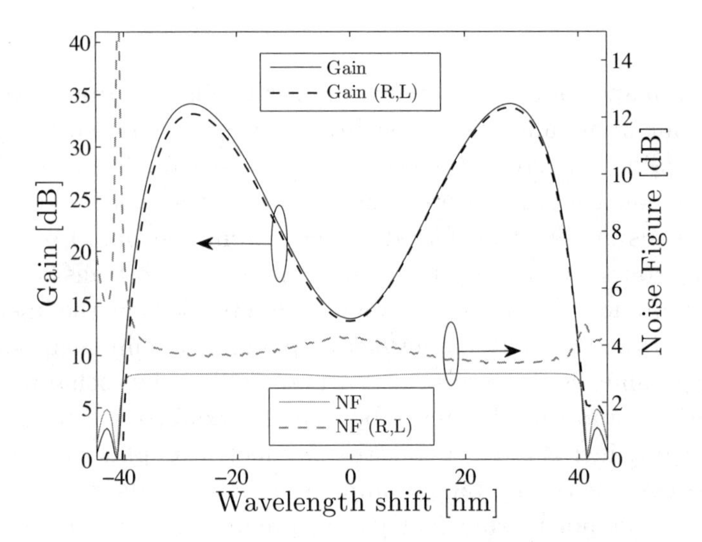

Figure 10.5: Predicted gain and noise figure for a phase insensitive amplifier, parameters are given in the main text. The solid lines are predicted under ideal operation i.e. no loss and no Raman gain, whereas the dashed curves, labeled (R,L) are obtained by taking Raman gain and fiber-attenuation into account - parameters given in the main text, using the method in [123].

In the case we consider here, i.e. the phase insensitive amplifier, the ideal noise figure, i.e. when neglecting depletion, loss and Raman scattering, etc. is

$$F = \frac{2G - 1}{G}. \tag{10.44}$$

In the high gain limit, $G \to \infty$, the noise figure approximates 3 dB, whereas for low gain, $G \to 1$, the noise figure approximates 0 dB.

As an example let us consider a 300 m long dispersion shifted highly nonlinear optical fiber. The fiber has a nonlinear strength γ of 11 $(\text{Wkm})^{-1}$, and a Raman peak gain coefficient of 5.77 $(\text{Wkm})^{-1}$. The group velocity of the fiber is zero at 1559 nm, with a slope of the group velocity dispersion of 0.03 ps/(nm^2 km), and an attenuation of 0.4 dB/km. A pump power of 1.4 W, located at 1560.7 nm, is assumed. Regarding the spontaneous Raman scattering, room temperature is assumed. Predictions for gain and noise figures are shown in Figure 10.5. The two sets of curves illustrate results obtained by neglecting loss and Raman gain, and by including loss and Raman gain, respectively.

From Figure 10.5 a maximum gain close to 35 dB is found at a wavelength shift of close to $\pm$ 25 nm from the pump wavelength. The impact on the gain due to the fiber background loss and Raman scattering is seen to be at maximum close to 1 dB. However, the impact of loss and especially Raman scattering is most significant on the noise figure which increases from the theoretical lower limit of 3 dB to 3.5 dB.

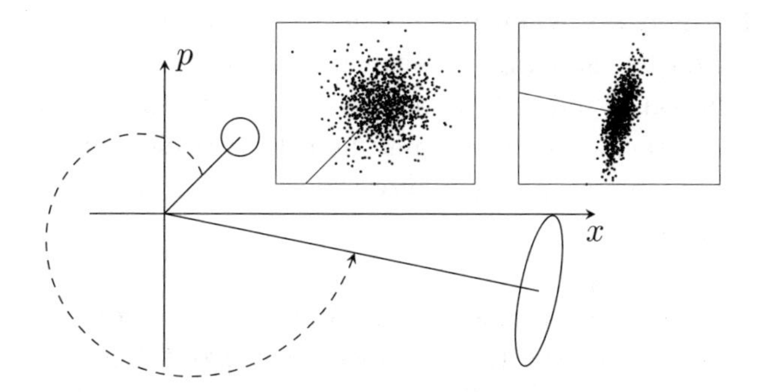

Figure 10.6: A constellation diagram showing the two quadratures x and p as discussed in Section 10.3.1 as they have evolved from the input - the circle, to the output - the ellipse, of a saturated parametric amplifier. The two insets illustrate a zoom of the ensembles of the signal.

As noted earlier in this section, the parametric amplifier may alter the variance in the two quadratures of a signal differently, i.e. even though $\delta x = \delta p$ at the input they may be different from each other at the output. This may happen when the amplifier is in depletion [123]. An illustration of this is shown in the constellation diagram in Figure 10.6. The figure shows the two quadratures of a signal. A coherent input state shows up as a circle, more accurately in the zoom an ensemble with $\delta x = \delta p$ in the first quadrant of the plot. After parametric amplification, the signal is amplified, evident by the ensemble shifted further away from the origo. However, the ensemble now has the shape of an ellipse, i.e. $\delta x = \delta p$, i.e. it is no longer a coherent state. This is due to the fact that the amplifier is saturated. For more details the reader is referred to [123].

From Figure 10.6 it is clear that the output state is not coherent since $\delta x \neq \delta p$. The figure also shows that the standard noise figure as used above may not be sufficient when describing the performance of a parametric amplifier. Finally, it is noted that states with $\delta x \neq \delta p$ may be so-called squeezed light, and are very attractive within quantum optics. However, this is outside the scope of this book and will not be discussed in further detail.

Impairments

The phase insensitive amplifier has many interesting characteristics, including the potential of wide bandwidth, low noise figure, and operation at any wavelength, only constrained by fiber group velocity dispersion as well as the availability of a pump laser. However, the amplifier is also challenging to work with. Some of the challenges are discussed briefly in the following.

Raman Scattering

To obtain significant four-wave mixing one has to use a strong pump, with the consequence that Raman scattering occurs simultaneously with four-wave mixing. This statement is valid, especially when Raman active material such as silica is used as a waveguide and when the spacing between pump, signal, and idler is close to the phonon energy of the waveguide materials. In relation to four-wave mixing, the advantage of Raman scattering is that it may help to transfer energy from the pump to the signal. However, the drawback is more severe, since Raman scattering results in spontaneous emission.

In this book, we include Raman scattering in the description of four-wave mixing with the aid of Chapter 7, where the necessary background of Raman scattering is provided.

In the Raman amplifier spontaneous emission is added to the signal while the signal is being amplified. When predicting the noise performance of a Raman amplifier it is customary simply to evaluate the amount of spontaneous emission added during amplification.

Raman scattering is usually described by equations in power. However, since four-wave mixing involves phase matching, the governing equations are in amplitudes and phase. The complete description of four-wave mixing including Raman scattering requires that the following terms are added to the propagation equations

Spontaneous emission on the signal This is dominated by a contribution from the pump.

$$\frac{\partial E_s^0}{\partial z} = \frac{g_R^{ps}}{2} |A_p|^2 (\eta_T + 1), \tag{10.45}$$

where η_T is the phonon population factor as discussed in Chapter 7. This is a Stokes process.

Spontaneous emission on the idler There are two contributions, one from the signal and one from the pump. The dominant one from the pump is

$$\frac{\partial E_i^0}{\partial z} = \frac{g_R^{ip}}{2} \frac{\omega_i}{\omega_s} |A_p|^2 \eta_T. \tag{10.46}$$

This is an anti-Stokes process.

In the above discussion of terms we have considered a single pumped amplifier, with a signal wavelength longer than the pump and an idler wavelength shorter than the pump. In addition, the pump is much stronger than the signal and idler at any position along the propagation.

Relative Intensity noise (RIN)

A signal, here in fact a pump beam, that fluctuates in time, is characterized by a quantity referred to as the relative intensity noise. More specifically, this quantity is defined as the ratio of the mean-square optical intensity noise to the square of the average optical power.

Since four-wave mixing requires the pump and signal to propagate simultaneously, and since the response time of the four-wave mixing process is dominated by the repose time of the intensity dependent refractive index, i.e. few fs, then any intensity fluctuations in the pump are transferred to the signal. This has been pointed out as one of the main contributions to noise in four-wave mixing [124] and [125].

The signal distortion caused by transfer of RIN depends upon the amplifier design, i.e. fiber length, pump power etc., more specifically to what degree the amplifier is depleted [126]. Further treatment is outside the scope of this book.

Dispersion fluctuations

The impact on four-wave mixing due to a group velocity dispersion that varies along the propagation may be significant. This is of course due to the fact that the energy transfer in the four-wave mixing process is directly determined by the phase matching $\Delta\beta$. The impact on gain and noise figure due to dispersion fluctuations in a parametric amplifier has been reported by several research groups for example [127] and [128].

The fluctuation in group velocity is a result of small variations in the fiber geometry along the length of the fiber. Even variations on a very small scale may cause fluctuations in the group velocity dispersion. This may be the wavelength at which the group velocity equals zero or the slope of the dispersion curve or both. To quantify this, E. Myslivets et al. have proposed an experimental method to characterize the group velocity as a function of position [129]. Alternatively, statistical data such as the mean and variance of the group velocity dispersion may also be obtained from simple measurements of the efficiency of four-wave mixing as a function of wavelength [130].

To address the challenges imposed by the fluctuations of the group velocity as a function of distance, B. P. P. Kuo and S. Radic [131] have proposed a fiber design that is invariant to fabrication fluctuations.

Polarization dependence

In the theoretical frame discussed in Section 10.2 we quickly assumed that the electric field was linearly polarized and remained so during propagation. However, it is without question that this is a crude approximation of a real transmission fiber due to the existence of birefringence in optical fibers. This is further emphasized when the frequencies taking part in the four-wave mixing are further and further separated in wavelength space. Consequently, waves that were initially co-polarized do not remain co-polarized, and the four-wave mixing efficiency is reduced. However, to circumvent this it has been suggested to pump the parametric amplifier using two pump wavelength [107] as for example illustrated in Figure 10.1.

Stimulated Brillouin scattering (SBS)

In accordance with Chapter 8, stimulated Brillioun scattering acts like a mirror to a narrow linewidth laser beam. This effect is detrimental to four-wave mixing since one often would like to use CW pump beams at power levels around 1 W or more. Unfortunately, Brillouin scattering limits the maximum power in a CW beam to power levels well below this threshold. Thus, since parametric amplifiers require pump power levels close to 1 W, Brillioun scattering has to be avoided. Fortunately, there are several ways by which that can be done. The method that was used in Figure 10.3, is based on the approach by Hansryd et al. [109]. In this approach the pump spectrum is broadened by applying a phase modulation. Alternatively, the properties of the fiber may also be altered. For example, in a simple approach by applying a temperature gradient [115] or a strain applied to the fiber [117][118]. Finally, as discussed in Chapter 8 one may also change the co-dopant materials of the fiber, for example by using aluminum rather than germanium to raise the refractive index of the core of the optical fiber [89]. These methods have the drawback that they also change the dispersion properties of the fiber.

10.4.2 Single pump–phase sensitive amplification (PSA)

One of the most interesting, if not the most interesting application of four-wave mixing, is to achieve so-called phase sensitive amplification. As compared to the phase insensitive amplifier discussed above, the phase sensitive amplifier provides the possibility of amplifying a signal with a given phase while attenuating the signal with other phases. This opens the possibility for optical signal regeneration or noise free amplification.

Input conditions

In a single pump configuration of a PSA, a pump and a signal as well as an idler need to be launched into the amplifier simultaneously. Besides having the correct frequency, the relative phase between the pump signal and idler requires accurate control.

The gain, or attenuation, as well as the noise performance of the PSA is governed by the total relative phase θ among the three interacting waves. In accordance with Eqn. (10.14) through Eqn. (10.15b) the governing phase is defined as $\theta = 2\phi_p - \phi_s - \phi_i$. To obtain amplification of a signal θ has to equal $-\pi/2$. In addition to the phases, the performance, both gain and noise performance of the amplifier, is controlled by the amplitude of the signal and idler [132] and[122]. For optimum performance both signal and idler have to be identical. Finally, the wavelength of the idler has to match the pump and signal such that there is energy conversation among the three interacting waves, i.e. $\omega_i + \omega_s = 2\omega_p$. One way to obtain an optimized idler is by launching the signal into a PIA, before it is launched to the PSA. Thereby one generates a copy of the signal, but shifted to the wavelength of the idler. By controlling the phase of the generated idler the three waves are now prepared to a state where they may be launched into the PSA. This method is described in [132] and [122].

Gain

The gain of the phase sensitive amplifier is defined as the output signal power relative to the launched signal power, i.e.

$$G(z) = \frac{\langle |A_s(z)|^2 \rangle}{\langle |A_s(0)|^2 \rangle}. \tag{10.47}$$

Tong et al. [122] have shown that in the limit of high PSA gain and under the assumption of equal signal and idler power at launch, the gain of the PSA is four times, or 6 dB higher than the gain of the PIA.

As an example Figure 10.7 illustrates predicted gain, and noise figures for a PSA. It is noted that since strong pumps are used, Raman gain also impacts the obtained gain. Consequently, the figure includes two graphs, one with Raman gain and one without. In addition to Raman scattering, the graph labeled (R,L) also includes loss. The parameters used are the same as used in Figure 10.5. The impact on the gain due to Raman scattering and loss is limited to around 1 dB. However, impact on the noise performance is more significant; we return to this below.

Noise figure

As mentioned, the noise performance of the PSA is very interesting. The reason is that in theory the amplifier may amplify a signal without adding noise, or when predicting the signal to noise ratio through the photon statistics, this may even be improved, by squeezing [133]. However, it has to be noted that to obtain the optimum performance, an idler, which is a copy of the signal, has to be generated and launched as the idler into the amplifier. Nevertheless, the impact of being able to amplify without adding noise is very important, and therefore the understanding of the PSA is extremely important.
In the phase sensitive amplifier, the signal to noise ratio may be defined as [112]

$$\mathrm{SNR} = \frac{\langle |A_s|^2 + |A_i|^2 \rangle^2}{V_{(|A_s|^2 + |A_i|^2)}}. \tag{10.48}$$

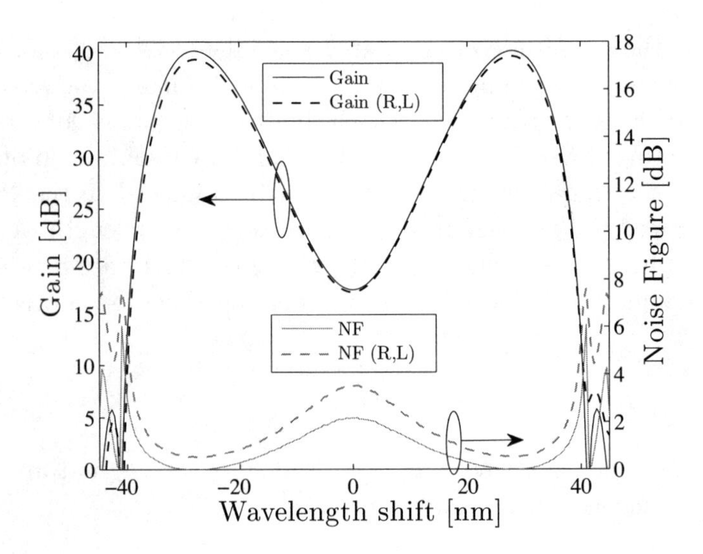

Figure 10.7: Low signal gain and noise figures for a phase sensitive amplifier - parameters are given in the main text, using the method described in [123].

Figure 10.7 illustrates the noise performance. As for the gain, Raman scattering also impacts the noise properties. As opposed to having a 0 dB noise figure, Raman scattering results in a noise figure now closer to 1 dB. Consequently, Raman scattering is even more detrimental for the noise properties than for the gain, since the ideal noise free amplification is hindered by Raman scattering.

In relation to the noise performance it is further noted that other noise sources impact the final noise performance of the PSA, for example fluctuations in the pump transferred to the signal [134]. In 2012 Tong et al. [122] demonstrated experimentally a low noise amplifier with a noise figure of 1.1. dB. The deviation from 0 dB is explained by the aforementioned other sources of noise.

It is noted that if the signal and only the signal is characterized this does in fact experience a 3 dB improvement from input to output of the PSA, when the amplifier noise figure is 0 dB. However, it is important to remember that to measure this it is essential that the PSA is operated with a phase controlled copy of the signal at the idler. Thus, to make the amplifier the input needs to be copied first, which at best causes a 3 dB reduction in signal to noise ratio of the signal. In addition, a redundant copy of the signal is transmitted together with the desired signal.

10.4.3 Two pump amplifiers

Both the phase insensitive amplifier and the phase sensitive amplifier discussed above, were single pumped amplifiers. For various reasons , including the purpose of broadening the gain spectrum, to make a more flat gain spectrum, to make the gain polarization independent, and to reduce the impact of dispersion fluctuations, it has been suggested to use dual pump configurations as shown in the right-hand side illustration of figure 10.1 as well as in Figure 10.2.

In 2002 McKinstrie and coworkers demonstrated some of the potential of dual pumped parametric amplifiers [135]. In a detailed study they demonstrate configurations capable of producing exponential gain over tens of THz. In addition, they also point to some of the potential impairments such as fluctuations of fiber parameters within fibers used as the amplifier fiber.

Marhic and coworkers showed in 2002 that the gain of a dual pumped parametric amplifier may be made polarization independent by using two orthogonally polarized pumps [107]. They demonstrate a flat - polarization independent - gain of 15 dB over a 20 nm bandwidth.

As pointed out by McKinstrie in [135], one of the challenges in parametric amplifiers is fluctuations of the group velocity dispersion along the fiber length. In 2004 Yaman and coworkers discussed the impact of such fluctuations in a dual pumped parametric amplifier [128]. The team show that the gain spectrum of the parametric amplifier becomes non-uniform. However, they also point to one potential solution, simply by reducing the wavelength separation between the pumps.

As discussed above, the dual pumped parametric amplifier has many promising characteristics. One of these was discussed in 2008 by Boggio and coworkers [136]. They demonstrated a parametric amplifier providing gain over 81 nm. The amplifier was a 2 pump configuration where each pump wavelength was located symmetrically around the zero dispersion wavelength of the fiber, which was measured to be 1561.9 nm, with a dispersion slope of 0.025 ps/(nm^{2}km). One pump wavelength was at 1511.29 nm while the other was at 1613.85 nm. Approximately 1 W of pump power was used in each of the two pump beams. The fiber in their experiment was 350 m long with nonlinear strength of 14 (Wkm)$^{-1}$.

Gain and noise performance

The gain delivered from an amplifier pumped by two lasers each at different wavelengths and each launching the pump power $P_p(z=0)$.

$$G = 1 + \frac{\left[\frac{2\gamma P_p(z=0)}{g}\sinh(gL)\right]^2}{1 + \frac{P_s}{P_{\text{sat}}}\left[\frac{2\gamma P_p(z=0)}{g}\sinh(gL)\right]^2}, \tag{10.49}$$

where $g = \sqrt{(2\gamma P_p(z=0))^2 - (\kappa/2)^2}$ and $\kappa = 2\gamma P_p(z=0) + \Delta\beta$, and $P_p(z=0)$ is the power launched from each of the two pumps. P_{sat} is the maximum power that

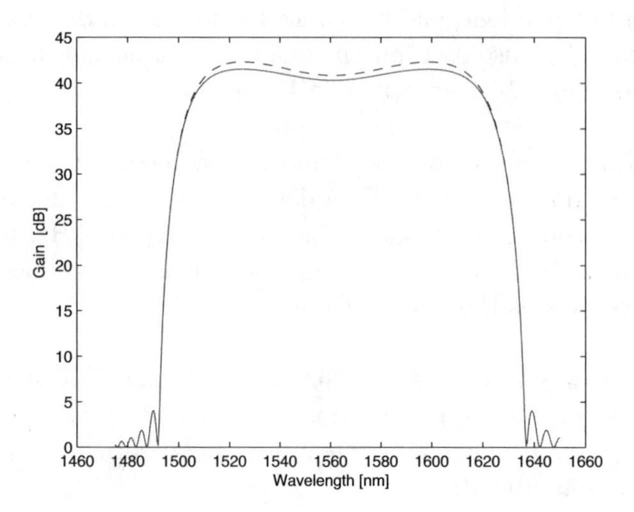

Figure 10.8: Two pump FOPA top curve undepleted (bottom curve including depletion) from [137].

may be coupled to the idler given by

$$P_{\text{sat}} = \begin{cases} \frac{1}{6}(\frac{\Delta\beta}{\gamma} - P_s) + P_p(z=0) & , \quad -6P_p(z=0) < \frac{\Delta\beta}{\gamma} < P_s \\ -\frac{1}{2}(\frac{\Delta\beta}{\gamma} - P_s) + P_p(z=0) & , \quad P_s < \frac{\Delta\beta}{\gamma} < 2P_p(z=0). \end{cases} \tag{10.50}$$

If the pump wavelengths are symmetrically located around the zero dispersion wavelength, the linear phase mismatch $\Delta\beta$ equals

$$\Delta\beta \approx \frac{\beta_4}{12}\left[(\omega_s - \omega_{ZD})^4 - (\omega_p - \omega_{ZD})^4\right], \tag{10.51}$$

where β_n is the nth-order derivative of the propagation constant with respect to frequency, evaluated at the zero dispersion frequency ω_{ZD}. ω_p is any of the two pump frequencies. However, if the two pump wavelengths are not symmetrically located around the zero dispersion wavelength, one has to make a larger effort to find the linear phase mismatch.

To illustrate the gain profile of a dual pumped parametric amplifier, we consider a 250 m long fiber with a nonlinear strength of $\gamma = 11.5$ (Wkm)$^{-1}$, using two pumps each launching a pump power of $P_p(z=0) = 1$ W into the fiber, one pump with a wavelength at 1505.5 nm and the other at the wavelength 1621.0 nm. Fig. 10.8 illustrates the predicted gain versus wavelength for this example. The wavelength at which the group velocity dispersion equals zero is: 1560.5 nm, while $\beta_3 = 2.49 \cdot 10^{-41}$ s^3/m, and $\beta_4 = 8.82 \cdot 10^{-56}$s^4/m.

The results in Figure 10.8 were obtained analytically by H. Steffensen et al. [137]. They also showed, semi-analytical solutions of the gain when considering depletion.

To achieve accurate solutions, one has to rely on numerical solutions of equations similar to Eqn. (10.8) through Eqn. (10.10), but extended to include four different frequencies, see for example [3].

As with any other optical amplifier, the noise properties are as important as the gain properties. In fact the noise properties of the dual pump parametric amplifier may be even more interesting than the single pump parametric amplifier. The reason is that energy may be transferred between signal and idler rather that between strong pumps and signals. This may enable improved noise performance since for example coupling of RIN from a strong pump to the signal may be avoided.

In 2004 McKinstrie and coworkers showed the quantum noise properties of a dual pumped parametric amplifier [111]. As one of the results of their comprehensive study they show that noise figures close to 3 dB for the dual pumped amplifier may be obtained and noise figures close to 0 dB for a frequency converter may be obtained.

10.5 Other applications

The topic of parametric amplification has attracted more and more interest within recent years. This has been spurred by demonstrations of various applications and the development of high power fiber lasers and specifically tailored highly nonlinear, dispersion engineered optical fibers. The FOPA is an attractive device since it not only provides phase insensitive amplification but also phase sensitive amplification, wavelength conversion, and in addition has an almost instantaneous response as well. Some of the recent demonstrations of applications are highlighted below.

In addition to amplification as discussed above, several other applications of four-wave mixing have been described. These include multi casting [138] [139], signal processing [140], wavelength conversion, optical sampling, and regeneration. The latter three are briefly highlighted below. Finally, it is noted that four-wave mixing plays an essential role in supercontinuum generation, see for example [64].

Wavelength conversion

Parametric amplifiers have great potential as devices for wavelength converters, since a signal by nature is converted to an idler through propagation through the parametric fiber amplifier.

J.M. Chavez Boggio and coworkers demonstrated in 2008 a conversion over 730 nm [141]. In their experiment a 15 m long highly nonlinear fiber, with very low fourth-order dispersion, was used. The nonlinear strength of the fiber was 11.5 $(\text{Wkm})^{-1}$, the wavelength of zero dispersion of the fiber was at 1582.8 nm, and the dispersion slope was 0.027 ps/(nm^{2}km). The third- and fourth-order dispersion coefficients were 0.038 ps^3/km and $1.4 \cdot 10^{-5}$ ps^4/km, respectively. To achieve the wavelength conversion a pulsed pump was used. The peak power of each pump pulse was 200 W. In one example a signal at 1312.6 nm was converted to 1999 nm with a conversion efficiency of 30 dB.

Optical sampling

Another promising application of parametric processes in optical fibers is to perform waveform sampling of a signal. To this end, in principle, a wavelength converter is designed, with the only purpose of generating an idler, which is a sample of the signal. The power generated in the idler is proportional to the power of the signal. P. Andrekson and coworkers have demonstrated this approach and have shown high resolution optical waveform sampling, including phase information in constellation diagrams of a 640 Gb/s data stream [142] and [143].

Regeneration

As a final example of a promising application of four-wave mixing, regeneration has recently been demonstrated. This is of particular interest because of application of advanced modulation formats such as phase shift keyed modulation formats has proven superior transmission properties and enabled record high capacity transmission. However, one of the limiting factors in such high capacity communication links is signal distortion due to the nonlinear phase shift that is accumulated during transmission. Due to intensity fluctuations of the signal, a phase shift keyed signal is distorted significantly since the induced nonlinear phase shift fluctuates in accordance with the intensity of the signal. However, by employing a method to reduce the intensity fluctuations, the phase distortion may be reduced, leading to a further improved transmission capacity if phase shift keyed signals are used.

In 2008 C. Peucheret and coworkers [144] demonstrated experimentally that a parametric amplifier may be used as a limiter and consequently mitigate the accumulated phase penalties. In [144] the dynamic range of a 40 Gb/s return to zero differential phase shift keyed signal was enhanced. An optical signal-to-noise ratio penalty of 3.5 dB was reduced to 0.2 dB by using a single pumped FOPA with a 22 dB small signal gain. The improvement in the optical signal-to-noise ratio was experimentally found through bit error rate measurements.

10.6 Summary

- In four-wave mixing four waves mutually interact through the intensity dependent refractive index. Two out of the four frequencies may be identical in degenerate four wave mixing. As opposed to many other nonlinear processes ideally no energy is exchanged with the material that the light propagates through. Consequently, the processes are referred to as parametric processes.
- The propagation is described through three coupled equations. In a single mode optical fibers these reduce to Eqn. (10.25)

$$\frac{dA_p}{dz} = i\tilde{\Gamma}\left[\{|A_p|^2 + 2(|A_s|^2 + |A_i|^2)\} A_p + 2A_s A_i A_p^* \exp(i\Delta\beta z)\right], \quad (10.25a)$$

$$\frac{dA_s}{dz} = i\tilde{\Gamma}\left[\{|A_s|^2 + 2(|A_i|^2 + |A_p|^2)\} A_s + A_i^* A_p^2 \exp(-i\Delta\beta z)\right], \quad (10.25b)$$

$$\frac{dA_i}{dz} = i\tilde{\Gamma}\left[\{|A_i|^2 + 2(|A_s|^2 + |A_p|^2)\} A_i + A_s^* A_p^2 \exp(-i\Delta\beta z)\right]. \quad (10.25c)$$

- As opposed to for example Raman amplifiers or other typical optical amplifiers, the parametric amplifiers have a nearly instantaneous response time. In addition, four-wave mixing requires that the pump and signal propagate in the same direction, consequently fluctuations in the pump may couple to the signal and cause fluctuations in the signal.
- The process may be used to obtain amplification, wavelength conversion or more complicated functionalities as optical regeneration. The amplification may be phase insensitive as known in many other amplifiers or it may be phase sensitive. The phase sensitive process may in fact be used to turn an amplifier into an attenuator.
- The operation wavelength of the amplifier and its bandwidth are determined by the dispersion properties together with the nonlinearity of the optical fiber used.
- One of the interesting properties of the parametric amplifier is that, ideally, the process is free of spontaneous emission, since no energy reservoir is associated with the process. Noise figures down to even zero dB may be obtained see Figure (10.7). However, in typical configurations and materials, that process of four-wave mixing is associated with Raman scattering, which may happen spontaneously.

10.6 Summary

Appendix A

Tensors

Tensors are mathematical tools used to characterize properties of physical systems. Tensors are simply arrays of numbers, or functions of for example time and space. A tensor may be defined at a single or multiple points, or it may vary continuously from point-to-point. A tensor may be transformed according to certain rules under a change of coordinates. The tensor concept may be illustrated by some examples. In the simplest case a tensor consists of a single number, for example the mass of a body. In this case the tensor is referred to as a tensor of order zero, or a zero-rank tensor, or simply a scalar.

The next simplest tensor is the tensor of order one, a first-rank tensor, a vector. Just as tensors of any order, it may be defined at a point in space, or it may vary continuously from point-to-point, thereby defining a vector field. In ordinary three dimensional space, a vector has three components (contains three numbers, or three functions of position). In general, a vector (tensor of order one) in an n-dimensional space, has n components. A vector can be visualized as being written in a column.

An example of a vector field is provided by the description of an electric field in space. The electric field at any point requires more than one number to characterize since it has both a magnitude and acts along a definite direction. Generally, both the magnitude and the direction of the field vary from point-to-point.

Next are tensors of order two, second-rank tensors, these are often referred to as matrices. The components of a second-order tensor can be written as a two dimensional array, with rows and columns.

An example of a second-rank tensor is the so-called inertia matrix of an object. For three dimensional objects, it is a $3 \times 3 = 9$ element array that characterizes the behavior of a rotating body. An other example is a gyroscope. The response of a gyroscope to a force along a particular direction (described by a vector). Such response is generally re-orientation along some other direction different from that of the applied force or torque. Thus, rotation must be characterized by a mathematical entity more complex than either a scalar or a vector; namely, a tensor of order two.

There are yet more complex phenomena that require tensors of even higher order. An example is the susceptibility that relates the induced polarization vector of a material to the applied electric field vector.

A.1 Notation

In general, tensors are simply arrays of numbers, or functions, that transform between different coordinate systems according to transformation rules that are described in the following.

Vectors, first-rank tensors, are characterized with a column of numbers given by a_i, whereas matrices, second-rank tensors, are characterized with rows and columns of numbers given by a_{ij}. An nth-rank tensor is characterized by numbers with n indices.

We use tensors to relate vectors for physical measures to each other, for example the induced polarization vector in a material originating from an applied electric field vector. In the simplest case the linear induced polarization vector is related to the electric field through the first-order susceptibility $\chi^{(1)}$. The relation is simply

$$\mathbf{P} = \varepsilon_0 \chi^{(1)} \mathbf{E}, \tag{A.1}$$

where $\chi^{(1)}$ is a second-rank tensor, and where $\mathbf{P}$ and $\mathbf{E}$ are vectors of the induced polarization and the electric field. Since the induced polarization vector $\mathbf{P}$ and the electric field vector $\mathbf{E}$ in Cartesian coordinates has three components, an x, y, and z-component, $\chi^{(1)}$ contains $3^2 = 9$ elements

$$\chi^{(1)} = \left\{ \begin{array}{ccc} \chi_{xx}^{(1)} & \chi_{xy}^{(1)} & \chi_{xz}^{(1)} \\ \chi_{yx}^{(1)} & \chi_{yy}^{(1)} & \chi_{yz}^{(1)} \\ \chi_{zx}^{(1)} & \chi_{zy}^{(1)} & \chi_{zz}^{(1)} \end{array} \right\}. \tag{A.2}$$

The so-called second-order induced polarization originates from the electric field 'mixing' with the induced polarization and is characterized by a third-rank tensor $\chi^{(2)}$. The induced polarization is given by

$$(\mathbf{P})^{(2)} = \varepsilon_0 K \chi^{(2)} \mathbf{E}\mathbf{E}, \tag{A.3}$$

where the double dot indicates that Eqn. (A.3) is not a simple dot product but a product involving a tensor. The third-rank tensor is defined through its $3^3 = 27$ elements, i.e.

$$\chi^{(2)} = \left\{ \begin{array}{ccccccccc} \chi_{xxx}^{(2)} & \chi_{xyy}^{(2)} & \chi_{xzz}^{(2)} & \chi_{xyz}^{(2)} & \chi_{xzy}^{(2)} & \chi_{xzx}^{(2)} & \chi_{xxz}^{(2)} & \chi_{xxy}^{(2)} & \chi_{xyx}^{(2)} \\ \chi_{yxx}^{(2)} & \chi_{yyy}^{(2)} & \chi_{yzz}^{(2)} & \chi_{yyz}^{(2)} & \chi_{yzy}^{(2)} & \chi_{yzx}^{(2)} & \chi_{yxz}^{(2)} & \chi_{yxy}^{(2)} & \chi_{yyx}^{(2)} \\ \chi_{zxx}^{(2)} & \chi_{zyy}^{(2)} & \chi_{zzz}^{(2)} & \chi_{zyz}^{(2)} & \chi_{zzy}^{(2)} & \chi_{zzx}^{(2)} & \chi_{zxz}^{(2)} & \chi_{zxy}^{(2)} & \chi_{zyx}^{(2)} \end{array} \right\}. \tag{A.4}$$

Often a single coordinate of the vector $\mathbf{P}^{(2)}$ is often quoted as

$$(P_i)^{(2)} = \varepsilon_0 K \sum_{jk} \chi^{(2)}_{ijk} E_j E_k = \varepsilon_0 K \left(\chi^{(2)}_{xxx} E_x E_x + \chi^{(2)}_{xyy} E_y E_y + \chi^{(2)}_{xzz} E_z E_z + \right. \tag{A.5}$$

$$\left. \chi^{(2)}_{xyz} E_y E_z + \chi^{(2)}_{xzy} E_y E_z + \chi^{(2)}_{xzx} E_z E_x + \chi^{(2)}_{xxz} E_x E_z + \chi^{(2)}_{xxy} E_x E_y + \chi^{(2)}_{xyx} E_y E_x \right),$$

and the summation is often omitted and one simply writes

$$(P_i)^{(2)} = \varepsilon_0 K \chi^{(2)}_{ijk} E_j E_k, \tag{A.6}$$

where it is implied that one has to evaluate the sum over all possible values of the repeated indices, this is referred to as the Einstein notation.

A.1.1 Symmetry and antisymmetry

Any tensor T can be written as the sum of a symmetric S and an antisymmetric (or skew symmetric) tensor A. In the case of a second-rank tensor an element in the tensor may be written as

$$T_{ab} = \frac{1}{2}\left[T_{ab} + T_{ba}\right] + \frac{1}{2}\left[T_{ab} - T_{ba}\right], \tag{A.7}$$

where the first square bracket represents the symmetric tensor and the second square bracket represents the antisymmetric tensor.

A symmetric tensor is a tensor that is unchanged by changing any two indices of the tensor, i.e.

$$S_{x_1,\ldots,x_i,\ldots,x_j,\ldots,x_n} = S_{x_1,\ldots,x_j,\ldots,x_i,\ldots,x_n}, \tag{A.8}$$

whereas an antisymmetric tensor is a tensor that changes sign when two indices are interchanged, i.e.

$$A_{x_1,\ldots,x_i,\ldots,x_j,\ldots,x_n} = -A_{x_1,\ldots,x_j,\ldots,x_i,\ldots,x_n}. \tag{A.9}$$

A.1.2 Change of basis

Consider two coordinate systems described in Cartesian coordinates, an unprimed basis with a vector expressed as $\mathbf{x}$ and a primed basis with the transformed vector expressed as $\mathbf{x}'$. The coordinate systems are rotated with respect to each other and the relation between the coordinates described by the 3×3 transformation matrix $\underline{\underline{R}}$ as

$$\mathbf{x}' = R\mathbf{x} \qquad x'_\alpha = R_{\alpha\beta} x_\beta \left(= \sum_\beta R_{\alpha\beta} x_\beta \right), \tag{A.10}$$

where the vector $\mathbf{x} = (x, y, z)^T$ and $\mathbf{x}' = (x', y', z')^T$ are two column vectors. In general, there are two types of rotations

- $\det(\underline{\underline{R}}) = 1$; Right-handed systems keep being right-handed, and left-handed keep being left-handed - proper rotations.

- $\det(\underline{\underline{R}}) = -1$; Right-handed systems are transformed into left-handed and vice versa - improper rotations.

Both the rotations are characterized by the fact that they preserves length. Thus a vector $\mathbf{x}$ in the unprimed system and a vector $\mathbf{x}'$ in the primed system have the same length. In this book we are primarily concerned with proper rotations.

Using these facts the inverse transformation between the coordinate systems is expressed as

$$\mathbf{x} = \underline{\underline{R}}^{-1}\mathbf{x}' \qquad x_\beta = R^{-1}_{\alpha\beta}x'_\alpha \left(= \sum_\alpha R^{-1}_{\alpha\beta}x'_\alpha \right). \tag{A.11}$$

The electric field $\mathbf{E}$ and the polarization $\mathbf{P}$ are both physical quantities that may be expressed in regular Cartesian coordinates hence they can both be transformed between two coordinate systems (x, y, z) and (x', y', z') using the above. Thus, in new primed coordinates the electric field, $\mathbf{E}'$, is related to the electric field in the unprimed coordinates, $\mathbf{E}$, through

$$\mathbf{E}' = \underline{\underline{R}}\mathbf{E}, \tag{A.12}$$

and similarly for the induced polarization

$$\mathbf{P}' = \underline{\underline{R}}\mathbf{P}. \tag{A.13}$$

Using these transformation rules, we now derive the form of the susceptibilities in rotated coordinate frames.

From the transformation rule for the induced polarization and using the standard form as previously expressed, we have for the first-order polarization in the primed coordinate system

$$\begin{aligned} \mathbf{P}' = \underline{\underline{R}}\mathbf{P} &= \underline{\underline{R}}\left(\varepsilon_0 \int_{-\infty}^{\infty} \chi^{(1)}\mathbf{E}d\omega\right) \\ &= \underline{\underline{R}}\left(\varepsilon_0 \int_{-\infty}^{\infty} \chi^{(1)} \left(\underline{\underline{R}}^{-1}\mathbf{E}'\right) d\omega\right). \end{aligned} \tag{A.14}$$

By now introducing

$$\chi^{(1)'} = \underline{\underline{R}}\chi^{(1)}\underline{\underline{R}}^{-1} \quad \Leftrightarrow \quad \chi^{(1)'}_{\mu\alpha} = R_{\mu u}R_{\alpha a}\chi^{(1)}_{ua}, \tag{A.15}$$

where the latter equation only bold when $\underline{\underline{R}}^{-1} = \underline{\underline{R}}^T$ the induced polarization in the primed system equals

$$\mathbf{P}' = \varepsilon_0 \int_{-\infty}^{\infty} \chi^{(1)'}\mathbf{E}'d\omega. \tag{A.16}$$

In a manner completely analogous to the first-order susceptibility, the transformation rule between the primed and unprimed coordinate systems can be obtained for the nth-order susceptibility tensor as

$$\chi^{(n)}_{\mu\alpha_1\cdots\alpha_n}{}' = R_{\mu u}R_{\alpha_1 a_1}\cdots R_{\alpha_n a_n}\chi^{(n)}_{ua_1\cdots a_n}. \tag{A.17}$$

For completeness it is noted that there are mathematical entities that occurs in physics generally consisting of multi-dimensional arrays of numbers or functions, like tensors, but that are NOT tensors. Most noteworthy are so-called spinors. Spinors differ from tensors in how the values of their elements change under coordinate transformations. For example, the values of the components of all tensors, regardless of order, return to their original values under a 360 degree rotation of the coordinate system in which the components are described. By contrast, the components of spinors change sign under a 360 degree rotation, and do not return to their original values until the describing coordinate system has been rotated through two full rotations = 720 degrees.

A.2 Pseudotensors

All tensors that we are interested in in this book are so-called polar tensors. As opposed to an axial tensor also referred to as pseudo-tensor, polar tensors transform such that the sign of their elements is unaltered by a transformation that change the handedness of the coordinate system. Pseudo-tensors or axial tensors have elements that change sign when the handedness of the coordinate system is changed by a transformation [35]. In some cases the tensor describing a physical property is not polar but axial. An example is the tensor describing optical activity.

Given a right-handed coordinate system with basis vectors $i_1 i_2 i_3$. With right-handed the thumb of the right-hand points to the direction of i_3 if we position our right hand so that our four fingers can rotate from i_1 to i_2.

A pseudotensor has 3^n components just like an axial or polar tensor but under coordinate change it transforms according to

$$\chi^{(n)}_{\mu\alpha_1\cdots\alpha_n}{}' = \Delta R_{\mu u}R_{\alpha_1 a_1}\cdots R_{\alpha_n a_n}\chi^{(n)}_{ua_1\cdots a_n}. \tag{A.18}$$

Δ equals one if the old and the new coordinate-systems have the same handedness (a proper transformation) whereas Δ equals minus one if the old and new coordinate-systems have different handedness (improper transformations).

From Yariv and Yeh [4] an example of a tensor that describes an optically active material is given by

$$G = \begin{pmatrix} 0 & G_{12} & G_{13} \\ -G_{12} & 0 & G_{23} \\ -G_{13} & -G_{23} & 0 \end{pmatrix}, \tag{A.19}$$

where we note that G is antisymmetric

There are several examples from physics of pseudotensors of rank one i.e. a pseudovector; most of these are related to the physical quantities defined by a vectorial crossproduct. However, in the following example, the Levi-Civita tensor is described.

Example A.1. *the Levi-Civita pseudotensor.*
By definition the Levi-Civita tensor is defined by

$$\epsilon_{ijk} = \mathbf{i}_i(\mathbf{i}_j \times \mathbf{i}_k), \tag{A.20}$$

that is

$$\begin{aligned} &\epsilon_{123} = \epsilon_{231} = \epsilon_{312} = 1 \\ &\epsilon_{213} = \epsilon_{132} = \epsilon_{321} = -1 \\ &\text{all others are zero.} \end{aligned} \tag{A.21}$$

The above definition holds in any coordinate system, and since it is indifferent, we have without stating this, assumed a standard right-handed coordinate system K. However, let us now consider ϵ_{jkl} in a left-handed coordinate system, i.e. $i_1' = i_1, i_2' = i_2$ and $i_3' = -i_3$ and the same origin. To transform the Levi-Civita tensor, we need the matrix that transforms between the original right-handed coordinate system K and the new left-handed coordinate-system K'. This is

$$\underline{\underline{R}} = \begin{pmatrix} 1 & 0 & 0 \\ 0 & 1 & 0 \\ 0 & 0 & -1 \end{pmatrix}. \tag{A.22}$$

It is noted that the determinant equals minus one, which means that it is an improper transformation. Using the transformation of polar tensors we find

$$\epsilon'_{123} = R_{1u}R_{2a_1}R_{3a_2}\epsilon_{ua_1a_2} = -1. \tag{A.23}$$

However, by definition

$$\epsilon'_{123} = i_1' \cdot (i_2' \times i_3') = 1. \tag{A.24}$$

Consequently ϵ_{jkl} does not transform as a tensor, this leads to the concept of a pseudotensor where the transformation into a new coordinatesystem is multiplied by the determinant. That is, the correct way of transforming the pseudotensor is

$$\epsilon'_{123} = \Delta R_{1u}R_{2a_1}R_{3a_2}\epsilon_{ua_1a_2} = 1. \tag{A.25}$$

— ■ —

Appendix B

Hamiltonian and polarization

B.1 Hamiltonian

In quantum mechanics, the total energy of a system, here the sum of the kinetic energies of all electrons and nuclei, plus their potential energy as for example the potential energy that exists between two neighboring nuclei, is described by an operator, the so-called Hamiltonian, here denoted $\hat{H}$, where the 'hat' symbolizes an operator. In relation to nonlinear optics, the energy of the system, the kinetic and the potential energy is impacted significantly as an intense electric field propagates through, and interacts with, a material. Thus it is fair to expect that the Hamiltonian may be divided into a 'ground state' Hamiltonian in the absence of an electric field plus an interaction Hamiltonian, describing the changes in the energy when the system is exposed to the electric field. However, it is also fair to assume that the Hamiltonian is a complicated operator since one would expect a significant coupling between the energy of the electrons and the nuclei.

In continuation of the above, the global Hamiltonian of a system is written as

$$\hat{H} = \hat{H}_0 + \hat{H}_{\text{int}}, \tag{B.1}$$

where $\hat{H}_0$ is the unperturbed Hamiltonian ($\mathbf{E} = 0$) and $\hat{H}_{int}$ is the interaction Hamiltonian that describe the interaction between the electric field and nuclei and electrons. Within the context of the **Born-Oppenheimer** approximation, $\hat{H} = \hat{H}_0^{\text{BO}} + \hat{H}_{\text{int}}^{\text{BO}}$, where $\hat{H}_0^{\text{BO}}$ is the unperturbed Hamiltonian and $\hat{H}_{\text{int}}^{\text{BO}}$ is the interaction Hamiltonian. The unperturbed Hamiltonian is the sum of kinetic energy of the nuclei, the coulomb interaction and the ground state energy

$$\hat{H}_0^{\text{BO}} = \hat{T}_N + \sum_{\alpha,\beta} \frac{e_\alpha e_\beta}{|\hat{\mathbf{r}}_\alpha - \hat{\mathbf{r}}_\beta|} + \hat{W}_{00}, \tag{B.2}$$

where $\hat{T}_N$ is the kinetic energy of the nuclei, $\sum_{\alpha,\beta} \frac{e_\alpha e_\beta}{|\hat{\mathbf{r}}_\alpha - \hat{\mathbf{r}}_\beta|}$ is the Coulomb interaction among nuclei of charges α and β with absolute charges e_α and e_β located at positions $\mathbf{r}_\alpha$ and $\mathbf{r}_\beta$. Finally, $\hat{W}_{00}$ is the ground state energy of the electrons.

Now, as an electric field is applied to the system, both the ground state energy of the electrons and the potential energy of the system are changed. Consequently, the interaction Hamiltonian has two contributions, one from the electron-field interaction and one from the nuclei-field interaction.

Electronic ground state energy The electron ground state energy $\hat{W}_0$ may be described by a time dependent perturbation as $\hat{W}_0 = \hat{W}_{00} - [\hat{\mu}_i E_i + \frac{1}{2}\hat{\alpha}_{ij} E_i E_j + \frac{1}{3}\hat{\beta}_{ijk} E_i E_j E_k + \cdots]V$, where $\hat{W}_{00}$ is the unperturbed electronic ground state energy, when no external macroscopic field is applied and $\hat{\mu}_i$ is the electronic dipole moment, $\hat{\alpha}_{ij}$ the electronic polarizability, $\hat{\beta}_{ijk}$ are higher-order hyperpolarizabilities, and V is the considered volume.

Interaction with the nuclei Interaction with the nuclei of charges e_α causes a change in the nucluei-field Hamiltonian ($\hat{H}_{\text{EN}} = -\sum e_\alpha \hat{r}_{i\alpha} E_i$)

From these two contributions the interaction Hamiltonian $\hat{H}_{\text{int}}^{\text{BO}}$ is

$$\begin{aligned} \hat{H}_{\text{int}}^{\text{BO}} &= -\sum e_\alpha \hat{r}_{i\alpha} E_i - \left[\hat{\mu}_i E_i + \frac{1}{2}\hat{\alpha}_{ij} E_i E_j + \frac{1}{3}\hat{\beta}_{ijk} E_i E_j E_k + \cdots\right] V \qquad \text{(B.3)} \\ &= \left[-\hat{m}_i E_i - \frac{1}{2}\hat{\alpha}_{ij} E_i E_j - \frac{1}{3}\hat{\beta}_{ijk} E_i E_j E_k + \cdots\right] V \\ &= \hat{v}_1 + \hat{v}_2 + \hat{v}_3, \end{aligned}$$

where $\hat{m}_i = \sum_\alpha e_\alpha \hat{r}_{i\alpha}/V + \hat{\mu}_i$ which is the sum of interaction of nuclei of charge e_α with macroscopic field and the electron field interaction.

The total effective Hamiltonian is the sum of the unperturbed system Hamiltonian $\hat{H}_0^{\text{BO}}$, Eqn. (B.2) and the interaction Hamiltonian $\hat{H}_{\text{int}}^{\text{BO}}$, Eqn. (B.3).

B.2 Polarization operator

The polarization density is the vector field that describes the density of permanent or induced electric dipole moments in a dielectric material. The total dipole moment is the sum of the electronic dipole moment and the dipole moment of the nuclei. Within the BO approximation the ***polarization density operator*** $\hat{p}_i$ is the sum of an electronic contribution and the dipole moment density of the nuclear charges.

The electronic contribution Since the electrons are assumed to follow the applied electric field instantaneously, the electrons occupy their ground state where they have an electric dipole moment, $\hat{M}_i^{el}$. This is expressed in terms of the energy of the ground state, $\hat{W}_0$, by $\hat{M}_i^{el} = -\partial \hat{W}_0/\partial E_i$, where the energy of the ground state $\hat{W}_0$ is described above as $\hat{W}_0 = \hat{W}_{00} - \{\hat{\mu}_i E_i + \frac{1}{2}\hat{\alpha}_{ij} E_i E_j + \frac{1}{3}\hat{\beta}_{ijk} E_i E_j E_k + \cdots\}V$. From this, the dipole moment operator may be expanded in terms of the macroscopic field, and written as $\hat{M}_i^{el} = \{\hat{\mu}_i + \hat{\alpha}_{ij} E_j + \hat{\beta}_{ijk} E_j E_k + \cdots\}V$.

The nuclear contribution The dipole moment operator associated with the sum of instantaneous positions of all nuclei of individual charges e_α and positions $r_{i\alpha}$ is given by $\sum_\alpha e_\alpha \hat{r}_{i\alpha}/V$.

The dipole moment density operator is the sum of these two contributions, i.e.

$$
\begin{aligned}
\hat{p}_i &= \frac{\sum_\alpha e_\alpha \hat{r}_{i\alpha} - \frac{\partial \hat{W}_0}{\partial E_i}}{V} \\
&= \hat{m}_i + \hat{\alpha}_{ij} E_j + \hat{\beta}_{ijk} E_j E_k + \hat{\gamma}_{ijkl} E_j E_k E_l + \cdots \\
&= \hat{p}_{i0} + \hat{p}_{i1} + \hat{p}_{i2} + \hat{p}_{i3} + \cdots ,
\end{aligned}
\tag{B.4}
$$

where $\hat{m}_i = \sum_\alpha e_\alpha \hat{r}_{i\alpha}/V + \hat{\mu}_i(r_{i\alpha})$ has been introduced. $\hat{m}_i$ describes the electric dipole in the absence of an applied electric field. It is noted that as opposed to the Hamiltonian operator, which is a scalar, the dipole density operator is a vector quantity.

Appendix C

Signal analysis

This appendix describes some signals that are important, for example in the analysis of communication signals, in Section C.1 in the time domain and in Section C.2 in the frequency domain. Signals should be understood in a broad sense since they are simply functions in the time and frequency domain used to describe for example a signal or a material response.

It is noted that there are numerous textbooks within the topic of signal analysis and mathematical treatment of signals, see for example [145]. Focusing on the topics covered in this book, we review some basic principles and signal analysis tools in the following.

C.1 Time domain signals

A time domain signal may be either an applied electric (or magnetic) field, i.e. one of the vectorial components in the field, or functions that may be applicable to describe the material response, i.e. one of the tensor elements used in the response tensor, even though we call it a signal.

Below, follows examples of response functions that are used in this book. This is followed by a description of some typical signals.

Example C.1. *The Dirac delta function*
In relation to a material response, an often used approximation is to assume that the material response is instantaneous, i.e. the induced polarization is simultaneous with the electric field. Mathematically this is described by using the *Dirac delta function*, denoted $\delta(t)$. The Dirac delta function is defined by its assigned properties, i.e.

$$\delta(t) = 0 \quad , \quad t \neq 0, \tag{C.1}$$

and

$$f(0) = \int_{-\infty}^{\infty} f(t)\delta(t)dt. \tag{C.2}$$

where $f(t)$ is a continues function at $t = 0$. The Dirac delta function is not a function in the usual sense that it may be described explicitly. However, it has some well defined characteristics including

$$\int_{-\infty}^{\infty} \delta(t)dt = 1, \tag{C.3}$$

which together with Eqn.(C.1) has the consequence that the delta function must be an infinitely high and infinitely thin spike at $t = 0$.

The delta function has several approximations. Here only one example will be given, the Gaussian function

$$\delta(x) \lim_{a\to\infty} \frac{1}{a\sqrt{\pi}} \exp\left(-x^2/a^2\right). \tag{C.4}$$

When convolving a function $f(t)$ with a delayed Dirac delta function, the result is a time delayed version of the function $f(t)$

$$f(t) * \delta(t-T) = \int_{-\infty}^{\infty} f(\tau)\delta(t-T-\tau)d\tau = f(t-T), \tag{C.5}$$

where $*$ represents the convolution. The result in Eqn. (C.5) is very convenient when describing a sampled signal or a periodic signal, as demonstrated in the following.

—— ■ ——

Example C.2. *Linear response function*
Another important response corresponds to a damped oscillation. In [30] this is given by

$$g(t) = \frac{\tau_1^2 + \tau_2^2}{\tau_1\tau_2} \exp\left(-t/\tau_2\right) \sin\left(t/\tau_1\right), \tag{C.6}$$

where τ_1 and τ_2 are two material parameters and $t > 0$ is a time delay. This response function is often used in relation to Raman, i.e. in Chapter 7. The ratio in front of the product between the exponential and the sinusoid ensures that the response is normalized, i.e. when integrated from 0 to ∞ the integral equals 1. We return to this example later in this appendix when we discuss the Lotentz response in Section C.3.

—— ■ ——

Example C.3. *Birefringence*
Assume that a material is birefringent but has an instantaneous response. This means that the response functions are delta functions but with different values for

each tensor element. If we constrain this example to a linear material, then the response tensor reduces to a diagonal matrix

$$R(t) = \left\{ \begin{array}{ccc} a\delta(t) & 0 & 0 \\ 0 & b\delta(t) & 0 \\ 0 & 0 & c\delta(t) \end{array} \right\}. \tag{C.7}$$

Note, the strength of the birefringence is included through the parameters a, b, and c. Consequently, the response in Eqn. (C.7) is not normalized. If the three parameters are all different, the material is called biaxial, whereas if two of the three are identical, the material is referred to as uniaxial. If the refractive index is independent of the direction the material is isotropic.

— ■ —

Regarding examples of signals in the time domain, described through $f(t)$ where t is the time, we typically use a Gaussian signal $f(t) = a\exp(-(t/T_0)^2)$, where a is the amplitude of the signal and T_0 describes the width of the signal, a soliton $f(t) = a\,\mathrm{sech}(bt)$, where a is the amplitude and b describes the width of the signal, a continuous signal, i.e. a constant for all times t, a delta function as described above or a square signal that takes the amplitude a for times t between $\pm T_0/2$ and equals zero otherwise, i.e.

$$f(t) = \left\{ \begin{array}{ll} a & -T_0/2 \leq t \leq T_0/2 \\ 0 & \text{otherwise.} \end{array} \right. \tag{C.8}$$

Finally, a signal that is important in communication and often used when making numerical simulations (not the least because discrete Fourier transforms indirectly assume strictly periodic signals) is a periodic signal as described in the following.

Sampled signal

Let's consider a periodically sampled signal, this is given by

$$g_s(t) = f(t) \sum_n \delta(t - nT_s), \tag{C.9}$$

where T_s is the time between two samples. This signal corresponds well to the situation when one performs numerical simulations.

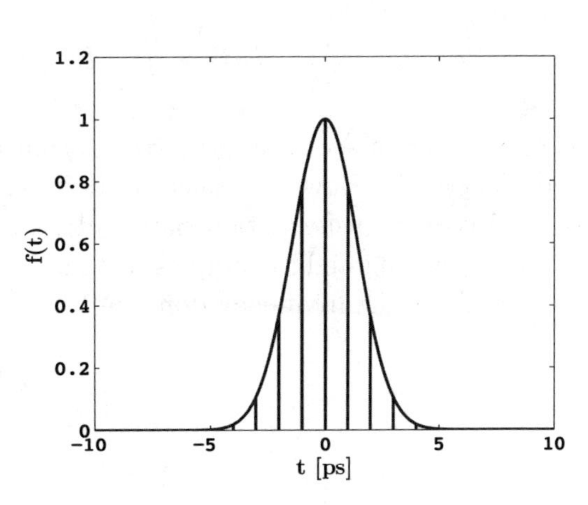

Figure C.1

Example of a sampled signal: $f(t) = \exp(-(t/T_s)^2)$, $T_s = 2$ ps.

Periodic signal
A periodic signal is obtained when

$$g_p(t + T_p) = g_p(t), \qquad (C.10)$$

where T_p is the period. This may also be written as a convolution between the signal describing one period of the signal and a periodic delta function

$$g_p(t) = f_p(t) * \sum_{n=-\infty}^{\infty} \delta(t - nT_p), \qquad (C.11)$$

where $f_p(t)$ is one period of the original signal and T_p is the period of the signal.

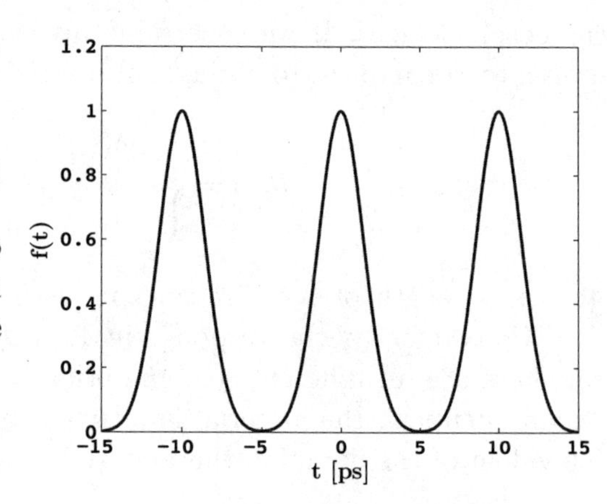

Figure C.2
Example of a periodic signal:
$f_p(t) = \exp(-(t/T_s)^2)$ for $|t| < 5$ ps
$f_p(t) = 0$ for $|t| > 5$ ps
$T_p = 10$ ps.

C.2 Frequency domain signals

To translate back and forth between the time and frequency domain we use the Fourier transformation

$$E(\omega) = \frac{1}{2\pi} \int_{-\infty}^{\infty} E(t) \exp(i\omega t) dt, \qquad (C.12)$$

with the inverse relation

$$E(t) = \int_{-\infty}^{\infty} E(\omega) \exp(-i\omega t) d\omega. \qquad (C.13)$$

There are different sets of Fourier transformation pairs; for example where to include the prefactor $\frac{1}{2\pi}$. However, once a set is chosen it is important to stick to the chosen Fourier transformation pair and its corresponding transformation rules. In this section we highlight examples of some useful Fourier transformation rules, and some signals in the frequency domain.

C.2.1 Review of basic Fourier transformation

When working with delta functions, some useful transformation rules are

$$\begin{array}{llcl} & \textit{Time domain} & \leftrightarrow & \textit{Frequency domain} \\ \text{A delta signal in time} & \delta(t) & \leftrightarrow & 1, \\ \text{A delta signal in freq.} & 1 & \leftrightarrow & 2\pi\delta(\omega), \\ \text{Sinusoids} & \exp(-i\omega_0 t) & \leftrightarrow & 2\pi\delta(\omega-\omega_0). \end{array} \tag{C.14}$$

The Fourier transformation of sums of delta functions are very relevant in nonlinear optics.

- A Fourier series:

$$\sum_{k=-\infty}^{\infty} a_k \exp(-ik\omega_0 t) \leftrightarrow \sum_{k=-\infty}^{\infty} 2\pi a_k \delta(\omega - k\omega_0). \tag{C.15}$$

- A periodic impulse train:

$$\sum_{n=-\infty}^{\infty} \delta(t - nT_s) \leftrightarrow \sum_{k=-\infty}^{\infty} \frac{2\pi}{T_s}\delta(\omega - k\omega_s), \tag{C.16}$$

where $\omega_s = 2\pi/T_s$.

Useful transformation rules. Let $g_1(t)$ have the Fourier transform $G_1(\omega)$ and $g_2(t)$ have the Fourier transform $G_2(\omega)$

$$\begin{array}{llcl} \text{Differentiation} & \frac{d^n g(t)}{dt^n} & \leftrightarrow & (-i\omega)^n G(\omega), \\ \text{Shift of time axis} & g(t+t_0) & \leftrightarrow & G(\omega)\exp i\omega t_0, \\ \text{Shift of frequency axis} & g(t)\exp(-i\omega_0 t) & \leftrightarrow & G(\omega+\omega_0), \\ \text{Convolution time} & g_1(t)*g_2(t) & \leftrightarrow & G_1(\omega)G_2(\omega), \\ \text{Convolution freq} & g_1(t)g_2(t) & \leftrightarrow & G_1(\omega)*G_2(\omega). \end{array} \tag{C.17}$$

Example C.4. *Spectrum of a Gaussian signal*
An envelope of an optical pulse is given by

$$f(t) = \exp\left[-(t/T_0)^2\right]. \tag{C.18}$$

Using the definition of the Fourier transformation given in Eqn. (C.12), we find the envelope in the frequency domain. In the Gaussian case this is

$$F(\omega) = (T_0/(2\sqrt{\pi}))\exp\left[-(\omega/\omega_0)^2\right], \tag{C.19}$$

where $\omega_0 = 2/T_0$. It is worth to note that this demonstrates that the Fourier transform of a Gaussian is another Gaussian.

— ■ —

Example C.5. *Spectrum of a Hyperbolic Secant, i.e. a soliton*
The Fourier transform of a soliton equals

$$\text{sech}(bt) \longleftrightarrow \frac{\pi}{b}\,\text{sech}\left(\frac{\pi\omega}{2b}\right). \tag{C.20}$$

— ■ —

Example C.6. *Spectrum of a sampled signal*

A sampled signal is

$$g_s(t) = f(t)\sum_n \delta(t - nT_s), \tag{C.21}$$

where T_s is the time between two samples. The spectrum of this signal is

$$G_s(\omega) = F(\omega) * \sum_{k=-\infty}^{\infty} \frac{2\pi}{T_s}\delta(\omega - k\omega_s). \tag{C.22}$$

This signal correspond well to numerical simulations, since this often involves discretizations of the signal in the time domain.

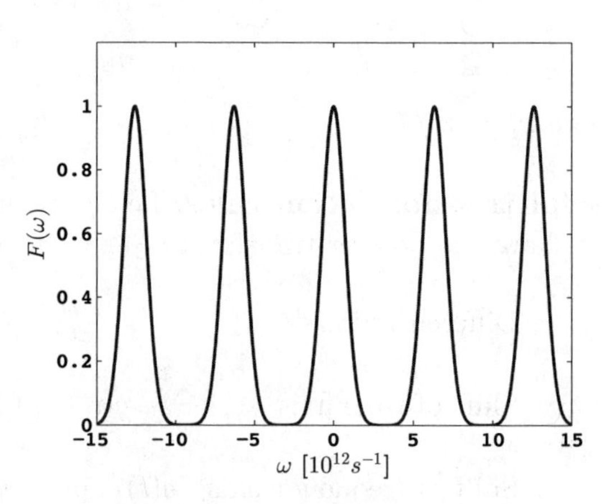

Figure C.3
Section of sampled signal in Figure C.1:
$F(\omega) = \exp\left(-(\omega/\omega_0)^2\right)$, $\omega_0 = 1$ ps^{-1}.

— ■ —

Example C.7. *Spectrum of a periodic signal*

A periodic signal is

$$g_p(t) = f(t) * \sum_{n=-\infty}^{\infty} \delta(t - nT_p), \tag{C.23}$$

where T_p is the time between two periods in the signal. To find its spectrum we use that the spectrum of a convolution of two functions equals the product of the two i.e. using Eqn.(C.15) and Eqn.(C.17)

$$G_p(\omega) = F(\omega) \sum_{k=-\infty}^{\infty} \frac{2\pi}{T_p} \delta(\omega - k\omega_0), \tag{C.24}$$

where $\omega_0 = 2\pi/T_p$. This also correspond to situations known from performing numerical simulations where a spetrum is sampled.

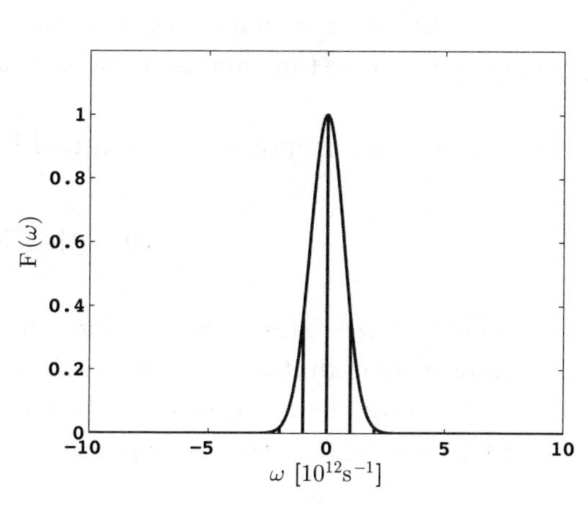

Figure C.4
Spectrum of sampled signal in Figure C.2.

——— ■ ———

Combining the last two examples, i.e. the spectrum of a, in the time domain periodically sampled signal, gives a periodic spectrum. That is, when performing numerical calculations of a signal in the time domain, this is treated as samples, consequently the spectrum becomes periodic with the sampling frequency as the periodicity. If the signal in the time domain furthermore is periodic, then the spectrum in the frequency domain consists of discrete points. Consequently, both in the time an frequency domain the signal is discrete (sampled) and periodic.

C.3 Models for linear material response

There exist several different models for the linear material response, dependent on the material that is being considered. Examples include the Lorentz model for solids like dielectrics and crystals, the Drude model for metals and the Debye models for liquids. Our main interest within this book is on solids; an introduction is given in Chapter 1. However, for completeness the following section outlines the main difference between different response functions. Several sources exist on the topic for example [146] and [147].

Lorentz

The Lorentz model is widely used to describe the susceptibility of solids, and we have also used this model in the main text, for example in the introduction in Chapter 1, when describing a charge (an electron) connected to a nuclei of a atom with a spring of spring constant and a damping force.

The general time response of a damped harmonic oscillator is

$$\frac{\omega_p^2}{\nu_0}\sin(\nu_0 t)e^{-\nu t/2}\Theta(t), \tag{C.25}$$

where $\Theta(t)$ is the Heaviside step function which is zero for $t < 0$ and one for $t \geq 0$. It is noted that other definitions of the Heaviside step function exist for example where $H(0) = 1/2$. Using the relation between the time and frequency domain response shown in Chapter 3, the frequency response, using the Lorentz model, is

$$\int_{-\infty}^{\infty}\frac{\omega_p^2}{\nu_0}\sin(\nu_0 t)e^{-\nu t/2}e^{i\omega t}\Theta(t)dt = \frac{\omega_p^2}{\nu_0^2 + (\nu/2)^2 - \omega^2 - i\omega\nu}. \tag{C.26}$$

If we introduce $\omega_0^2 = \nu_0^2 + (\nu/2)^2$ the result may be recognized from Chapter 1.

As described in the main text, more specifically in Chapter 1, Eqn. (1.12) and the surrounding text, it is the relative permittivity that is important for propagation. In the case of the Lorentz model the relative permittivity is

$$\varepsilon_r = 1 + \omega_p^2\frac{\omega_0^2 - \omega^2}{(\omega_0^2 - \omega^2)^2 + (\omega\nu)^2} + \omega_p^2\frac{i\nu\omega}{(\omega_0^2 - \omega^2)^2 + (\omega\nu)^2}. \tag{C.27}$$

That is the real part

$$\varepsilon' = 1 + Re[\chi] = 1 + \frac{\omega_p(\omega_0^2 - \omega^2)}{(\omega_0^2 - \omega^2)^2 + \gamma^2\omega^2}, \tag{C.28}$$

and the imaginary part

$$\varepsilon'' = 1 + Im[\chi] = \frac{\omega_p^2\gamma\omega}{(\omega_0^2 - \omega^2)^2 + \gamma^2\omega^2}. \tag{C.29}$$

Drude

The model is a special case of the Lorentz model where the spring constant is assumed zero, i.e. $k = 0$. This model is often used to describe optical properties of metals, where electrons are free charges. The time response of the Drude model is

$$\frac{\omega_p^2}{\nu}\left(1 - e^{-\nu t}\right)\Theta(t). \tag{C.30}$$

From this the frequency domain response is

$$\chi^{(1)} = \frac{\omega_p^2}{\nu} \frac{1}{i\omega - \nu}, \tag{C.31}$$

and the corresponding relative permittivity is

$$\varepsilon_r = 1 - \frac{\omega_p^2}{\nu} \left[\frac{\nu}{\omega^2 + \nu^2} + i \frac{\omega}{\omega^2 + \nu^2} \right]. \tag{C.32}$$

That is, the real

$$\varepsilon' = 1 - \frac{\omega_p^2}{\omega^2 + \nu^2}, \tag{C.33}$$

and imaginary part

$$\varepsilon'' = \frac{-\omega_p^2 \nu}{\omega(\omega^2 + \nu^2)}. \tag{C.34}$$

Debye

This model is used to describe dielectric properties of fluids with permanent, or static, dipoles. It provides an ideal relaxation response of an ideal, noninteracting population of dipoles to an alternating external electric field.

The time response of the Debye model is

$$\beta e^{-t/\tau} \Theta(t). \tag{C.35}$$

This corresponds to the frequency domain response

$$\chi^{(1)} = \beta \frac{1}{i\omega - 1/\tau}, \tag{C.36}$$

and the relative permittivity

$$\varepsilon_r = 1 + \frac{\beta}{\tau} \frac{1}{\omega^2 + (1/\tau)^2} - i \frac{\beta\omega}{\omega^2 + (1/\tau)^2}. \tag{C.37}$$

That is, the real

$$\varepsilon' = \frac{\beta\tau}{1 + \omega^2\tau^2}, \tag{C.38}$$

and imaginary part

$$\varepsilon'' = \frac{\omega\tau^2\beta}{(1 + \omega^2\tau^2)}. \tag{C.39}$$

Modified Debye

As the last response function we mention, the modified Debye has a time response

$$\omega_p^2 t e^{-\nu t/2}\Theta(t), \tag{C.40}$$

and corresponding frequency domain response

$$\omega_p^2 \frac{1}{(\nu/2)^2 - \omega^2 - i\nu\omega}. \tag{C.41}$$

Using this, the relative permittivity is

$$\varepsilon_r = 1 + \omega_p^2 \frac{(\nu/2)^2 - \omega^2}{((\nu/2)^2 - \omega^2)^2 + (\nu\omega)^2} + \omega_p^2 \frac{i\nu\omega}{((\nu/2)^2 - \omega^2)^2 + (\nu\omega)^2}. \tag{C.42}$$

Appendix D

Generating matrices and susceptibility tensors

The main idea of this book is not to discuss crystals and their structures. Consequently we do not describe spatial symmetries from a very general point of view.

Depending on their degree of symmetry, crystals are commonly classified in seven **crystal systems** trichlinic (the least symmetrical), monoclinic, orthombic, tetragonal, hexagonal, trigonal and isometric (the most symmetric).

The seven systems are in turn divided into **point groups** according to their symmetry with respect to a point. There are thirty-two such point groups.

In crystallography the properties of a crystal are described in terms of the natural coordinate system provided by the crystal itself. The axes of this natural system are the edges of a unit cell. I.e. the basis vectors that define the natural coordinate system may not be of equal lengths nor be perpendicular to each other. The natural axes are referred to as a,b,c, and α,β,γ. The angles between b and c, c and a and a and b respectively as shown in the main text.

In relation to a rectangular coordinate system (x, y, z) which is used in theoretical treatment of electromagnetic problems, a standard is described in [36].

1. the z-axis is chosen parallel to the c-axis

2. the x-axis is chosen perpendicular to the c-axis in the ac-plane. pointing in the positive a direction

3. y is normal to the ac-plane pointing in the positive b-direction and forming a right-handed coordinate system with z and x.

The designation for any class of symmetry is made of one, two or three symbols . **The first symbol** refers to the principal axis, if any, of the crystal, indicating the type of symmetry of that axis and the existence of reflection planes perpendicular to that axis. **The second symbol** refers to the secondary axis on the crystal. **The third symbol** names the tertiary axis if such exists.

D.1 Generating matrices

Inversion $M_1 = \begin{pmatrix} -1 & 0 & 0 \\ 0 & -1 & 0 \\ 0 & 0 & -1 \end{pmatrix}$	fourfold inversion-rotation about X_3-axis $M_8 = \begin{pmatrix} 0 & -1 & 0 \\ 1 & 0 & 0 \\ 0 & 0 & -1 \end{pmatrix}$
Twofold rotation about X_3-axis $M_2 = \begin{pmatrix} -1 & 0 & 0 \\ 0 & -1 & 0 \\ 0 & 0 & 1 \end{pmatrix}$	threefold rotation about X_3-axis $M_9 = \frac{1}{2}\begin{pmatrix} -1 & -\sqrt{3} & 0 \\ \sqrt{3} & -1 & 0 \\ 0 & 0 & 2 \end{pmatrix}$
Reflection in X_1X_2 plane $M_3 = \begin{pmatrix} 1 & 0 & 0 \\ 0 & 1 & 0 \\ 0 & 0 & -1 \end{pmatrix}$	Threefold inversion-rotation about X_3-axis $M_{10} = \frac{1}{2}\begin{pmatrix} 1 & -\sqrt{3} & 0 \\ \sqrt{3} & 1 & 0 \\ 0 & 0 & -2 \end{pmatrix}$
Twofold rotation about X_1-axis $M_4 = \begin{pmatrix} 1 & 0 & 0 \\ 0 & -1 & 0 \\ 0 & 0 & -1 \end{pmatrix}$	Sixfold rotation about X_3-axis $M_{11} = \frac{1}{2}\begin{pmatrix} -1 & -\sqrt{3} & 0 \\ \sqrt{3} & 1 & 0 \\ 0 & 0 & 2 \end{pmatrix}$
reflection in X_2X_3-plane $M_5 = \begin{pmatrix} -1 & 0 & 0 \\ 0 & 1 & 0 \\ 0 & 0 & 1 \end{pmatrix}$	Sixfold inversion-rotation about X_3-axis $M_{12} = \frac{1}{2}\begin{pmatrix} -1 & -\sqrt{3} & 0 \\ \sqrt{3} & -1 & 0 \\ 0 & 0 & -2 \end{pmatrix}$
reflection in X_1X_3 plane $M_6 = \begin{pmatrix} 1 & 0 & 0 \\ 0 & -1 & 0 \\ 0 & 0 & 1 \end{pmatrix}$	threefold rotation about [111] direction $M_{13} = \begin{pmatrix} 0 & 0 & 1 \\ 1 & 0 & 0 \\ 0 & 1 & 0 \end{pmatrix}$
Fourfold rotation about X_3-axis $M_7 = \begin{pmatrix} 0 & -1 & 0 \\ 1 & 0 & 0 \\ 0 & 0 & 1 \end{pmatrix}$	Threefold inversion-rotation about [111]-direction $M_{14} = \begin{pmatrix} 0 & -1 & 0 \\ 0 & 0 & -1 \\ -1 & 0 & 0 \end{pmatrix}$

D.2 Susceptibility tensors

The following serves to demonstrate some examples of susceptibility tensors. The examples are chosen to demonstrate the general structure of some susceptibility tensors as well as to give some examples of tensors relevant for some materials.

D.2.1 Triclinic material

The linear susceptibility has nine elements and is written as

$$\chi^{(1)} = \begin{bmatrix} xx & xy & zx \\ xx & yy & yz \\ zx & yz & zz \end{bmatrix}. \tag{D.1}$$

The second-order susceptibility (class 1) is

$$\chi^{(1)} = \begin{bmatrix} xxx & xyy & xzz & xyz & xzy & xzx & xxz & xxy & xyx \\ yxx & yyy & yzz & yyz & yzy & yzx & yxz & yxy & yyx \\ zxx & zyy & zzz & zyz & zzy & zzx & zxz & zxy & zyx \end{bmatrix}, \tag{D.2}$$

and the third-order susceptibility $\chi^{(3)}$ has 81 independent non-zero elements, that we will not write in a form like $\chi^{(1)}$ or $\chi^{(2)}$.

D.2.2 Trigonal

Linear susceptibility

$$\chi^{(1)} = \begin{bmatrix} xx & 0 & 0 \\ 0 & xx & 0 \\ 0 & 0 & zz \end{bmatrix}, \tag{D.3}$$

second-order susceptibility (class 3m)

$$\chi^{(2)} = \begin{bmatrix} 0 & 0 & 0 & 0 & 0 & xzx & xxz & \overline{yyy} & \overline{yyy} \\ \overline{yyy} & yyy & 0 & xxz & xzx & 0 & 0 & 0 & 0 \\ zxx & zxx & zzz & 0 & 0 & 0 & 0 & 0 & 0 \end{bmatrix}, \tag{D.4}$$

where the bar is standard notation for a negative element. The third-order susceptibility $\chi^{(3)}$ has 27 non-zero elements of which 14 are independent $zzzz$

$$xxxx = yyyy = xxyy + xyyx + xyxy \begin{cases} xxyy = yyxx \\ xyyx = yxxy \\ xyxy = yxyx \end{cases}$$

$yyzz = xxzz$
$zzyy = zzxx$
$zyyz = zxxz$
$yzzy = xzzx$

$$yzyz = xzxz$$
$$zyzy = zxzx$$

$$xxxz = \overline{xyyz} = \overline{yxyz} = \overline{yyxz}$$
$$xxzx = \overline{xyzy} = \overline{yxzy} = \overline{yyzx}$$
$$xzxx = \overline{xzyy} = \overline{yzxy} = \overline{yzyx}$$
$$zxxx = \overline{zxyy} = \overline{zyxy} = \overline{zyyx}.$$

D.2.3 Isotropic material

An example of an isotropic material is glass. In general, isotropic material has 21 nonzero elements of which only 3 are independent

$$xxxx = yyyy = zzzz$$
$$yyzz = zzyy = zzxx = xxzz = xxyy = yyxx$$
$$yzyz = zyzy = zxzx = xzxz = xyxy = yxyx$$
$$yzzy = zyyz = zxxz = xzzx = xyyx = yxxy$$
$$xxxx = xxyy + xyxy + xyyx.$$

For further examples see [2]. It is emphasized that the tensors are defined in a cartesian coordinate system with axes along the direction of the crystallographic axes.

Appendix E

Transverse field distributions

E.1 Lasermodes

The fundamental mode of a laser is the so-called Gaussian beam

$$E^0 = E_0^0 \frac{w_0}{w_0(z)} \exp\left[-\frac{r^2}{w^2(z)}\right] \exp\left[ikz - \zeta(z) - r^2 \frac{ik}{2R(z)},\right] \tag{E.1}$$

where w_0 is the beam waist, which expands as the distance from the beam center z increases according to $w^2(z) = w_0^2 \left(1 + \left(\frac{z\lambda}{\pi w_0^2}\right)^2\right)$, and ζ a the Gouy phase term.

Depending on the design of the laser cavity, more solutions may exist to the wave-equation. Using Cartesian coordinates (x, y, z) the electrical field may be expressed

$$\begin{aligned} E^0 = {} & E_0^0 \frac{\omega_0}{\omega_0(z)} H_m\left(\sqrt{2}\frac{x}{\omega}\right) H_n\left(\sqrt{2}\frac{y}{\omega}\right) \exp\left[-r^2/\omega^2\right] \\ & \times \exp\left[ikz - \zeta(z) - r^2\left(\frac{1}{\omega(z)} + \frac{ik}{2R(z)}\right)\right], \end{aligned} \tag{E.2}$$

where ω, R and ϕ define the beam, while H_j is the Hermite polynomial of order j. The four Hermite polynomials to lowest order are

$$\begin{aligned} H_0(\xi) &= 1, \\ H_1(\xi) &= 2\xi, \\ H_2(\xi) &= 4\xi^2 - 2, \\ H_3(\xi) &= 8\xi^3 - 12\xi. \end{aligned} \tag{E.3}$$

In some cases the laser resonator invites circular cylindrical symmetry and the transverse electrical field may preferably be expressed in cylindrical coordinates (r, θ, z) using Lagurre polynomial functions rather than Cartesian coordinates and Hermite polynomials.

E.2 Fibermodes

The most common optical waveguide is the optical fiber, the simplest being the so-called step index fiber, where the refractive index changes as a step function at the core-cladding interface. In the following we use n_1 as the refractive index of the core and n_0 as the refractive index of the cladding, $n_1 > n_0$.

Unlike the ideal metallic waveguide the equations for the tangential components of electric field $\mathbf{E}$, and magnetic field $\mathbf{H}$ in an optical fiber are coupled. In general the modes in a cylindrical fiber can not be grouped into TE and TM guided waves. Only modes with no azimuthal variations separate into TE and TM modes. If the symmetry axis of the fiber is the z-axis, the modes with both E_z and H_z nonzero are known as EH and HE hybrid modes.

The normalized frequency, or the V-number of a fiber is defined as $V = kaNA$ where k is the wavenumber, a the radius of the core and NA the numerical aperture defined through the index contrast as $NA^2 = n_1^2 - n_0^2$. The cutoff frequency, i.e. the frequency under which a mode can not exist, is non-vanishing for all modes except the HE_{11} mode which has no cutoff frequency i.e. it always exist. The next mode has cutoff for $V = 2.405$. When more than one mode exists in the optical fiber, the fiber is referred to as a multi-mode fiber. The number of modes approximates $V^2/2$.

In the weakly-guiding approximation $(n_1 - n_0)/n_1 << 1$, the modes propagating in the fiber are linearly polarized, and denoted LP modes (the z-component of the electromagnetic field is very small in this approximation). These are characterized by two indices, m and n. The LP modes are combinations of the modes found when the exact theory of the waveguide is applied. The fundamental mode is the HE_{11} mode, which in the LP mode picture becomes the LP_{01} mode.

The first subscript m gives the number of azimuthal nodes in the electric field distribution, while the second subscript, n, gives the number of radial nodes. The zero field at the outer edge of the field distribution is counted as a node.

In a multi-mode fiber the different modes corresponding to the same frequency ω have different propagation constants. In addition, the propagation constant for each mode depends on the frequency ω. In a single mode fiber the propagation constant depends on the frequency, causing group velocity dispersion.

E.2.1 The LP_{01} mode

When a waveguide consists of homogeneous layers,as for example the step index fiber, the scalar wave equation apply for the electric and the magnetic field apply within each individual layer, i.e.

$$\nabla_{\perp}^2 \phi + (k^2 n_i^2 - \beta^2)\phi = 0, \tag{E.4}$$

where ϕ is any of the vector components of the electric or magnetic field. n_i is the refractive, $i = 1$ in the core and $i = 0$ in the cladding, a constant in each layer. To arrive at Eqn. (E.4) a solution to the waveequation in the form

$$E = \psi(r, \theta) e^{i(\beta z - \omega t)}. \tag{E.5}$$

has been assumed. ∇_t^2 is the transverse Laplace operator. In cylindrical coordinates the Laplace operator is

$$\frac{\partial^2}{\partial r^2} + \frac{1}{r}\frac{\partial}{\partial r} + \frac{1}{r^2}\frac{\partial^2}{\partial \theta^2}. \tag{E.6}$$

Furthermore, solutions must have separable radial and azimuthal dependence

$$\phi(r, \theta) = R(r) \left\{ \begin{array}{c} \cos(m\theta) \\ \sin(m\theta) \end{array} \right\}. \tag{E.7}$$

Inserting this into the wave equation together with the Laplacian operator, the radial dependence, i.e. the mode is found from

$$\frac{\partial^2 R}{\partial r^2} + \frac{1}{r}\frac{\partial R}{\partial r} - \frac{m^2}{r^2} R + (k^2 n_i^2 - \beta^2) R = 0. \tag{E.8}$$

This is a Bessel differential equation when $\beta < kn_i$ and a modified Bessel differential equation when $\beta > kn_i$. Applying the boundary conditions at infinity ($r \to \infty$) and at the center ($r = 0$) and continuity at the core-cladding interface, the modes are found as solutions from

$$u \frac{J_{l+1}(u)}{J_l(u)} = v \frac{K_{l+1}(v)}{K_l(v)}, \tag{E.9}$$

where J_l and K_l are the Bessel and modified Bessel function respectively. Both characterized by their-order l. The parameters

$$\begin{aligned} u^2 &= a^2(k^2 n_1^2 - \beta^2) \\ v^2 &= a^2(\beta^2 - k^2 n_2^2), \end{aligned} \tag{E.10}$$

where n_1 is the refractive index of the core and n_2 the refractive index of the cladding, as noted earlier. It is noted that $V^2 = u^2 + v^2 = k^2 a^2 (n_1^2 - n_2^2)$ defines the normalized frequency.

When solving the mode problem one knows the wavelength at which one would like to find the propagation constant β. In addition the geometry i.e. the core radius and the refractive index of the core and cladding also needs to be known. With the wavelength and the refractive index profile as known parameters one can solve the set of equations given by Eqn. (E.9) and the normalized frequency. Once this problem is solved, one know the propagation constant β, hence also u and v in Eqn. (E.10), and one can then find the electric field in the core and cladding from

$$\begin{aligned}
E_{x1} &= -i2B(\beta a/u)J_l(ur/a)\cos(l\theta), \\
E_{y1} &= 0, \\
E_{z1} &= B\left[J_{l+1}(ur/a)\cos((l+1)\theta) - J_{l-1}(ur/a)\cos((l-1)\theta)\right], \\
H_{x1} &= 0, \\
H_{y1} &= -i2Bn_1\eta(\beta a/u)J_l(ur/a)\cos(l\theta), \\
H_{z1} &= Bn_1\eta\left[J_{l+1}(ur/a)\sin((l+1)\theta) + J_{l-1}(ur/a)\sin((l-1)\theta)\right],
\end{aligned} \tag{E.11}$$

where $\eta = \sqrt{\varepsilon_0/\mu_0}$, and in the cladding,

$$\begin{aligned}
E_{x2} &= -i2B(\beta a/v)\frac{J_{l+1}(u)}{K_{l+1}(v)}K_l(vr/a)\cos(l\theta), \\
E_{y2} &= 0, \\
E_{z2} &= B\frac{J_{l+1}(u)}{K_{l+1}(v)}\left[K_{l+1}(vr/a)\cos((l+1)\theta) + K_{l-1}(vr/a)\cos((l-1)\theta)\right], \\
H_{x2} &= 0, \\
H_{y2} &= -i2Bn_2\eta(\beta a/v)\frac{J_{l+1}(u)}{K_{l+1}(v)}K_l(vr/a)\cos(l\theta), \\
H_{z2} &= Bn_2\eta\frac{J_{l+1}(u)}{K_{l+1}(v)}\left[K_{l+1}(vr/a)\sin((l+1)\theta) - K_{l-1}(vr/a)\sin((l-1)\theta)\right].
\end{aligned} \tag{E.12}$$

It is important to note that from the propagation constant β, one may define an effective refractive index n_{eff} through

$$\beta = \frac{2\pi}{\lambda}n_{\text{eff}}, \tag{E.13}$$

where λ is the vacuum wavelength for which the wave equation has been solved. The effective refractive index is the refractive index that a mode experiences and for a guided mode its value is between the refractive index of the core and the refractive index of the cladding, $n_0 < n_{\text{eff}} < n_1$.

Appendix F

The index ellipsoid

F.1 Background

Birefringence, or double refraction, is the decomposition of a ray of light into two rays, depending on the state of polarization of the light, see Figures F.1 and F.2. The two rays are referred to as the ordinary ray and the extraordinary ray. This effect only occurs if the structure of the material is anisotropic, i.e. directionally dependent, as opposed to isotropic. If the material has a single axis of anisotropy it is termed uniaxial, whereas material with more than one axis of anisotropy is referred to as biaxial [24].

When describing the linear refractive index of a material, a coordinate system can always be chosen in a way so that the linear susceptibility tensor reduces to a diagonal matrix. This is done by finding eigenvalues and eigenvectors [148]; this coordinate system defines the principal axis of the material.

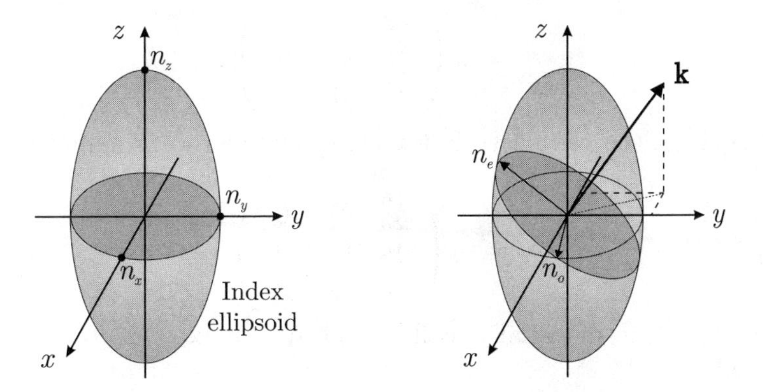

Figure F.1: When an optical signal propagates through a birefringent material, the electrical field is decomposed into two orthogonally polarized rays; the ordinary ray and the extraordinary ray.

F.2 The index ellipsoid

The propagation of optical fields in birefringent materials, can be evaluated through Maxwell equations to obtain the phase velocities and the vectors **D**, **H** and **E**. In anisotropic media the electric field vector and the dielectric displacement vector can be non-parallel although being linearly related. Often the index ellipsoid is used to describe the propagation of fields. The index ellipsoid is defined on the basis of the surface of constant energy density U

$$U = \frac{1}{2}\,\mathbf{E}\cdot\mathbf{D}. \tag{F.1}$$

Using that $\mathbf{D} = \varepsilon_0\varepsilon_r\mathbf{E}$ in the material, the permittivity is a diagonal matrix in the principal coordinate system, i.e. $D_i = \varepsilon_0\varepsilon_{ii}E_i$, where $i \in (x, y, z)$. Eqn. (F.1) can now be written as

$$U = \frac{1}{2\varepsilon_0}\left(\frac{D_x^2}{\varepsilon_{xx}} + \frac{D_y^2}{\varepsilon_{yy}} + \frac{D_z^2}{\varepsilon_{zz}}\right). \tag{F.2}$$

This is an ellipsoid in the coordinates (D_x, D_y, D_z), which gives the magnitude and direction of the displacement vector. Replacing $\mathbf{D}/\sqrt{2U}$ by $\mathbf{r} = (\tilde{x}, \tilde{y}, \tilde{z})$ gives

$$1 = \frac{\tilde{x}^2}{n_x^2} + \frac{\tilde{y}^2}{n_y^2} + \frac{\tilde{z}^2}{n_z^2}, \tag{F.3}$$

which defines the ellipsoid in the normalized coordinates $(\tilde{x}, \tilde{y}, \tilde{z})$ [4].

The dielectric tensor can always be reduced to a diagonal tensor, defining the principal axis of the material. Considering different groups of materials, in isotropic materials the index ellipsoid reduces to a sphere, meaning that all three tensor elements are equal

$$\varepsilon_i = \varepsilon_0\begin{pmatrix} n^2 & 0 & 0 \\ 0 & n^2 & 0 \\ 0 & 0 & n^2 \end{pmatrix}. \tag{F.4}$$

In uniaxial crystals the refractive index along two of the principal dielectric axes are equal

$$\varepsilon_u = \varepsilon_0\begin{pmatrix} n_o^2 & 0 & 0 \\ 0 & n_o^2 & 0 \\ 0 & 0 & n_e^2 \end{pmatrix}. \tag{F.5}$$

If further $n_e < n_o$ the material is called negative uniaxial, if $n_e > n_o$ it is called positive uniaxial.

In biaxial materials the refractive indices are all different along the three principal dielectric axes

$$\varepsilon_b = \varepsilon_0\begin{pmatrix} n_x^2 & 0 & 0 \\ 0 & n_y^2 & 0 \\ 0 & 0 & n_z^2 \end{pmatrix}. \tag{F.6}$$

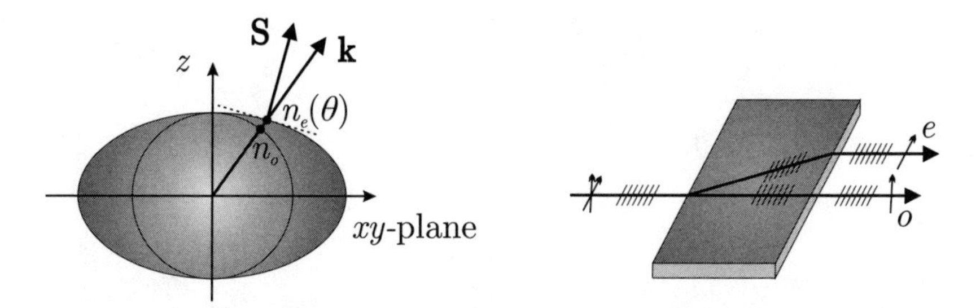

Figure F.2: The ordinary and the extraordinary index of refraction is found as the intersection of the wavevector and the circle and ellipse, respectively.

From the index ellipsoid the actual indices for the ordinary and extraordinary polarization can be found by considering a plane normal to the wavevector through the origin of the coordinate system, see Figure F.1. The intersection of the plane and the ellipsoid form an ellipse. The refractive index is found as the intersection of the wavevector and the circle (ordinary index) and the ellipse (extraordinary index), as illustrated in Figure F.2 for a positive uniaxial material.

The ordinary index of refraction in an uniaxial material is independent of the direction of propagation through the material. Defining the optical axis as the direction where the ordinary and extraordinary indices are identical, the extraordinary index of refraction, which depends on the angle θ, between the optical axis and the direction of propagation, can be calculated as

$$n_e(\theta) = \left(\frac{\cos^2(\theta)}{n_o^2} + \frac{\sin^2(\theta)}{n_e^2} \right)^{-1/2}. \tag{F.7}$$

In general the Pointing vector $\mathbf{S}$, and the wavevector $\mathbf{k}$, are non-parallel, as a result of the electric field vector $\mathbf{E}$, and the displacement vector $\mathbf{D}$ not being parallel, when the field is not propagating along one of the principal axes of the material. The Pointing vector is orthogonal to the surface of the ellipse, as seen from Figure F.2. Snell's law relates to the wavevector; however, the power flow is in the direction of the Pointing vector as indicated in Figure F.2. This effect is often referred to as beam walk-off. The angle ρ, between the wave vector and the Pointing vector is

$$\rho(\theta) = \pm \arctan\left(\left(\frac{n_o}{n_e} \right)^2 \tan(\theta) \right) \mp \theta, \tag{F.8}$$

where the upper and lower signs are for negative and positive uniaxial materials, respectively.

Appendix G

Materials commonly used in nonlinear optics

The purpose of this appendix is to provide data for typical examples of photonic materials used within nonlinear optics. The list of materials is not comprehensive but may hopefully be useful to gain insight and to get an overview of typical nonlinear photonic materials.

G.1 Lithium Niobate - $LiNbO_3$

Negative uniaxial crystal: $n_o > n_e$
Point group: 3m
Transparency range: 0.4 - 4.5 μm

Sellmeier parameters [149]:

Parameters	n_e	n_o
a_2	0.0983	0.1185
a_3	0.2020	0.2091
a_4	189.32	89.61
a_5	12.52	10.85
a_6	$1.32 \cdot 10^{-2}$	$1.97 \cdot 10^{-2}$
b_1	$2.860 \cdot 10^{-6}$	$7.941 \cdot 10^{-7}$
b_2	$4.700 \cdot 10^{-8}$	$3.134 \cdot 10^{-8}$
b_3	$6.113 \cdot 10^{-8}$	$-4.641 \cdot 10^{-9}$
b_4	$1.516 \cdot 10^{-4}$	$-2.188 \cdot 10^{-6}$

Sellmeier equation:

$$n_i^2 = a_1 + b_1 \cdot f + \frac{a_2 + b_2 \cdot f}{\lambda^2 - (a_3 + b_3 \cdot f)^2} + \frac{a_4 + b_4 \cdot f}{\lambda^2 - a_5^2},$$

where the wavelength is to be inserted in [µm]. The index i specifies the ordinary or extraordinary index. The temperature dependence is given as

$$f = (T - 24.5)(T + 570.82),$$

inserting T in degrees Celsius.

Typical values for the nonlinear coefficients at 1064 nm [150]:

$$\begin{aligned} d_{22} &= 2.1 \text{ pm/V}, \\ d_{31} &= 4.6 \text{ pm/V}, \\ d_{33} &= 25.2 \text{ pm/V}, \end{aligned}$$

where d_{ij} is defined in Chapter 4 in the main text.

$$\begin{pmatrix} (P_x^0)^{(2)} \\ (P_y^0)^{(2)} \\ (P_z^0)^{(2)} \end{pmatrix} = \begin{pmatrix} 0 & 0 & 0 & 0 & d_{15} & -d_{22} \\ -d_{22} & d_{22} & 0 & d_{15} & 0 & 0 \\ d_{31} & d_{31} & d_{33} & 0 & 0 & 0 \end{pmatrix} \begin{pmatrix} (E_x^0)^2 \\ (E_y^0)^2 \\ (E_z^0)^2 \\ 2E_y^0 E_z^0 \\ 2E_x^0 E_z^0 \\ 2E_x^0 E_y^0 \end{pmatrix}.$$

Effective nonlinear coefficients[151]:

$$\begin{aligned} d_{ooe} &= d_{31} \sin(\theta) - d_{22} \cos(\theta) \sin(3\phi), \\ d_{eoe} &= d_{oee} = d_{22} \cos^2(\theta) \cos(3\phi). \end{aligned}$$

Typical values for the electro optic coefficients at 633 nm [4]:

$$\begin{aligned} r_{13} &= 9.6 \text{ pm/V}, \\ r_{22} &= 6.8 \text{ pm/V}, \\ r_{33} &= 30.9 \text{ pm/V}, \\ r_{51} &= 32.6 \text{ pm/V}, \\ r_{12} &= 21.1 \text{ pm/V}, \end{aligned}$$

$$r = \begin{pmatrix} 0 & -r_{12} & r_{13} \\ 0 & r_{22} & r_{13} \\ 0 & 0 & r_{33} \\ 0 & r_{51} & 0 \\ r_{51} & 0 & 0 \\ -r_{22} & 0 & 0 \end{pmatrix}.$$

G.2 Potassium Titanyl Phosphate - KTP

Positive biaxial crystal
Point group: mm2
Transparency range: 0.35 - 4.5 μm

Sellmeier parameters [152] and [153]:

Parameters	n_x	n_y	n_z
a	3.0065	3.0333	3.3134
b	0.03901	0.04154	0.05694
c	0.04251	0.04547	0.05658
d	0.01327	0.01408	0.01682
e	0.1323	0.5014	0.3896
f	0.4385	2.0030	1.3332
g	1.2307	3.3016	2.2762
h	0.7709	0.7498	2.1151

Sellmeier equation:

$$n_i^2 = a_i + \frac{b_i}{\lambda^2 - c_i} - d_i \cdot \lambda^2,$$

where the wavelength is to be inserted in [μm]. The temperature dependence is gives as

$$\frac{dn_i}{dT} = e_i \cdot \lambda^{-3} - f_i \cdot \lambda^{-2} + g_i \cdot \lambda^{-1} + h_i.$$

Typical values for the nonlinear coefficients at 1064 nm [154]:

$$\begin{aligned} d_{15} &= 1.91 \text{ pm/V}, \\ d_{24} &= 3.64 \text{ pm/V}, \\ d_{31} &= 2.54 \text{ pm/V}, \\ d_{32} &= 4.35 \text{ pm/V}, \\ d_{33} &= 16.9 \text{ pm/V}, \end{aligned}$$

$$\begin{pmatrix} (P_x^0)^{(2)} \\ (P_y^0)^{(2)} \\ (P_z^0)^{(2)} \end{pmatrix} = \begin{pmatrix} 0 & 0 & 0 & 0 & d_{15} & 0 \\ 0 & 0 & 0 & d_{24} & 0 & 0 \\ d_{31} & d_{32} & d_{33} & 0 & 0 & 0 \end{pmatrix} \begin{pmatrix} (E_x^0)^2 \\ (E_y^0)^2 \\ (E_z^0)^2 \\ 2E_y^0 E_z^0 \\ 2E_x^0 E_z^0 \\ 2E_x^0 E_y^0 \end{pmatrix}.$$

Effective nonlinear coefficients [155]:

$$\begin{array}{ll} xy \text{ - plane :} & d_{eoe} = d_{oee} = d_{31}\sin^2(\phi) - d_{32}\cos^2(\phi), \\ yz \text{ - plane :} & d_{oeo} = d_{eoo} = d_{31}\sin(\theta), \\ xz \text{ - plane, } \theta < V_z : & d_{ooe} = d_{32}\sin(3\theta), \\ xz \text{ - plane } \theta > V_z : & d_{oeo} = d_{eoo} = d_{32}\sin(\theta). \end{array}$$

Typical values for the electro optic coefficients [58]:

$$r_{13} = 9.5 \text{ pm/V},$$
$$r_{23} = 15.7 \text{ pm/V},$$
$$r_{33} = 36.3 \text{ pm/V},$$
$$r_{51} = 7.3 \text{ pm/V},$$
$$r_{42} = 9.3 \text{ pm/V},$$

$$r = \begin{pmatrix} 0 & 0 & r_{13} \\ 0 & 0 & r_{23} \\ 0 & 0 & r_{33} \\ 0 & r_{42} & 0 \\ r_{51} & 0 & 0 \\ 0 & 0 & 0 \end{pmatrix}.$$

G.3 Third-order nonlinear materials

Two examples of third order nonlinear materials that have found extensive use is silica (SiO_2), within fiber optics, and silicon, within silicon photonics in lightwave circuitry. Both materials are cubic however, silica is furthermore isotropic as described in the main text.

Below, some characteristics are shown in the table, from [156]:

Material	n_0	$\chi^{(3)}$ (esu)	$n_2(cm^2/w)$
Silicon	3.4	$2.0 \cdot 10^{-10}$	$2.7 \cdot 10^{-14}$
Silica	1.47	$1.8 \cdot 10^{-14}$	$3.2 \cdot 10^{-16}$

To translate between SI units and esu units, [2] provide the following relation

$$\chi^{(n)}(SI) = \chi^{(n)}(esu)\frac{4\pi}{(10^{-4}c)^{n-1}}. \tag{G.1}$$

It is noted that the third order susceptibility may be complicated to measure and especially for materials as silicon since the tensor elements differ significantly. [157]

and [158] provide useful insight into how this can be done and also the relative magnitude among the different tensor elements.

For completeness the susceptibility for a few other interesting materials are listed below:

Material	n_0	$\chi^{(3)}$ (esu)	$n_2(cm^2/w)$
Lead bismuth gallate	2.3	$1.6{\cdot}10^{-12}$	$1.3{\cdot}10^{-14}$
Carbon disulfide	1.63	$2.2{\cdot}10^{-12}$	$3.2{\cdot}10^{-14}$
water	1.33	$1.8{\cdot}10^{-14}$	$4.1{\cdot}10^{-16}$
Air	1.0003	$1.2{\cdot}10^{-17}$	$5.0{\cdot}10^{-19}$

Bibliography

[1] R. W. Boyd. *Radiometry and the Detection of Optical Radiation.* John Wiley & Sons, New York, 1983.

[2] P. N. Butcher and D. Cotter. *The Elements of Nonlinear Optics.* Cambridge University Press, Cambridge, 1990.

[3] G. P. Agrawal. *Nonlinear Fiber Optics.* Academic Press, San Diego, 1995.

[4] A. Yariv and P. Yeh. *Optical waves in crystals.* J. Wiley and Sons inc., Hoboken New Jersey, USA, 2003.

[5] P. A. Franken, A. E. Hill, C. W. Peters, and G. Weinreich. "Generation of optical harmonics". *Phys. Rev. Lett.*, 7:118, 1961.

[6] A. L. Shawlow and C. H. Townes. "Infrared and optical masers". *Phys. Rev.*, 112:1940, 1958.

[7] T. Miya, Y. Terunuma, T. Hosaka, and T. Myashita. "Ultimate low-loss single-mode fibre at 1.55 μm". *Electron. Lett.*, 15(4):106–108, 1979.

[8] K. Okamoto. *Fundamentals of optical waveguides, Second Edition.* Academic Press, London, 2006.

[9] A. W. Snyder and J. D. Love. *Optical Waveguide Theory.* Kluwer Academic Publishers, Massachusetts, 2000.

[10] J. D. Jackson. *Classical Electrodynamics, Third Edition.* John Wiley & Sons, San Diego, 1999.

[11] H. A. Haus and D. A. B. Miller. "Attenuation of Cutoff Modes and Leaky Modes of Dielectric Slab Structures". *IEEE J. of Quantum Electronics*, QE-22(7):310–318, 1986.

[12] M. Born and E. Wolf. *Principles of Optics.* Pergamon Press, New York, sixth edition, 1980.

[13] S. Huard. *Polarization of Light.* John Wiley and Sons, 1997.

[14] P. W. Milonni and J. H. Eberly. *Lasers.* Wiley Interscience, New York, 1988.

[15] M. Fox. *Optical Properties of Solids.* Oxford University Press, Oxford, second edition, 2010.

[16] R. W. Boyd. "Order-of-magnitude estimates of the nonlinear optical susceptibility". *J. of Modern Optics*, 46(3):367–378, 1999.

[17] R. W. Hellwarth. "Third-order optical susceptibilities of liquids and solids". *Progress in Quantum Electr.*, 5, Part 1:1–68, 1977.

[18] P. Schwerdtfeger. *Computational Numerical and Mathematical Methods in Sciences and Engineering - Vol. 1 ATOMS, MOLECULES AND CLUSTERS IN ELECTRIC FIELDS Theoretical Approaches to the Calculation of Electric Polarizability*, edited by George Maroulis, Chapter 1. University of Patras, Greece.

[19] C. C. Wang. "Empirical Relation between the Linear and the Third-order Nonlinear Optical Susceptibilities". *Phys. Rev.*, B2:2045, 1970.

[20] T. Monro and H. Ebendorff-Heidepriem. "Progress in Microstructured Optical Fibers". *Annual Review of Materials Research*, 36:467–95, 2006.

[21] S. Sevincli, H. Henkel, C. Ates, and T. Pohl. "Nonlocal Nonlinear Optics in Cold Rydberg Gases". *Phys. Rev. Letters*, PRL 107:153001, 2011.

[22] M. Shen, Y. Y. Lin, C.-C. Jeng, and K.-K. Lee. "Vortex pairs in nonlocal nonlinear media". *J. of Optics*, 14:065204, 2012.

[23] W. Krolikowski, O. Bang, N. I. Nikolov, D. Neshev, J. Wyller, J. J. Rasmussen, and D. Edmundson. "Modulation Instability, solitons and beam propagation in spatially nonlocal nonlinear media". *J. of Optics B*, 6(5):S288, 2004.

[24] B. E. A. Saleh and M. C. Teich. *Fundamentals of Photonics, Second Edition.* Siley Series in Pure and Applied Photonics, Hoboken, NJ, 2007.

[25] R. Loudon. *The Quantum Theory of Light.* Oxford Science Publications, Clarendon Press, Oxford, second edition, 1983.

[26] R. W. Hellwarth, A. Owyoung, and N. George. "Origin of the nonlinear refractive index of liquid CCl_4". *Physical Rev. A*, 4:2342, 1971.

[27] A. Owyoung, R. W. Hellwarth, and N. George. "Intensity-Induced Changes in Optical Polarizations in Glasses". *Physical Rev. B*, 5:628, 1972.

[28] R. Hellwarth, J. Cherlow, and T Yang. "Origin and frequency dependence of nonlinear optical susceptibilities of glasses". *Phys. Rev. B*, 11(2):964 – 967, 1975.

[29] R. H. Stolen, J. P. Gordon, W. J. Tomlinson, and H. A. Haus. "Raman response function of silica-core fibers". *J. of Opt. Soc. Am. B.*, 6:1159, 1989.

[30] K. J. Blow and D. Wood. "Theoretical Description of Transient Stimulated Raman Scattering in Optical Fibers". *IEEE J. Quantum Electron.*, 25(12):2665 – 2673, 1989.

[31] G. E. Walrafen and P. N. Krishnan. "Model analysis of the Raman spectrum from fused silica optical fibers". *Appl. Optics*, 21(3):359 – 350, 1982.

[32] P. D. Maker, R. W. Terhune, and C. M. Savage. "Intensity-dependent changes in the refractive index of liquids". *Physical Rev. Letters*, 12:507, 1964.

[33] R. W. Boyd. *Nonlinear Optics.* Academic Press, New York, 2007.

[34] D. Hollenbeck and C. D. Cantrell. "Multi-vibrational-mode model for fiber-optic Raman gain spectrum and response function". *J. Opt. Soc. Am. B*, 19(12):2886–2892, 2002.

[35] R. F. Tinder. *Tensor properties of solids, phenomelogical development of the tensor properties of crystals.* Morgan and Claypool, 2008.

[36] Committee. "Standards on Pizoelectric Crystals". *Proceedings of the IRE*, December:1378–1394, 1949.

[37] E. Hartmann. *An Introduction to Crystal Physics, description of the physical properties of crystals.* Univ. College Cardiff Press, electronic edition, Cardiff, Wales, 2001.

[38] D. E. Sands. *Introduction to Crystallography.* Dover Publications, Inc., Mineola, New York, 1993.

[39] Q. Lin, O. J. Painter, and G. P. Agrawal. "Nonlinear optical phenomena in silicon waveguides: Modeling and applications". *Opt. Express*, 15:16604, 2007.

[40] J. M. Manley and H. E. Rowe. "Some general properties of nonlinear elements - Part 1. general energy relations". *Proc. IRE*, 44:904–913, 1956.

[41] Condren and Corzine. *Diode lasers and photonic integrated circuits.* Wiley series in microwave and optical engineering, New York, 1995.

[42] S. V. Afshar and T. M. Monro. "A full vectorial model for pulse propagation in emerging waveguides with subwavelength structures Part 1: Kerr ninlinearity". *Optics Express*, 17:2298, 2009.

[43] M. D. Turner, T. M. Monro, and S. V. Afshar. "A full vectorial model for pulse propagation in emerging waveguides with subwavelength structures Part 1: Stimulated Raman Scattering". *Optics Express*, 17:11565, 2009.

[44] F. Poletti and P. Horak. "Description of ultrashort pulse propagation in multimode optical fibers". *J. Opt. Soc. Am. B*, 25(10):1645, 2008.

[45] F. Poletti and P. Horak. "Dynamics of femtosecond supercontinuum generation in fibers". *Optics Express*, 17(8):6134, 2009.

[46] M. Kolesik and J. V. Moloney. "Nonlinear optical pulse propagation simulation: From Maxwell's to unidirectional equations". *Physical Reveiw*, E70:036604, 2004.

[47] M. E. V. Pedersen, J. Cheng, C. Xu, and K. Rottwitt. "Transverse Field Dispersion in the Generalized Nonlinear Schrodinger Equation: Four Wave Mixing in a Higher Order Mode Fiber". *J. of Lightwave Technol.*, 31:3425, 2013.

[48] J. A. Buck. *Optical fibers.* John Wiley and Sons, second edition, 2004.

[49] T. H. Mainman. "Stimulated Optical Radiation in Ruby". *Nature*, 187:493–494, 1960.

[50] V. G. Dmitriev, G. G. Gurzadyan, and D. N. Nikogosyan. *Handbook of Nonlinear Optical Crystals.* Springer Series in Optical Sciences, third edition, 1999.

[51] R. L. Byer. "Quasi-phasematched nonlinear interactions and devices". *Journal of Nonlinear Optical Physics & Materials*, 6:549–592, 1997.

[52] M. Yamada, N. Nada, M. Saitoh, and K. Watanabe. "First order quasiphase matched linbo 3 waveguide periodically poled by applying an external field for efficient blue second harmonic generation". *Appl. Phys. Lett.*, 62:435–436, 1993.

[53] J. Janousek, S. Johansson, P. Tidemand-Lichtenberg, S. Wang, J. L. Mortensen, P. Buchhave, and F. Laurell. "Efficient all solid-state continuous-wave yellow-orange light source". *Opt. Express*, 13:1188–1192, 2005.

[54] C. Pedersen, E. Karamehmedovic, J. S. Dam, and P. Tidemand-Lichtenberg. "Enhanced 2D-image upconversion using solid-state lasers". *Opt. Express*, 23:20885–20890, 2009.

[55] J. S. Dam, C. Pedersen, and P. Tidemand-Lichtenberg. "Room temperature mid-IR single photon spectral imaging". *Nat. Phot.*, 6:788–793, 2012.

[56] W. Brunner and H. Paul. "Theory of Optical Parametric Amplification and Oscillation". *Progress in Optics*, 15:3–73, 1977.

[57] G. D.Boyd and D. A. Kleinman. "Parametric Interaction of Focused Gaussian Light Beams". *J. Appl. Phys.*, 39:3539, 1968.

[58] W. P. Risk, T. R. Gosnell, and A. V. Nurmikko. *Compact Blue-Green Lasers.* Cambridge University Press, 2003.

[59] V. Delaubert, M. Lassen, D. R. N. Pulford, H.-A. Bachor, and C. C. Harb. "Spectral noise figure of Er^{3+}-doped fiber amplifiers". *IEEE Photon. Technol. Lett.*, 2:208, 1990.

[60] A. Penzkofer, A. Laubereau, and W. Kaiser. "High intensity Raman interactions". *Prog. in Quant. Electr.*, 6:55, 1982.

[61] C. Kittel. *Introduction to Solid State Physics.* J Wiley and Sons inc., Hoboken New Jersey, USA, 8th edition.

[62] N. Shibata, M. Horigudhi, and T. Edahiro. "Raman spectra of binary high-silica glasses and fiber containing GeO_2, P_2O_5 and B_2O_3". *Non-Crystalline Solids*, 45:115, 1981.

[63] A. Pasquarello and R. Car. "Identification of Raman Defect Lines as Signatures of Ring Structures in Vitreous Silica". *Phys. Rev. Lett.*, 80:5145, 1998.

[64] J. M. Dudley and J. R. Taylor. *Supercontinuum Generation in Optical Fibers.* Cambridge University Press, New York, 2010.

[65] A. Hasegawa. "Numerical study of optical soliton transmission amplified periodically by the stimulated Raman process". *Appl. Opt.*, 23:3302, 1984.

[66] K. Rottwitt and H. D. Kidorf. "A 92-nm bandwidth Raman amplifier". *Proc. Optical Fiber Commun. Conf.*, San Jose, California, USA:Paper PD6, 1998.

[67] K. Rottwitt, J. H. Povlsen, A. Bjarklev, O. Lumholt, B. Pedersen, and T. Rasmussen. "Noise in Distributed Erbium-Doped Fibers". *IEEE Photon. Technol. Lett.*, 5:218, 1993.

[68] J. Bromage, K. Rottwitt, and M. E. Lines. "A method to predict the Raman gain spectra of germanosilicate fibers with arbitrary index profiles". *IEEE Photon. Technol. Lett.*, 14:24–26, 2002.

[69] J. Kani, M. Jinno, and K. Oguchi. "Fibre Raman amplifier for 1520 nm band WDM transmission". *Electron. Lett.*, 34:1745, 1998.

[70] C. R. S Fludger, V. Handerek, and R. J. Mears. "Pump to Signal RIN Transfer in Raman Fiber Amplifiers". *J. Lightwave Technol.*, 19:1140, 2001.

[71] H. Kidorf, K. Rottwitt, M. Nissov, M. Ma, and E. Rabarijaona. "Pump Interactions in a 100-nm Bandwidth Raman Amplifier". *IEEE Photon. Technol. Lett.*, 11:530, 1999.

[72] K. Rottwitt, M. Nissov, and F. Kerfoot. "Detailed analysis of Raman amplifiers for long-haul transmission". *Proc. Optical Fiber Commun. Conf.*, San Jose, California, USA:Paper TuG1, 1998.

[73] J. Bromage, C. H. Kim, R. M. Jopson, K. Rottwitt, and A. J. Stentz. "Dependence of double Rayleigh backscatter noise in Raman amplifiers on gain and pump depletion". *Proc. Optical Amplifiers and their Applications Conf.*, Stresa, Italy, 2001.

[74] K. Rottwitt and A. J. Stentz. In *Optical Fiber Telecommunications Vol. IVA, Chapter 5.* Academic Press, San Diego, 2002.

[75] E. Desurvire, M. J. F. Digonnet, and H. J. Shaw. "Theory and Implementation of a Raman Active Fiber Delay Line". *J. Lightwave Technol.*, LT-4(4):426–443, 1986.

[76] S. Popov, E. Vanin, and G. Jacobsen. "Influence of polarization mode dispersion value in dispersion-compensating fibers on the polarization dependence of Raman gain". *Opt. Lett.*, 27(10):848, 2002.

[77] Q. Lin and G. P. Agrawal. "Polarization mode dispersion-induced fluctuations during Raman amplification in optical fibers". *Opt. Lett.*, 27(24):2194–2196, 2002.

[78] Q. Lin and G. P. Agrawal. "Statistics of polarization-dependent gain in fiber-based Raman amplifiers". *Opt. Lett.*, 28(4):227–229, 2003.

[79] Q. Lin and G. P. Agrawal. "Vector theory of stimulated Raman scattering and its application to fiber-based Raman amplifiers". *J. Opt. Soc. Am. B.*, 20(8):1616–1631, 2003.

[80] J. Bromage. "Raman Amplification for Fiber Communications Systems". *J. Lightwave Technol.*, 22(1):79–93, 2004.

[81] R. H. Stolen. "Polarization effects in fiber Raman and Brillouin lasers". *IEEE J. Quantum Electron.*, QE-15:1157, 1979.

[82] E. Son, J. Lee, and Y. Chung. "Gain variations of Raman amplifier in birefringent fiber". *Proc. Optical Fiber Commun. Conf.*, Atlanta, Georgia, USA, 2003.

[83] T. B. Hansen and A. D. Yaghjian. *Plane-Wave Theory of Time-Domain Field, Near-Field scanning applications.* IEEE Press, New York, 1999.

[84] I. Bar-Joseph, A. A. Friesem, E. Lichtman, and R. G. Waarts. "Steady and relaxation oscillations of stimulated Brillouin scattering in single-mode optical fibers". *J. Opt. Soc. Am. B.*, 2(10):1606, 1985.

[85] J. R. Ott, M. E. V. Pedersen, and K. Rottwitt. "Self-pulsation threshold of Raman amplified Brillouin fiber cavities". *Optics Express*, 17:16166, 2009.

[86] A. Yariv. *Quantum Electronics.* John Wiley & Sons, New York, 1989.

[87] D. Cotter. “Observation of stimulated Brillouin scattering in low loss silica fibre at 1.3μm”. *Electron. Lett.*, 18:495, 1982.

[88] R. W. Tkach, A. R. Chraplyvy, and R. M. Derosier. “Spontaneous Brillouin Scattering for Single-mode optical-fibre characterisation”. *Electron. Lett.*, 22:1011, 1986.

[89] L. Grüner-Nielsen, S. Dasgupta, M. D. Mermelstein, D. Jakobsen, S. Herstrøm, M. E. V. Pedersen, E. L. Lim, S. Alam, F. Parmigiana, D. Richardson, and B. Palsdottir. “A silica based highly nonlinear fibre with improved threshold for stimulated Brillouin scattering”. *Proc ECOC2010*, page Tu.4.D.3, 2010.

[90] R. G. Smith. “Optical power handling capacity of low loss optical fibers as determined by stimulated Raman and Brillouin scattering”. *Appl. Optics*, 11:2489, 1972.

[91] M.-J. Li, X. Chen, J. Wang, S. Gray, A. Liu, J. A. Demeritt, A. B. Ruffin, A. M. Crowley, D. T. Walton, and L. A. Zenteno. “Al/Ge co-doped large mode area fiber with high SBS threshold”. *Optics Express*, 15:8290, 2007.

[92] X. Chen, M.-J. Li, and A. Liu. “Ge-doped laser fiber with suppressed stimulated Brillouin scattering”. *Proc. OSA/OFC/NFOEC*, pages OMG3–5465197, 2010.

[93] T. Nakanishi, M. Tanaka, T. Hasegawa, M. Hirano, T. Okuno, and M. Onishi. “Al_2O_3-SiO_2 Core Highly Nonlinear Dispersion-shifted Fiber with Brillouin Gain Suppression Improved by 6.1 dB”. *Proc. ECOC'06*, page Th4.2.2, 2006.

[94] E. P. Ippen and R. H. Stolen. “Stimulated Brillouin scattering in optical fibers”. *Appl. Phys. Lett.*, 21(11):539, 1972.

[95] T. R. Parker, M. Farhadiroushan, V. A. Handerek, and A. J. Rogers. “Temperature and strain dependence of the power level and frequency of spontaneous Brillouin scattering in optical fibers”. *Optics Letters*, 22:787, 1997.

[96] Y. Okawachi, M. S. Bigelow, J. E. Sharping, Z. Zhu, A. Schweinber, D. J. Gauthier, R. W. Boyd, and A. L. Gaeta. “Tunable All-Optical Delay via Brillouin Slow Light in an optical fiber”. *Physical Review Letters*, 94:153902, 2005.

[97] M. Suzuki, I. Morita, N. Edagawa, S. Yamamoto, H. Taga, and A. Akiba. “Reduction of Gorgon-Haus timing jitter by periodic dispersion compensation in soliton transmission”. *Electron. Letters*, 31:2027, 1995.

[98] N. J. Smith, K. J. Blow F. M. Knoxand N. J. Doran, and I. Bennion. “Enhanced power solitons in optical fibres with periodic dispersion management”. *Electron. Letters*, 32(1):54, 1996.

[99] Gradshteyn and Ryzhik. *Tables of integrals, series and products.* corrected and enlarged edition, Academic Press, 1980.

[100] H. Haken, H. C. Wolf, and W. D. Drewer. *The Physics of Atoms and Quanta.* Springer-Verlag, New York, 1990.

[101] J. N. Elgin and S. M. J. Kelly. "Spectral modulation and the growth of resonant modes associated with periodically amplified solitons". *Opt. Lett.*, 18(10):787–789, 1993.

[102] K. J. Blow and N. J. Wood. "Average soliton dynamics and the operation of soliton systems with lumped amplifiers". *IEEE Photon. Technol. Lett.*, 3(4):369 – 371, 1991.

[103] K. Rottwitt, B. Hermann, J. H. Povlsen, and J. N. Elgin. "Adiabatic soliton transmission at very high bit rates". *J. Opt. Soc. Am. B*, 12(7):1307–1310, 1995.

[104] L. F. Mollenauer and K. Smith. "Demonstration of soliton transmission over more than 4000 km in fiber with loss periodically compensated by Raman gain". *Opt. Lett.*, 13(8):675–677, 1988.

[105] C. J. McKinstrie and J. P. Gordon. "Field Fluctuations Produced by Parametric Processes in Fibers". *IEEE J. of Selected Topics in Quantum Electronics*, 18(2):958, 2012.

[106] C. J. McKinstrie and S. Radic. "Parametric amplifiers driven by two pump waves with dissimilar frequencies". *Optics Letters*, 27(13):1138, 2002.

[107] K. K. Y. Wong, M. E. Marhic, K. Uesaka, and L. G. Kazovsky. "Polarization-Independent Two-Pump Fiber Optical Parametric Amplifier". *IEEE Photon. Technol. Lett.*, 14:911, 2002.

[108] R. H. Stolen and J. E. Bjorkholm. "Parametric amplification and frequency conversion in optical fibers". *IEEE J. Quantum Electron.*, 18:1062, 1982.

[109] J. Hansryd, P. A. Andrekson, M. Westlund, J. Li, and P. Hedekvist. "Fiber-Based Optical Parametric Amplifiers and Their Applications". *IEEE J. of Selected Topics in Quantum Electronics*, 8:506, 2002.

[110] C. Gerry and P. Knight. *Introductory Quantum Optics.* Cambridge University Press, Cambridge, UK, 2005.

[111] C. J. McKinstrie, S. Radic, and M. G. Raymer. "Quantum noise properties of parametric amplifiers driven by two pump waves". *Optics Express*, 12(21):5037, 2004.

[112] P. L. Voss, G. Kahraman, G. Köprülü, and P. Kumar. "Raman-noise-induced noise-figure quantum limits for $\chi^{(3)}$ nondegenerate phase-sensitive amplification and quadrature squeezing". *J. Opt. Soc. Am. B*, 23:598, 2006.

[113] E. Desurvirre. *Erbium Doped Fiber Amplifiers, Principles and Applications.* Academic Press, San Diego, 1994.

[114] J. B. Coles, B. P. P. Kuo, N. Alic, S. Moro, C. S. Bres, J. M. C. Boggio, P. A. Andrekson, M. Karlsson, and S. Radic. "Bandwidth-efficient phase modulation techniques for Stimulated Brillouin Scattering suppression in fiber optic parametric amplifiers". *Optics Express*, 18:18138, 2010.

[115] J. Hansryd, F. Dross, M. Westlund, P. A. Andrekson, and S. N. Knudsen. "Increase of the SBS Threshold in a Short Highly Nonlinear Fiber by Applying a Temperature Distribution". *J. of Lightwave Technol.*, 19:1691, 2001.

[116] M. R. Lorenzen, D. Noordegraaf, C. V. Nielsen, O. Odgaard, L. G. Nielsen, and K. Rottwitt. "Suppression of Brillouin scattering in fibre-optical parametric amplifier by applying temperature control and phase modulation". *Electron. Lett.*, 45(2):125, 2009.

[117] R. Engelbrecht, M. Mueller, and B. Schmauss. "SBS shaping and suppression by arbitrary strain distributions realized by a fiber coiling machine". *Proc. IEEE/LEOS Winter Topicals*, Paper WC1.3:248, 2009.

[118] R. Engelbrecht. "Analysis of SBS Gain Shaping and Threshold Increase by Arbitrary Strain Distributions". *J. of Lightwave Technol.*, 32:1689, 2014.

[119] A. Yeniay, J. M. Delavaux, and J. Toulouse. "Spontaneous and Stimulated Brillouin Scattering Gain in Optical Fibers". *J. of Lightwave. Technol.*, 20:1425, 2002.

[120] K. Shiraki, M. Ohashi, and M. Tateda. "SBS Threshold of a fiber with a Brillouin Frequency Shift Distribution". *J. of Lightwave. Technol.*, 14:50, 1996.

[121] M. E. Marhic, C. H. Hsia, and J. M. Jeong. "Optical amplification in a nonlinear fiber interferometer". *Electron. Lett.*, 27:210, 1991.

[122] Z. Tong, C. Lundström, P. A. Andrekson, M. Karlsson, and A. Bogris. "Ultralow Noise, Broadband Phase-Sensitive Optical Amplifiers, and Their Applications". *IEEE J. of selected topics in quantum electronics*, 18:1016, 2012.

[123] S. M. M. Friis, K. Rottwitt, and C. J. McKinstrie. "Raman and loss induced quantum noise in depleted fiber optical parametric amplifiers". *Optics Express*, 21:29320, 2013.

[124] P. L. Voss and P. Kumar. "Raman-noise-induced noise-figure limit for $\chi^{(3)}$ parametric amplifiers". *Optics Letters*, 29:445, 2004.

[125] P. Kylemark, P. O. Hedekvist, H. Sunnerud, M. Karlsson, and P. A. Andrekson. "Noise Characteristics of Fiber Optical Parametric Amplifiers". *J. of LightWave. Technol.*, 22:409, 2004.

[126] V. Cristofori, Z. Lali-Dastjerdi, T. Lund-Hansen, C. Peucheret, and K. Rottwitt. "Experimental investigation of saturation effect on pump-to-signal intensity modulation transfer in single-pump phase-insensitive fiber parametric amplifiers". *J. Opt. Soc. Am. B*, 30:884, 2013.

[127] P. Velanas, A. Bogris, and D. Syvridis. "Impact of Dispersion Fluctuations on the Noise Properties of Fiber Optic Parametric Amplifiers". *J. Lightwave Technol.*, 24:2171, 2006.

[128] F. Yaman, Q. Lin, S. Radic, and G. P. Agrawal. "Impact of Dispersion Fluctuations on Dual-Pump Fiber-Optic Parametric Amplifiers". *IEEE Photon. Technol. Lett.*, 16:1292, 2004.

[129] E. Myslivets, N. Alic, and S. Radic. "High Resolution Measurements of Arbitrary Dispersion Fibers: Dispersion Map Reconstruction Techniques". *J. of Lightwave. Technol.*, 28(23):3478, 2010.

[130] L. S. Rishøj, A. S. Svane, T. Lunh-Hansen, and K. Rottwitt. "Quantitative evaluation of standard deviations of group velocity dispersion in optical fibre using parametric amplification". *Electron. Lett.*, 50:199, 2014.

[131] B. P. P. Kuo and S. Radic. "Highly nonlinear fiber with dispersive characteristic invariant to fabrication fluctuations". *Optics Express*, 20:7716, 2012.

[132] Z. Tong, C. Lundström, P. A. Andrekson, C. J. McKinstrie, M. Karlsson, D. J. Blessing, E. Tipsuwannakul, B. J. Puttnam, and H. Toda amd L. Grüner-Nielsen. "Towards ultrasensitive optical links enables by lo-noise Phase-Sensitive Amplifiers". *Nature Photonics*, 5:430, 2011.

[133] M. D. Levenson, R. M. Shelby, and S. H. Perlmutter. "Squeezing of classical noise by nondegenerate four-wave mixing in an optical fiber". *Optics Letters*, 10:514, 1985.

[134] Z. Tong, A. Bogris, M. Karlsson, and P. A. Andrekson. "Full characterization of the signal and idler noise figure spectra in single-pumped fiber optical parametric amplifiers". *Optics Express*, 18:2884, 2010.

[135] C. J. McKinstrie, S. Radic, and A. R. Chraplyvy. "Parametric amplifiers driven by two pump waves". *IEEE J. of. Selected Topics in Quantum Electronics*, 8(3):538, 2002.

[136] J. M. Chavez Boggio, C. Lundström, J. Yang, H. Sunnerud, and P. A. Andrekson. "Double-pumped FOPA with 40 dB flat gain over 81 nm bandwidth". *ECOC*, paper Tu.3.B.5, Brussels, 2008.

[137] H. Steffensen, J. R. Ott, K. Rottwitt, and C.J. McKinstrie. "Full and semi-analytical analysis of two-pump parametric amplification with pump deletion". *Optics Express*, 19:6648, 2011.

[138] Q. Lin, C. F. Marki, C. J. McKinstrie, R. Jobson, J. Ford, G. P. Agrawal, and S. Radic. "40-Gb/s Optical Switching and Wavelength Multicasting in a Two-Pump Parametric Device". *IEEE Photonics Technol. Lett.*, 17:2376, 2005.

[139] C.-S. Bres, A. O. J. Wiberg, B. P. P Kuo, N. Alic, and S. Radic. "Wavelength Multicasting of 320 gb/s challen self-seeded parametric amplifier". *IEEE Photonics Technol. Lett.*, 21:1002, 2009.

[140] C.-S. Bres, A. O. J. Wiberg, B. P. P Kuo, J. M.Chavez-Boggio, C. F. Marki, N. Alic, and S. Radic. "Optical demultiplexing of 320 Gb/s to 8 × 40 gb/s in single parametric gate". *J. Lightwave Technol*, 28:434, 2010.

[141] J. M. Chavez Boggio, J. R. Windmiller, M. Knutzen, R. Jiang, C. Bres, N. Alic, B. Strossel, K. Rottwitt, and S. Radic. "730 nm optical parametric conversion from near to short-wave infrared band". *Optics Express*, 16:5435, 2008.

[142] P. A. Andrekson and M. Westlund. "Nonlinear optical fiber based high resolution all-optical waveform sampling". *Laser and photonics reviews*, 1:231, 2007.

[143] M. Westlund, P. A. Andrekson, H. Sunnerud, J. Hansryd, and J. Li. "High-performance optical-fiber nonlinearity based optical waveform monitoring". *J. of Lightwave Technol.*, 23(6):2012, 2005.

[144] C. Peucheret, M. Lorenzen, J. Seoane, D. Noordegraaf, C. V. Nielsen, L. Grüner-Nielsen, and K. Rottwitt. "Amplitude Regeneration of RZ-DPSK Signals in Single-Pump Fiber-Optic Parametric Amplifiers". *IEEE Photonics Technol. Lett.*, 21:872, 2009.

[145] E. Kreyszig. *Advanced Engineering Mathematics.* J. Wiley and Sons, Inc., Hoboken New Jersey, USA, 10th edition, 2011.

[146] C. F. Bohren and D. R. Huffman. *Absorption and Scattering of Light by Small Particles.* John Wiley and Sons, New York, 1998.

[147] G. Kristensson, S. Rikte, and A. Sihvola. "Mixing formulas in the time domain". *JOSA A*, 15(5):1411, 1998.

[148] A. Yariv. *Optical electronics.* Rinehart and Winston Inc., Orlando, Florida, 3rd edition, 1985.

[149] O. Gayer, Z. Sacks, E. Galun, and A. Arie. "Temperature and wavelength dependent refractive index equations for MgO-doped congruent and stoichiometric $LiNbO_3$". *Appl. Phys. B*, 91:343–348, 2008.

[150] I. Shoji, T. Kondo, A. Kitamoto, M. Shirane, and R. Ito. "Absolute scale of second-order nonlinear-optical coefficients". *J. Opt. Soc. Am. B*, 14:2268–2294, 1997.

[151] J. E. Midwinter and J. Warner. "The effect of phase matching method and of uniaxial crystal symmetry on the polar distribution of second-order non-linear optical polarization". *Brit. J. Appl. Phys.*, 16:1135–1142, 1965.

[152] K. Kato. "Parametric oscillation at 3.2 μm in KTP pumped at 1.064 μm". *IEEE J. QE*, 27:1137 – 1140, 1991.

[153] K. Kato. "Temperature Insensitive SHG at 0.5321 μm in KTP". *IEEE J. QE*, 28:1974 – 1976, 1992.

[154] H. Vanherzeele and J. D. Bierleini. "Magnitude of the nonlinear-optical coefficients of KTiOPO4". *Opt. Lett.*, 17:982–984, 1992.

[155] V. G. Dmitriev and D. N. Nikogosyan. "Effective nonlinearity coefficients for three-wave interactions in biaxial crystals of mm2 point group symmetry". *Opt. Comm.*, 95:173–182, 1993.

[156] R. W. Boyd and G. L. Fisher. *Nonlinear Optical Materials.* Elsevier Science Ltd., Encyclopedia of Materials: Science and Technology, ISBN:0-08-0431526.

[157] Y. Wang, Chia-Yu Lin, A. Nikolaenko, V. Raghunathan, and E. O. Potma. "Four-wave mixing microscopy of nanostructures". *Advances in Optics and Photonics*, 3:1–52, 2011.

[158] G. R. Meredith, B. Buchalter, and C.Hanzlik. "Third order susceptibility determination by third harmonic generation. ii". *The Journal of Chemical Physics*, 78(3):1543, 1983.

Index

K

L

M

N

O